KEYWORDS IN TECHNICAL AND PROFESSIONAL COMMUNICATION

Foundations and Innovations in Technical and Professional Communication

Series Editor: Lisa Melonçon

Series Associate Editor: Sherena Huntsman

The Foundations and Innovations in Technical and Professional Communication series publishes work that is necessary as a base for the field of technical and professional communication (TPC), addresses areas of central importance within the field, and engages with innovative ideas and approaches to TPC. The series focuses on presenting the intersection of theory and application/practice within TPC and is intended to include both monographs and co-authored works, edited collections, digitally enhanced work, and innovative works that may not fit traditional formats (such as works that are longer than a journal article but shorter than a book).

The WAC Clearinghouse and University Press of Colorado are collaborating so that these books will be widely available through free digital distribution and low-cost print editions. The publishers and the series editors are committed to the principle that knowledge should freely circulate and have embraced the use of technology to support open access to scholarly work.

Other Books in the Series

Jason C. K. Tham (Ed.), *Keywords in Design Thinking: A Lexical Primer for Technical Communicators & Designers* (2022)

Kate Crane and Kelli Cargile Cook (Eds.), *User Experience as Innovative Academic Practice* (2022)

Joanna Schreiber and Lisa Melonçon (Eds.), *Assembling Critical Components: A Framework for Sustaining Technical and Professional Communication* (2022)

Michael J. Klein (Ed.), *Effective Teaching of Technical Communication: Theory, Practice, and Application* (2021).

KEYWORDS IN TECHNICAL AND PROFESSIONAL COMMUNICATION

Edited by Han Yu and Jonathan Buehl

The WAC Clearinghouse
wac.colostate.edu
Fort Collins, Colorado

University Press of Colorado
upcolorado.com
Denver, Colorado

The WAC Clearinghouse, Fort Collins, Colorado 80523

University Press of Colorado, Denver, Colorado 80203

ISBN 978-1-64215-192-3 (PDF) | 978-1-64215-193-0 (ePub) | 978-1-64642-501-3 (pbk.)

DOI 10.37514/TPC-B.2023.1923

Produced in the United States of America

Library of Congress Cataloging-in-Publication Data

Names: Yu, Han, 1980– editor. | Buehl, Jonathan, editor.
Title: Keywords in technical and professional communication / edited by Han Yu and Jonathan Buehl.
Description: Fort Collins, Colorado : The WAC Clearinghouse ; Denver, Colorado : University Press of Colorado, [2023] | Series: Foundations and innovations in technical and professional communication | Includes bibliographical references and index.
Identifiers: LCCN 2023037179 (print) | LCCN 2023037180 (ebook) | ISBN 9781646425013 (pbk.) | ISBN 9781642151923 (PDF) | ISBN 9781642151930 (ePub)
Subjects: LCSH: Communication of technical information. | Technology—Terminology.
Classification: LCC T10.5 .K174 2023 (print) | LCC T10.5 (ebook) | DDC 603—dc23/eng/20240126
LC record available at https://lccn.loc.gov/2023037179
LC ebook record available at https://lccn.loc.gov/2023037180

Copyeditor: Meg Vezzu
Designer: Mike Palmquist
Series Editor: Lisa Melonçon
Series Associate Editor: Sherena Huntsman

The WAC Clearinghouse supports teachers of writing across the disciplines. Hosted by Colorado State University, it brings together scholarly journals and book series as well as resources for teachers who use writing in their courses. This book is available in digital formats for free download at wac.colostate.edu.

Founded in 1965, the University Press of Colorado is a nonprofit cooperative publishing enterprise supported, in part, by Adams State University, Colorado State University, Fort Lewis College, Metropolitan State University of Denver, University of Alaska Fairbanks, University of Colorado, University of Denver, University of Northern Colorado, University of Wyoming, Utah State University, and Western Colorado University. For more information, visit upcolorado.com.

Land Acknowledgment. The Colorado State University Land Acknowledgment can be found at https://landacknowledgment.colostate.edu.

Contents

Thematically Organized Contents

Types of Technical and Professional Communication

Technical and Professional Communication Practices

Features of Technical and Professional Communication

Acknowledgments

The concept for this collection began in 2019 when Han was chatting with Phil Nel—her colleague at Kansas State University, renowned children's literature scholar, and co-editor of *Keywords for Children's Literature*. That day, Phil and Han just happened to talk about the keywords essay genre and how the field of technical and professional communication might benefit from having a keywords collection. Thank you, Phil, for that catalyzing conversation!

After thinking about the scope and potential significance of the project, Han was very interested but immediately knew that she needed at least one skilled co-editor to make it happen. Thankfully, Jonathan said yes! It was an easy decision for Jonathan to make, for he often wished for a resource similar to *Keywords in Writing Studies* when teaching graduate courses in technical and professional communication. And from previous projects, he knew Han to be both an astute editor *and* a great collaborator.

Still, this collection would have remained just an idea if not for the support and work of our 44 fellow contributors. We thank you for your confidence in this project and in us as editors, your willingness to work with guidelines that challenged your typical writing processes, and the knowledge and insight you brought to your individual contributions. Working with so many contributors—and working through the COVID-19 pandemic—meant inevitable delays and complications. We thank our contributors for their patience and perseverance and for celebrating each milestone with us as the project slowly took shape.

We were delighted when Johndan Johnson-Eilola and Stuart Selber agreed to write the foreword for this volume, and their reflective work provides a shrewd opening frame for the collection. Likewise, we are indebted to Kristen Moore and Lauren Cagle for suggesting the inclusive citation audit, which both greatly informed the editorial process and served as the basis for their afterword written with Nicole Lowman. Their analysis and suggestions to authors led to more inclusive references in the final draft and a better acknowledgement of our debt to marginalized and underrepresented scholars.

We want to thank Lisa Melonçon, Series Editor of Foundations and Innovations in Technical and Professional Communication, for her enthusiasm for the project and for arranging informative and critically constructive peer reviews. We are also grateful for the work of the anonymous reviewers; their comments and suggestions improved every chapter.

Our publisher Mike Palmquist and his colleagues at the WAC Clearinghouse are superb professionals to work with. They made the production process of the book as smooth for us as possible, and we are so grateful for their labor.

Finally, we'd like to acknowledge our students, past and present, whose questions, passions, needs, and interests inspired our work on this volume. We hope it helps you as you navigate the terminological complexity of technical and professional communication.

Foreword: Technical and Professional Communication

Johndan Johnson-Eilola
Clarkson University

Stuart A. Selber
Pennsylvania State University

Every field of study is a moving target, challenging its members—researchers, practitioners, teachers, and students—to find ways of taking stock of knowledge claims and current practices in order to assess the state of play and imagine what the future might hold for their work. For technical and professional communication, a field aligned historically with the arts and sciences of discourse, keywords are an insightful location for development and analysis because we understand language to be constitutive of our being in the world. Language isn't the only thing that helps construct reality—consider our increasing interest in material matters—but articulating keywords helps us to take a useful snapshot in time of the field's ongoing development. Language is always open to interpretation and reinterpretation, but this quality can be seen as a feature with positive effects. As such, the keywords in this book are meant to invite discussion and debate, raise questions, and aid both reflection and invention, not pin down some absolute sense of central aspects of our professional domain.

This foreword itself functions as a keyword entry for technical and professional communication. We build on the entry for technical communication that Carolyn Rude wrote in 2015 for *Keywords in Writing Studies*. Rude traced the modern history of technical and professional communication, focusing mainly on practice and theory since the 1970s. She considered developments in U.S. culture that have moved the field in various new directions, new rhetorics for understanding what technical and professional communication is and does, growth and expansion of our research agendas, challenges of professional legitimization, expanded capabilities that new technologies have afforded to both technical and professional communicators and users, and more. Understandably for a short piece about an entire field, Rude pitched the discussion at the broadest possible level, tracing general contours and outlining some of the main accents of technical and professional communication as an evolving area of study. We encourage readers of this volume to read or revisit her keyword essay for another valuable starting point, and to think of it as something of a companion piece to our own, for we begin where she left off by considering the nature of change in our current period.

DOI: https://doi.org/10.37514/TPC-B.2023.1923.1.2

The final paragraph of Rude's essay raises the specter of a field collapsing on itself by expanding outward repeatedly in ways that confound coherence: "But because of considerable changes in practice and the term's divergent meanings, technical communication may become most interesting as an artifact of history" (2015, p. 168). Rude is referencing several consequential realities here—among them, technical and professional communication is a contested term with a wide variety of different theories, models, and emphases; many people produce technical and professional communication, not just technical and professional communicators, including workers in a growing array of affiliated fields; technical and professional communication content can be mutable and parasitic, living inside products, larger systems, and networks that users can configure and reconfigure, affecting the content; and the tools of technical and professional communication can automate certain types of production tasks. Although these realities aren't exactly new, they've become amplified and intensified in recent years, further complicating questions of boundaries, identities, and exclusions. We don't believe the field is in danger of becoming irrelevant or anachronistic, but we agree with Rude that the future depends on articulating and delivering on comprehensible research agendas. Our view is that those research agendas should attend to realities like those listed above. We consider them to be a key facet of the rhetorical contexts for technical and professional communication today.

Context isn't a separate keyword in this collection because discussions of context permeate rhetorical treatments in all of the chapters. But we want to focus on context because many of the realities of our current period are either a product of the growing complexity of sociotechnical structures and processes or a reflection of our growing awareness of complexity in consequential settings. We submit that seeing, understanding, and managing complexity in the contexts of our professional domain should be a defining objective for the field and a (not the) productive path forward for researchers, practitioners, teachers, and students. In the complex contexts we're imagining, it can prove difficult to pin down meaning, determine cause and effect, assign agency, and gauge how power is exercised and negotiated. In addition, such contexts are dynamic and fluid, changing over time, and can produce unintended consequences that become preconditions for future action. Complexity is a characteristic that is interwoven with the technical, the professional, and the communicative, affecting the full spectrum of our concerns.

If disambiguating complex contexts is a complicated and confounding task, the field must still find ways to make sense of them in order to work productively and responsibly in seeking solutions to domain problems. For this keyword entry, we want to offer one view of the field by characterizing the complex contexts that promise to be particularly salient to the future of technical and professional communication. To reiterate, on some level the realities in these contexts have been with us for some time now, but in recent years they've become more intensified and more integrated into the settings of everyday practice, growing complexity but also encouraging us to see complexity that was always there but not really recognized adequately. Although the aspects we discuss are intertwined in various ways, we separate them for analytic purposes, constructing a set of themes or topics for thinking about the complex contexts of technical and professional communication. In these contexts, the technical, the professional, and the communicative are bound up in interdisciplinarity, ambiguity, mutability, intertextuality, and interconnectivity. The result is a dynamic scene for the production, reception, and circulation of knowledge that's as challenging, interesting, and engaging as any problem-solving landscape.

In terms of interdisciplinarity, technical and professional communicators, by definition and by their increasingly expanding roles in a variety of settings, invariably function at the nexus of multiple fields, and so we must thrive on and even nurture any and all approaches that add value to integrative work. Technical and professional communicators have long been responsible for learning the fundamentals of other disciplines, but our endeavors now require a much richer and much more diverse set of practices and perspectives. Consider a single keyword in this volume: *Documentation* by David Farkas (we will reference this hallmark area throughout the rest of our foreword, but we could have selected any keyword entry as an example, for all of them have evolved in complex ways with time). We can trace the transformation of this term from an almost incidental offshoot of the primary activity, computer programming, into the wide variety and range of activities encompassed in the term today. Early UNIX documentation, both print and online, was written primarily by two UNIX programmers in the late 1960s at the direction of their manager. Today, documentation encompasses concerns from genre, social media, intercultural communication, ethics, social justice, editing, and plain language, to name just some of the other relevant keywords. This array of related areas, many disciplines in their own right, may seem daunting and too much to contemplate or reasonably consider applying. But navigating these areas, and bringing them to bear on specific problems in complex contexts, is a strength and major contribution of our field. Our field is a connective tissue that assembles aspects from many other disciplines into a coherent, working whole for users of technical and professional communication.

A second reality that characterizes the state of contemporary technical and professional communication is a growing appreciation for the complexity of the concept of ambiguity. Historically, eliminating ambiguity in written language has been discussed as one key strategy for achieving clarity, which often serves as a measure of excellence for our information products. In writing documentation, for example, we have encouraged technical and professional communicators to prefer the active voice ("Attach part A to part B") to passive voice ("Part A is attached to part B") because the active voice signals agency more directly and clearly: The human or nonhuman entity performing or experiencing the action is in the subject position of the sentence. In other words, in the passive version of the sentence, is part B already attached to part A, or does the user need to attach it? In many situations, preferring the active voice continues to be useful advice for helping to reduce ambiguity in technical and professional communication. However, the field has also come to understand that ambiguity is actually a property of language (and of technology) that cannot be eliminated or controlled completely on the production side of the equation. Consider a sentence we use with our students to make this point in a basic way: "I decided on the boat." Although the grammatical pattern of this sentence is a very simple one—subject-verb-direct object—there are at least two possible meanings one can draw from the very same words in the very same sequence. In order for a technical or professional communicator to encourage the appropriate interpretation in a specific situation, they will need to craft additional content that guides meaning making in the right direction. The point is that language always includes a surplus of meaning and that we should invent and emphasize strategies for contending with this surplus. We would add, also, that a benefit of the technology-as-text metaphor is that it can attune us to ambiguity in the design of technical systems (researchers in affiliated fields such as human-computer interaction account for such ambiguity in work in the area of "interpretive flexibility"). Because ambiguity is a property of language and technology and not just a problem to be solved, we're left wondering if the field might come to think of it as a positive resource to be leveraged in complex contexts. Exploring this topic could open a useful avenue for future research.

In addition to the paradigmatic nature of ambiguity, mutability brings syntagmatic complexity to technical and professional communication. The post-structuralist turn in communication in general has moved beyond the simple sender-receiver model towards a more textured and open-ended (albeit less stable) system in which meaning remains in constant flux. In one way of thinking, technical and professional communication would not be possible without the slippage of signification that allows a specific person to insert themselves, for example, into a sentence in a user manual or screenshot in online help to be translated into the working interface. Although research in areas such as contextual theory and design thinking has shown that the meaning of a sentence in a piece of documentation can shift around based on the complex, often messy contexts in which particular users work, we're beginning to see the mutability of content itself as a consequential affordance in technical and professional communication environments. Users of instructional videos on various streaming services can rate a video and search by user ratings, recontextualize a video by embedding its code in another website, add notes to help others interpret the instructions and navigate the video, filter notes to see only those added by other users, leave comments, add or suggest tags, post responses, see websites that link to a video, and flag inappropriate material. The ability to produce, use, and reinterpret metadata contributes to the construction of meaning as an active, collaborative process. This process is doubly collaborative in user-generated systems such as Wikimedia, where online help is under constant revision, positioning users as authors and editors of crowd-sourced documentation. If technical and professional communication seems more fraught with uncertainty than it was in the past, that recognition also tells us that meaning making was never really that simple in the first place.

In a closely related shift, the texts that technical and professional communicators work with today enact intertextuality, not just philosophically but functionally. While texts have always gestured to, cited, quoted, and echoed other texts, the introduction of hypertext links foreshadowed a fragmentation, circulation, and reassemblage of texts. We can also see precursors of this shift in technologies such as single-sourcing, which separated content from form, enabling technical and professional communicators to produce, for example, online help, reference sheets, and printed manuals from the same document database. Taking this practice to a new level, technical and professional communicators now build texts from pre-existing parts. Like programmers, they work with code and pattern libraries, templates, stock art, and other resources, transforming and combining them in novel ways. Although copy/paste has been with us for decades, building a document with substantial amounts of text (verbal, visual, and aural) from other sources is a relatively new practice. The production of a simple online tutorial might be built on top of a content management system, use a third-party cascading style sheet theme tweaked to conform to the technical or professional communicator's organizational style guide, be augmented with third-party plug-ins to offer features such as feedback forms, include edited and revised versions of text descriptions from the original product specification, and be illustrated with Creative Commons-licensed images of users at computers and icons licensed from The Noun Project. As this example illustrates, the distance from text to text today can easily collapse, no longer an intertextual pointer but now an adoption, an inclusion, an assemblage. Because traditional approaches to plagiarism fail to address this phenomenon in complex contexts, we're really just beginning to grapple with assessing and teaching intertextual practices.

Our final theme or topic, interconnectivity, reflects the reality that complex contexts have many interconnected parts, which interact to produce relationships, dynamics, and effects. We already mentioned that technical and professional communicators work with a variety of interconnected fields and texts, and that meaning making is interconnected with numerous aspects of interpretation, experience, and environment. In addition, however, the interconnections themselves are enmeshed in larger webs of affiliation; these larger webs link the technical, the professional, and the communicative in intricate and consequential ways. The field now understands that our processes and products are not isolated from organizational, social, and political conditions and challenges. In fact, technical and professional communication often finds itself at odds with its own interconnected complexity: balancing expediency with responsibility. The organizational style guide that directs a documentation specialist to tweak their cascading style sheet will also account for industry standards and genre conventions. A user constantly prompted to fix grammatical errors by their word processor may prioritize surface-level correctness over rhetorical effectiveness. Even "correctness" contains unseen assumptions about race, class, work, and more. Likewise, the existence of online help is loaded with powerful issues ranging from intercultural communication (Are non-English speakers relegated to using English-only online help or are localized versions available?) to agency (Are users empowered to work effectively or just quickly?) to pedagogy (Does the online help integrate or separate the why and the how?). Pulling at any strand within the complex weave of technical and professional communication tugs at both macro-level and micro-level concerns and realities.

The rest of the entries in this keyword volume continue to paint the complex picture of technical and professional communication as we know and understand it today. Some entries consider various aspects of interdisciplinarity, ambiguity, mutability, intertextuality, and interconnectivity, at times using alternative terms with a different set of connotations, while others employ additional terms to characterize the growing complexity of sociotechnical structures and processes or our growing awareness of complexity in consequential settings. In acknowledging and characterizing complexity rather than simply trying to solve it, we're advancing what we consider to be one useful stance for addressing future work in technical and professional communication. To repeat ourselves, we submit that seeing, understanding, and managing complexity should be a defining objective and productive path forward for researchers, practitioners, teachers, and students. This volume is a vital source of support and inspiration for this critical enterprise.

References

Rude, C. (2015). Technical communication. In P. Heilker & P. Vandenberg (Eds.), *Keywords in writing studies* (pp. 165–168). Utah State University Press. https://doi.org/10.7330/9780874219746.c033.

KEYWORDS IN TECHNICAL AND PROFESSIONAL COMMUNICATION

Introduction

Han Yu
Kansas State University

Jonathan Buehl
The Ohio State University

> *Even if any given terminology is a reflection of reality, by its very nature as a terminology it must be a selection of reality; and to this extent it must function also as a deflection of reality.*
>
> — Kenneth Burke, "Terministic Screens"

In 1966, the Society of Technical Writers and Publishers, Inc. (STWP) and the Carnegie Library of Pittsburgh published *An Annotated Bibliography on Technical Writing, Editing, Graphics and Publishing: 1950 to 1965*. This "aid to those with a general interest in technical writing" and "guide to those seeking specific information" (Philler et al., 1966, p. i) provides a remarkable snapshot of technical writing as a field in the middle of the 20th century. Its thousands of annotations, representing hundreds of scholarly and trade publications, cover pieces on issues unique to the period (e.g., "Soviet Scientific and Technical Propaganda") as well as issues we continue to grapple with today (e.g., "What is Technical Writing?").

To organize their 2,000 annotations, the editors used the "aid of a computer" to create a permuted title index—an indexing strategy that sorts works by key terms within their titles instead of purely alphabetically. The resulting list presents scannable clusters of related works (see Figure 1). We introduce the example of this bibliography and its information management strategies to highlight two points.

First, the key terms of the field we now call technical and professional communication (TPC) have a rich history that is worth both documenting and updating. The 1966 bibliography (covering 2,000 works published from 1950 to 1965) and a 1983 sequel (Carlson et al., 1983), which covers 2,700 works published from 1966 to 1980, provide synoptic views of the terms that mattered to the profession during this key thirty-year span. Genre types (e.g., manual, report, proposal) and key contexts (e.g., business, engineering) are among the most frequently indexed terms of both volumes. But even less frequent terms can tell us something about the development of the field and its concepts. For example, forms of the word *rhetoric* appear only five times in the 1966 bibliography but 45 times in the 1983 bibliography, which suggests the increasing importance of rhetoric as a framing concept for the field. Keywords related to gender are almost nonexistent in the 1966 bibliography—just one indexed work recommending technical communication as a good career path for women chemistry majors. By 1980, instances of terms indexing works on gender representation, equity, and discrimination (e.g.,

DOI: https://doi.org/10.37514/TPC-B.2023.1923.1.3

sexism and phrases including *women*), though not abundant, were at least present (e.g., the cluster of titles including *women* and *women's liberation* in Figure 2).

Furthermore, considering the terms that these index pages do not include is also instructive. For example, although we know from historical research that people of diverse backgrounds worked as technical communicators during this period (see Malone, this volume), that diversity is not reflected in the bibliographies' key terms. Only one work in the 1980 bibliography (and no works in the 1966 volume) was sorted by a keyword related to race and technical communication—a 1975 presentation titled "Language Engineering for Black Managers." Thus, like all attempts to provide a synoptic view of a field, these indexes function as what Kenneth Burke (1966) called "terministic screens." They simultaneously reflect some aspects of reality while deflecting others.

Although the rise of electronic bibliographic databases has made book-length bibliographies largely obsolete, other synoptic works can serve as similar terminological markers for the field of TPC. For example, the contents and alternate table of contents describing the works anthologized in Johndan Johnson-Eilola and Stuart A. Selber's (2004) *Central Works in Technical Communication* provide a snapshot of terms central to the academic discipline of technical communication as it flourished and evolved in the 1980s, 1990s, and early 2000s: *history*, *theory*, *ethics*, *power*, *pedagogy*, *collaboration*, *genre*, *gender*, *visual*, *usability*, etc. As we approach the twenty-year anniversary of that important anthology, it is time to revisit those central concepts and to consider other emergent terms.

Recent large-scale analyses of TPC publications have begun to do that work. For example, Ryan Boettger and Erin Friess (2020) conducted a content analysis of 672 articles published in technical communication journals between 1996 and 2017 to identify content and authorship patterns. Most relevant for our work, they coded each article with a "primary topic" content category. The fifteen core conceptual categories they identified—assessment, collaboration, communication strategies, comprehension, design, diversity, editing and style, genre, professionalization, knowledge and information management, pedagogy, research design, rhetoric, technology, and usability and user experience—overlap with many of the terms we identified as central concerns through our own content analysis of article abstracts and keywords.

More recently, Stephen Carradini's (2021) corpus analysis of 1,593 TPC abstracts examined word frequencies to identify shifts in technical communication research topics between 2000 and 2017. These include a shift in focus from print communication to digital communication, expanding boundaries of the technical communication field, and affirming its core identity. Key terms that emerged from Carradini's study included both well-established central concepts—such as *ethics* and *rhetoric*—as well as more recently emergent but nonetheless central TPC terms—such as *content management*, *social justice*, *user experience*, and *social media*. Unsurprisingly, our analysis identified similar terms, but the goal of our project is to move from identification of central terms (both old and new) to documentation of their multiple, nuanced, and sometimes contested uses.

PERMUTED TITLE INDEX

```
ING SCIENTIST AND ENGINEER--A REPORT ON THE FIRST AAAS- STWP JOINT PROGRAM.      THE ROLE OF THE WORK A-0850
                                                  ABBREVIATIONS.                                        A-0909
             ON SYMBOLS FOR UNITS OF MEASUREMENT (ABBREVIATIONS).                                       A-1043
                SYMBOLS FOR UNITS OF MEASUREMENT (ABBREVIATIONS).                                       A-1044
                            IRE STANDARDS ON ABBREVIATIONS.                                             A-0785
 GUIDES TO TECHNOLOGYS NEW TONGUE (ACRONYMS AND ABBREVIATIONS).                                         A-1413
IONS.                                             ABBREVIATIONS OF RUSSIAN SCIENTIFIC SERIAL PUBLICAT A-0760
                                                  ABBREVIATIONS DICTIONARY.                             B-1585
                               COMMON ELECTRONIC ABBREVIATIONS.                                         A-0652
                                         JOURNAL ABBREVIATIONS.                                         A-0704
L-TYPE PUBLICATIONS.           MILITARY STANDARD ABBREVIATIONS FOR USE ON DRAWINGS AND IN TECHNICA    B-1972
                                       TECHNICAL ABBREVIATIONS AND CONTRACTIONS IN ENGLISH.             A-0209
                                             THE ABC OF LANGUAGES AND LINGUISTICS.                      B-1747
                                             THE ABC OF STYLE.                                          B-1897
                                             THE ABECEDARIAN BOOK (VOCABULARY).                         B-1895
FECTIVE IN HELPING STUDENTS IMPROVE THEIR WRITING ABILITIES (EDUCATION). /S TEACHING APPROACH MORE EF A-0463
                HELP WANTED-- ENGINEERS, MUST BE ABLE TO WRITE.                                         A-0146
MATION OF I/ SUGGESTED FILING SYSTEM FOR KEEPING ABREAST OF TECHNICAL LITERATURE AND GENERAL INFOR    A-1236
                               TECHNICAL WRITING ABROAD.                                                A-0628
                           INFORMATION RETRIEVAL (ABSTRACTS).                                           A-0096
 TIME TO READ ALL HE NEEDS TO READ--A SOLUTION (ABSTRACTS).            FOR THE MAN WHO DOESNT HAVE   A-1261
FOR THE PREPARATION ANC PUBLICATION OF SYNOPSES (ABSTRACTS).                                   GUIDE   A-0811
                                  THE WRITING OF ABSTRACTS.                                             A-1409
OU WRITE TECHNICAL REPORTS POLISH YOUR TITLE AND ABSTRACT.                                       IF Y A-1368
                      WRITING BETTER TITLES AND ABSTRACTS.                                              B-1871
                             A SCRUTINY OF THE ABSTRACT.                                                A-0360
                                                  ABSTRACT OF THE TECHNICAL REPORT.                     A-1115
TION.                  THE ROLE OF SUMMARIES AND ABSTRACTS IN THE METABOLISM OF SCIENTIFIC INFORMA    A-0914
           CA PERIODICALS LIST CLIMBS (CHEMICAL ABSTRACTS).                                             A-1104
    TRAINING OF PATENT ABSTRACTORS FOR "CHEMICAL ABSTRACTS."                                            A-0958
          THE AUTOMATIC CREATION OF LITERATURE ABSTRACTS.                                               A-1268
               STYLE AND SPEED IN PUBLISHING ABSTRACTS.                                                 A-1335
                                           WATER ABSTRACTS.                                             A-1486
                    SOME READER REACTIONS TO ABSTRACT-BULLETIN STYLE.                                   A-1400
  CRITERIA FOR ACCEPTABLE ABSTRACTS--A SURVEY OF ABSTRACTERS INSTRUCTIONS.                              A-1307
                                         PHYSICS ABSTRACTING.                                           A-0433
                                      SCIENTIFIC ABSTRACTING AND INDEXING SERVICES.                     A-1028
                             TRAINING OF PATENT ABSTRACTORS FOR "CHEMICAL ABSTRACTS."                   A-0958
                        CRITERIA FOR ACCEPTABLE ABSTRACTS--A SURVEY OF ABSTRACTERS INSTRUCTIONS.        A-1307
                                       USAGE AND ABUSAGE--A GUIDE TO GOOD ENGLISH.                      B-1603
                                         USE AND ABUSE OF STATISTICS.                                   B-1695
                                                  ACCENTS IN COLD TYPE PREPARATION (PRINTING).          A-0051
(LETTER WRITING).                                 ACCENTUATE THE POSITIVE-- ELIMINATE THE NEGATIVE    A-0409
UCTIONS.                            CRITERIA FOR ACCEPTABLE ABSTRACTS--A SURVEY OF ABSTRACTERS INSTR    A-1307
                                        IS THERE ACCEPTED SCIENTIFIC JARGON.                            A-0135
                                            FAST ACCESS TO SYSTEM TECHNICAL INFORMATION (FASTI).        A-0921
                                  A PROGRAM FOR ACCREDITING TECHNICAL WRITERS.                          A-0761
                                          TODAY, ACCURACY DEMANDS THE FIRST PERSON.                     A-1119
                                TECHNICAL MANUAL ACCURACY AND ERROR LIABILITY.                          A-0725
         A DOMINANT NEED IN THE NATIONS SPACE-- ACCURATE TECHNICAL REPORTING (NASA).                    A-0822
                      PSYCHOLOGICAL ASPECTS OF ACHIEVING MEANING.                                       A-0885
                   TECHNICAL WRITING-- EASILY ACQUIRED SKILL.                                           A-0181
                                                  ACQUISITION OF SPECIAL MATERIALS.                     B-1999
                                     DATEC-- DATA ACQUISITION AND TECHNICAL EVALUATION CONCEPT.         A-0136
            GUIDES TO TECHNOLOGYS NEW TONGUE (ACRONYMS AND ABBREVIATIONS).                              A-1413
   HOW TO WRITE A REPCRT THAT WILL BE READ AND ACTED UPON.                                              A-1260
                                                  ACTIVITIES IN A PROPOSAL CONTROL CENTER.              A-0901
 NATIONAL SCIENCE FOUNDATION IN SCIENCE INFORMATION ACTIVITIES.                      THE ROLE OF THE  A-1465
       CURRENT TRENCS IN TECHNICAL INFORMATION ACTIVITIES IN THE DEPARTMENT OF DEFENSE.                 A-0298
      FUNDING TECHNICAL EDITING AND PUBLISHING ACTIVITIES.                                              B-1824
                                         SOCIETY ACTIVITIES AND TECHNICAL PUBLICATION.                  A-1288
 CASE HISTORY ON ADMINISTRATION COORDINATING THE ACTIVITY OF A TECHNICAL PUBLICATIONS GROUP.        A A-0306
IBRARIES.                                         ADAPTING TECHNICAL WRITING TO THE OBJECTIVES OF L   A-0332
                                    HIGH POLYMER ADJECTIVE SYNTHESIS.                                   A-0673
C AND TECHNICAL LIBRARIES--THEIR ORGANIZATION AND ADMINISTRATION.                           SCIENTIFI B-1994
                                                  ADMINISTRATION OF TECHNICAL EDITORS (TRAINING).       A-1113
                                                  ADMINISTRATION OF TECHNICAL INFORMATION GROUPS.       A-1401
                                                  ADMINISTRATION OF THE TECHNICAL PUBLICATIONS GROUP.   A-0981
HNICAL PUBLICATIONS GROUP.       A CASE HISTORY ON ADMINISTRATION COORDINATING THE ACTIVITY OF A TEC   A-0306
RMING REPORT WRITING FRCM CHORE TO BENEFIT (I) (ADMINISTRATIVE PROGRESS REPORTS).          TRANSFO   A-1493
                                                  ADMINISTRATIVE COMMUNICATION.                         A-0987
                                                  ADMINISTRATIVE COMMUNICATION.                         B-1790
         PRESENTATION OF TECHNICAL INFORMATION-- ADMINISTRATIVE ASPECT.                                 A-0097
            WRITTEN COMMUNICATIONS FOR BUSINESS ADMINISTRATORS.                                         B-1773
          EDITING REPORTS FOR THE SEMITECHNICAL ADMINISTRATOR.                                          A-0069
                                            MUCH ADO ABOUT HYPHENS.                                     A-0324
                           MICROFICHE STANDARDS ADOPTED.                                                A-0177
ACTICING PHYSICIAN IN STEP WITH MODERN SCIENTIFIC ADVANCES. /MEDICAL WRITERS PART IN KEEPING THE PR   A-1231
ACTICING PHYSICIAN IN STEP WITH MODERN SCIENTIFIC ADVANCES. /MEDICAL WRITERS PART IN KEEPING THE PR   A-1352
OTE BUSINESS AND INDUSTRIAL USES OF TECHNOLOGICAL ADVANCES.                  HOW PUBLICATIONS PROM   A-0630
OF INFORMATION.                        MICRO-VUE-- ADVANCED DEVELOPMENT IN THE PRESENTATION AND USES   A-1388
          MANUAL STANDARDIZATION PROGRAM REACHES ADVANCED PLANNING STAGE.                               A-0426
                                             THE ADVERB "SURE" (USAGE).                                 A-0292
TIONS OF THE TECHNICAL EDITOR TO THE READER AND ADVERTISER.                                   OBLIGA  A-1046
                                   PRODUCTION IN ADVERTISING.                                           B-1889
                       BRING BACK PLAIN ENGLISH (IN ADVERTISING).                                       A-0114
   TECHNICAL WRITING TO SELL PUBLIC RELATIONS AND ADVERTISING.                                          A-1034
                    THE TECHNICAL WRITERS ROLE IN ADVERTISING AND PUBLIC RELATIONS.                     A-0516
                                  BACKGROUNDS IN ADVERTISING AND SALES.                                 A-0637
                                                  ADVERTISING IN THE PRINTED MEDIA.                     B-1800
          PRODUCT ENGINEERING ARTICLES CAN BE ADVERTISING DIAMONDS.                                     A-0500
                                                  ADVERTISING LAYOUT AND TYPOGRAPHY.                    B-1843
                               THE TECHNIQUE OF ADVERTISING PRODUCTION.                                 B-1804
MON TERMINOLOGIES FOR PRESS, PRINT, BROADCAST, FILM ADVERTISING, AND COMMUNICATIONS RESEARCH. / OF COM B-1662
                                      INDUSTRIAL ADVERTISING.                                           A-0144
OPYWRITER SHOULD KNOW ABOUT CHANGES IN INDUSTRIAL ADVERTISING.           WHAT TODAYS INDUSTRIAL C    A-0718
/AND BOOKS IN PRINTING, PAPER, PUBLISHING, PRINTED ADVERTISING, AND THEIR CLOSELY RELATED INDUSTRIE/ B-1840
```

Figure 1. The first page of the permuted title index in An Annotated Bibliography on Technical Writing, Editing, Graphics and Publishing: 1950 to 1966. *Frequently repeated terms on this page include forms of the words* abbreviations, abstracts, *and* administration.

JOB~ THE ART OF CUT AND PASTE OR WHAT	WE REALLY DON'T WANT TO KNOW ABOUT OUR	A-4383
ANS" AND GOOD "GRAMMARIANS": WHAT CAN	WE SAY ABOUT ALL THIS HE/SHE, (S)HE BU..	A-4368
~ CAN	WE SELL OUR SERVICES (PUBLICATIONS)	A-3653
LARIZATION - AT LEAST WE'LL KNOW WHERE	WE STAND ON USAGE~ STOP COMPROMISE NO..	A-2604
TING AND THE TECHNICAL AUTHOR: WHERE	WE STAND TODAY ~ AUTOMATED EDI	A-2136
MULTIMILLION-DOLLOR QUESTION: HOW SHALL	WE TEACH HIM? OR, HOW DOES HE LEARN..	A-4029
~ WHAT SHALL	WE TELL THEM?	A-4530
~ WHAT KIND OF SENTENCE SHOULD	WE WRITE?	A-3879
ECHNICAL WRITERS AND EDITORS ABOUT HOW	WE WRITE ~ WHAT DEAFNESS TELLS T	A-4200
CTION) ~ PLATE	WEAR FROM ABRASION (PRINTING AND REPRODU	A-2947
SI) ~ WILL AMERICAN GIRLS	WEAR SIZE 90 BIKINIS IN 1975? (METRIC,	A-2682
CTION) ~ WORLDS LONGEST	WEB OFFSET PRESS (PRINTING AND REPRODU	A-3367
~ HOW TO SWITCH A NEWSPAPER TO	WEB OFFSET (PRINTING) (I)	A-3401
~	WEB VERSUS SHEETFED PRINTING (JOURNALS)	A-2587
ERS ~ EDITORS	WED TO TERMINALS - NOW RESENT TYPEWRIT	A-2538
ED COORDINATE INDEX ON MAGNETIC TAPE~	WEIGHTED TERM SEARCH - A COMPUTER PR..	A-3032
COSTS ~ BASIC	WEIGHTS - THE GUIDE TO ESTIMATING PAPER	A-3453
OURSE ~ DEVELOPING A	WELL-ROUNDED TECHNICAL COMMUNICATION C	A-4447
~ HOW GENERAL ELECTRIC HALFTONE COSTS	WERE CUT	A-2985
ING WITHIN AEROSPACE INDUSTRIES OF THE	WEST~ OUTLOOK FOR TECHNICAL WRITING R..	A-3587
RGANIZATIONS ~	WESTERN ELECTRIC DATA DESIGN - SUPPORT O	A-4421
~ DATA MANAGEMENT AT	WESTINGHOUSE	A-3507
RINTING AND REPRODUCTION)~ ALCOHOL AND	WETTING AGENTS IN DAMPENING SYSTEMS (P	A-2899
~ SO YOU'RE A SUPERVISOR - NOW	WHAT?	A-2041
LARGE COMPUTERS: A SHOPPER'S GUIDE~	WHAT'S AVAILABLE FOR FORMATTING TEXT ON	A-4427
~	WHAT'S IN A WORD	A-2247
SITION?" ~	WHAT'S SO SACRED ABOUT JUSTIFIED COMPO	A-2182
ICAL EDITOR AS TEACHER: HOW TO EXPLAIN	WHAT'S WRONG AND WHY ~ THE TECHN	A-4322
SIDER AN INDUSTRIAL MOTION PICTURE)~	WHENEVER YOU'RE READY, C.B. (WHEN TO CON	A-2159
~ GIACOMO VENTRESCA, MOONSHINE	WHISKEY, AND THE PASSIVE VOICE	A-2701
~ COLORFUL SLIDES FROM BLACK AND	WHITE ART - AND MORE	A-2160
N PRODUCTION OF JOURNALS - A BLACK AND	WHITE CASE ~ STANDARDIZATION I	A-2735
YOUNG PROFESSIONAL ~	WHITHER THOU GOEST: MANAGEMENT AND THE	A-2128
P ~ MAKING	WHOOPIE: A PROFESSIONAL WRITER LEARNS W	A-4239
NOMIC NECESSITY OR HOW TO SURVIVE AS A	WILD DUCK OF THE BUSINESS WORLD~ CREA..	A-4008
E AMONG THE NOMADS OF THE PUBLICATIONS	WILDERNESS (EMPLOYMENT) ~ LIF	A-3634
AL AND RELATED TECHNICAL LITERATURE OF	WILDLIFE CONSERVATION ~ CHEMIC	A-3006
TECHNICAL COMMUNICATION~ FOSTERING THE	WILLING SUSPENSION OF DISBELIEF: SIMU..	A-4299
~ HOW THE TECHNICAL WRITER CAN	WIN A SMALL PROPOSAL	A-3645
~ CHARTS AND GRAPHS QUICK AS A	WINK	A-2989
~ BIOS CAN MAKE YOUR PROPOSAL A	WINNER	A-2235
COORDINATOR/EDITOR AND THE STRAWMAN: A	WINNING PROPOSAL COMBINATION ~ THE	A-4070
~ DESIGNING	WINNING PROPOSALS	A-4379
(PUBLIC SPEAKING) ~	WINSTON CHURCHILL - A STUDY IN ORATORY	A-2361
RES (INFORMATION RETRIEVAL)~ USE OF THE	WISWESSER LINE NOTATION FOR DETERMIN..	A-3277
CONNECTIVITY MATRIX DERIVED FROM THE	WISWESSER NOTATION (INFORMATION STORAG..	A-3039
STORAGE AND RETRIEVAL)~ CONVERSION OF	WISWESSER NOTATION TO A CONNECTIVITY..	A-3026
AND TRAINING) ~ I AM NO	WOLF, AND YOU ARE NOT RABBITS (EDUCATION	A-2566
~ THE	WOMAN MANAGER: AN INSIDE PICTURE	A-4529
~ WORKING	WOMEN AND EQUAL EMPLOYMENT OPPORTUNITY	A-4074
RELATED JOURNALS - IMPLICATIONS FOR	WOMEN~ EDITORIAL PRACTICES OF PSYCHIA..	A-2550
~ THE STATUS OF	WOMEN IN COMMUNICATIONS	A-2202
THE SHIFT TO TECHNICAL COMMUNICATION BY	WOMEN IN MANAGEMENT ~	A-4509
AND CAMARADERIE ~ TOWARD	WOMEN'S ADVANCEMENT: POLITICS, TACTICS,	A-4516
E SINGLE OFFICE - OR, GRUBBLY PUBS MEETS	WOMEN'S LIB ~ SEX AND TH	A-2200
ICATION ~ THE IMPACT OF	WOMEN'S LIBERATION ON TECHNICAL COMMUN	A-4137
~ HOW TO WRITE REPORTS THAT	WON'T BE IGNORED	A-2808
~ A STYLE GUIDE FOR THE FUTURE: YOU	WON'T LIKE IT	A-4045
~ THE	WONDERFUL WORLD OF TECHNICAL ART	A-4092
~ WHAT'S IN A	WORD	A-2247
~	WORD BUILDING	B-3712
~	WORD COUNT - A BETTER YARDSTICK (COSTS)	A-2416
ND TECHNOLOGY~ IN THE BEGINNING WAS THE	WORD - INFORMATION NEEDS AND USES IN..	A-3955
~ WHAT IS IN A	WORD (LANGUAGE)	B-3710
~ JOHN DEERE MINIMUM -	WORD OPERATOR'S MANUALS	A-4487
~ THE PROS AND CONS OF CENTRALIZED	WORD PROCESSING	A-2135
~ WHO NEEDS A SMART TYPEWRITER?	WORD PROCESSING)	A-2526
~ TEXT EDITING	WORD PROCESSING)	A-2738
~ CHOOSING STORAGE MEDIA FOR	WORD PROCESSING	A-4415
APPLICATION OF TECHNOLOGICAL ADVANCES IN	WORD PROCESSING ~	A-4534
~ THE TRAUMA OF CONVERSION TO	WORD PROCESSING	A-4559
N SYSTEMS	WORD PROCESSING AND AUTOMATED PUBLICATIO	B-4693
L COMMUNICATION TOOL ~	WORD PROCESSING AS AN EFFECTIVE TECHNICA	A-4414
~ PRODUCTION STANDARDS IN A	WORD PROCESSING CENTER	A-2798
OLLED EXPERIMENT TO ASSESS CENTRALIZED	WORD PROCESSING ~ CONTR	A-2412
~ CRITERIA FOR SELECTING	WORD PROCESSING EQUIPMENT	A-4423
ATIONS DEPARTMENT ~ EVALUATING	WORD PROCESSING FOR A TECHNICAL PUBLIC	A-4150
STAFF ~ TEXT	WORD) PROCESSING FOR THE PROFESSIONAL	A-2790
AL STRUCTURE FOR THE MANAGEMENT OF THE	WORD PROCESSING FUNCTION~ AN ORGANIZA..	A-4528
ONS ENVIRONMENT ~	WORD PROCESSING IN A TECHNICAL PUBLICATI	A-4242
FACILITY ~	WORD PROCESSING IN A TECHNICAL RESEARCH	A-4520
ORATE PROFIT ~	WORD PROCESSING - NEW APPROACH TO CORP	A-2646
CATIONS ~ IMPACT OF	WORD PROCESSING ON ENGINEERING COMMUNI	A-2464
EDUCATION, SELECTION, AND TRAINING OF	WORD PROCESSING PERSONNEL ~	A-2511

Figure 2. An index page from An Annotated Bibliography on Technical Writing, Editing, Graphics and Publishing: 1965 to 1980 *(Carlson et al., 1983). Frequently repeated terms on this page include* Wiswesser line notation, women, *and* word processing. *Incongruously, the same index page listing works on serious issues faced by women in technical communication also includes an overtly sexist title about the potential adoption of the metric system in the United States: "Will American Girls Wear Size 90 Bikinis in 1975?"*

The second point we want to make by introducing examples from earlier bibliographies is to highlight the practical and methodological problems of organizing and accessing the keywords of a field like TPC. As Figure 1 demonstrates, the permuted title index is both helpful and problematic as an information management strategy. Through repetition, one gets a sense of some of the important terms (e.g., *abstracts*, *advertising*). However, as the example shows, such indexes can also be muddled by repeated terms that are not all that key. For example, *ABC* is treated as a keyword when it is a mere stylistic flourish deployed in more than one title. In other cases, the same term might be used in multiple ways; for example, the permutations of *program* lists computer programs, organizational initiatives, and academic programs interchangeably in the same section of the index. In still other cases, important terms might be represented but not necessarily be "key" terms with broad appeal. For example, works on *Wiswesser line notation* (one of the key terms in Figure 2) would have only been relevant for technical writers working with technical chemistry texts. Finally, the permuted title index (like any other term-based search strategy) is an insider's tool that is most useful when an information seeker knows which terms to search, whereas newcomers to a discipline often need guidance for understanding both the concepts and the complexities represented by key terms. Part of entering a field involves learning which terms matter and how those terms are used.

This volume, *Keywords in Technical and Professional Communication*, attempts to address both the need to document the evolving terminological complexity of TPC and the needs of newcomers unfamiliar with its key terms, though we use a different genre than the bibliography or anthology to do so—the keyword essay collection. The remainder of this introductory chapter explains the history and purpose of this genre, describes why we felt the 21st-century discipline of TPC needed a keyword essay collection, and documents how we, as editors, selected keywords and contributing authors for this volume.

What Is a Keyword Essay Collection? Why Does TPC Need One?

The keyword essay collection has emerged as a unique academic genre composed of short essays that discuss the multiple and sometimes conflicting uses of words central to a discipline. Examples include *Keywords for American Cultural Studies* (Burgett & Hendler, 2014), *Keywords in Writing Studies* (Heilker & Vandenberg, 2015), and *Keywords for Latina/o Studies* (Vargas et al., 2017). These and other keyword collections owe their origin to *Keywords: A Vocabulary of Culture and Society*, the first keyword collection, which was first published in 1976 by British cultural studies scholar Raymond Williams. In the introduction to that book, Williams recounts the personal motivation behind the book, which is worth quoting at length:

> In 1945, after the ending of the wars with Germany and Japan, I was released from the Army to return to Cambridge. University term had already begun, and many relationships and groups had been formed. It was in any case strange to travel from an artillery regiment on the Kiel Canal to a Cambridge college. I had been away only four and a half years, but in the movements of war had lost touch with all my university friends. Then, after many strange days, I met a man I had worked with in the first year of the war, when the formations of the 1930s, though under pressure, were still active. He too had just come out of the Army. We talked eagerly, but not about the past. We were too much preoccupied with this new and strange world around us. Then we both said, in effect simultaneously: 'the fact is, they just don't speak the same language.' (Williams, 1976/1983, p. 11)

After a mere four and a half years, people were no longer speaking the same language, which frustrated and intrigued Williams. The word *culture*, for example, took on shifting and nebulous meanings: Previously, it was used in teashops and similar places to denote social superiority or in artistic circles to refer to writing poems, working in theaters, and other expressive activities. Four years later, it was used to describe both the formation of values in the study of literature and a particular way of life, like "American culture" (Williams, 1976/1983). In an effort to help himself (and others) grapple with these shifting vocabularies, Williams started to collect what he later called keywords and to write short essays to document the genealogies of their usage. With each word, Williams covers centuries of evolving, divergent, and sometimes contested meanings, replete with specific examples and contexts. His approach has been replicated by numerous other authors and editors, and indeed a WorldCat search for "keywords in" and "keywords for" returns dozens of titles published since 2000 alone. However, despite its disciplinary history and terminological traditions, our field of TPC does not have its own keywords collection attending to its unique disciplinary context.

Williams' *Keywords* and many of its contemporary successors are situated in the disciplinary fields of literary and cultural studies. In fields more closely related to TPC, two collections have been published by the same editors: *Keywords in Composition Studies* (Heilker & Vandenberg, 1996) and *Keywords in Writing Studies* (Heilker & Vandenberg, 2015). These are two excellent collections delineating issues related to writing and composing, but they do not reflect the precise interests of TPC—a field with links to academia and industry, to the sciences as well as the humanities.

Keywords in Composition Studies has its "focus on the academic text, the writing student, and the classroom" (Heilker & Vandenberg, 2015, p. xii), a focus reflected by keyword choices such as *academic discourse*, *basic writing*, and *freshman English*. This focus diverges from TPC's interests in the workplace and in non-academic communication contexts. *Keywords in Writing Studies* is Paul Heilker and Peter Vandenberg's response to the changing nature of composition studies. As they acknowledged, the 1980s' social turn and the late 1990s' public turn forcefully demonstrated that "writing in universities is only a small slice of writing that goes on elsewhere in the world" (Bazerman, qtd. in Heilker & Vandenberg, 2015, p. xii). During these changes, classroom writings interacted with social practices, linguistic and cultural differences became central concerns, and methodological and theoretical approaches were increasingly plural (Heilker & Vandenberg, 2015). *Keywords in Writing Studies* captures these changes and shifts, as reflected in keyword choices such as *citizen*, *identity*, and *multilingual/ism*. Most interesting to us among the new keyword essays is "Technical Communication," which was authored by Carolyn Rude (2015).

In her essay, Rude skillfully introduced technical communication's history, major genres, key organizations, practices, curricular programs, and research domains. Each of these is a shifting landscape of convergences and contentions, but in the limited space of a single short essay, complexities had to be excluded, flattened, or rapidly glossed over. As we reviewed this informative essay, we could not help but think that the field of TPC needed not just one essay but its own keyword collection.

Now, it is not our intention here to define or redefine "TPC" vis-à-vis "writing studies." We merely hope to demonstrate that the field of TPC has considerable depth, width, complexities, and nebulousness on/in its own terms. Given decades of development and processes of professionalization, it has accumulated its own share of thorny keywords that are well worth documenting and unpacking.

Like other keyword essays, the essays in this volume are studies of words that "are ritually invoked or provocatively redefined," words that "anchor course titles, cue manuscript reviewers, situate *curricula vitae*, ping research-alert notifications, and tag conference panels" (Dryer, 2019, p. 214). In pithy essays, the origins of these words are examined, examples of usages are offered, and multiple and conflicting meanings are acknowledged.

It is also important to note that essays in this collection are not comparable to dictionary entries. Dictionaries attempt to close down, to fix the meanings of words and offer agreed-upon, clear, and consistent definitions. Keyword essays attempt to open up the meanings of words, to emphasize that meanings are always in flux, and to celebrate the different (but also overlapping) meanings of words as they are used in varied social, cultural, and disciplinary circles. As Heilker and Vandenberg (1996) put it, clear and consistent definitions are often "secured not by a sacred illumination, but through a process of forgetting, neglecting, denying" (p. 2). The alternative, and more productive and promising, approach, is to "listen openly, generously, and carefully" to a word's "many, layered voices, echoes, and overtones, especially the dissonant ones" (Heilker & Vandenberg, 2015, p. xvi). We *are* aware that the very attempt to portray these varied voices runs the risks of valuing some voices and devaluing others, but, as in any reflective attempt, we must start somewhere.

Like previous keyword books, our collection features an eclectic, carefully selected list of terms. In previous books, writers and editors often relied on their tacit knowledge of a discipline to arrive at their lists. Raymond Williams (1976/1983), for example, selected words that, as he put it, virtually forced themselves on his attention because the problems of their meaning were bound up with the problems they were used to discuss (p. 15). Bruce Burgett and Glenn Hendler (2014), editors of *Keywords for American Cultural Studies*, selected words whose meanings and debates are central to shaping the study of culture and society. Heilker and Vandenberg (2015) used two overarching impressions to select words for *Keywords in Writing Studies*: words that are part of their disciplinary parlance and words that are highly contested.

For our collection, we employed a more formalized, data-driven process to arrive at our list—a process that seemed to us more in keeping with the rigorous but eclectic methods and methodologies of TPC. We began by conducting a corpus analysis to identify words that are frequently used in field publications, and we followed that analysis with a survey that crowdsourced input from field professionals (details below in "Our Methods"). To borrow from Dylan Dryer (2019), doing so allows us to combine the quantitative account of a discipline, something that is "broad and flat," with the impressionistic account, something that is "deep and narrow" (p. 215).

However, we do not pretend that our process is disinterested or impartial, as we elaborate in "Our Methods." Indeed, as we made sense of our data, we did not proceed purely statistically but also interpretively, drawing upon our knowledge developed as members of the discipline. Some terms were relabeled, broadened, or narrowed. For example, the term *markup language* was replaced with *structure*—a term that can cover issues related to markup languages as well as other issues related to the material presentation of information across technologies. As we made decisions, we also considered both the field's history and emerging trends because we envision a future-oriented collection, a collection that not only captures words that are and have been frequently used in the discipline but also words that will be or should be.

Our ultimate decisions on what to include and exclude are fraught with problems (how can they not be given our necessarily localized and partial positionalities), and our readers may well disagree with those decisions. Indeed, the results of our survey already hinted at diverse opinions, with some participants believing, for example, that terms such as *feminism* are not unique/central to technical communication, while others applauded its inclusion and advocated for more counter-hegemonic terms.

Precisely because of these disagreements and partialities, we hope that this collection will be followed up with later efforts to document new keywords. Raymond Williams (1976/1983) intentionally included blank pages in his keyword book to signify that "the inquiry remains open" and that he "will welcome all amendments, corrections and additions" (p. 26). While we do not have blank pages in this collection, we share the same sentiment. Echoing Burgett and Hendler, editors of *Keywords for American Cultural Studies*, we invite readers "to revise, reject, and respond to the essays that do—and do not—appear in this publication, to create new clusters of meaning among them, and to develop deeper and richer discussions of what a given term does and can mean when used in specific local and global contexts" (2014, p. 5).

Our Methods

We used a two-phase process to arrive at the keywords included in this collection. In Phase 1, we conducted a corpus analysis of peer-reviewed journals in the field. This phase included two sub-phases: In Phase 1.1, article abstracts were analyzed using word clouds, which was itself a multi-step process; in Phase 1.2, journal- and author-provided keywords were analyzed using Microsoft Excel. In Phase 2, we surveyed TPC faculty, graduate students, and practitioners for their perspectives to help support and validate our Phase 1 findings. These phases and steps are summarized in Figure 3 and detailed in the following sections.

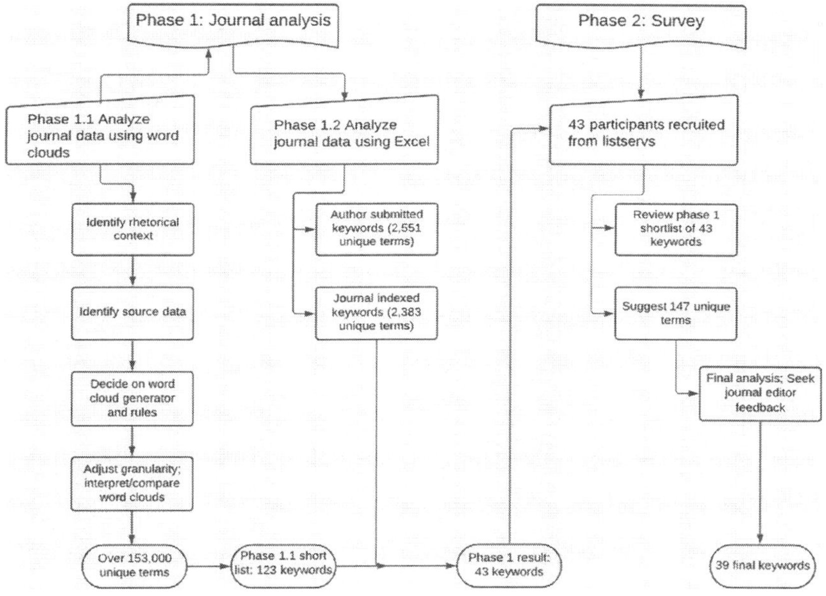

Figure 3. An overview of our methods for selecting the keywords for this collection.

Phase 1.1. Use Word Clouds to Extract Keywords from Article Abstracts

Word clouds, also known as text clouds or tag clouds, are visual representations of textual data. Started as web-based visualizations of keywords (or "tags") that categorize user-contributed online content, word clouds are now used as a general tool to mine source texts (Steinbock et al., 2007). Using visual attributes such as colors and font sizes, word clouds highlight terms that are most frequently used in a source text, giving readers an immediate summary of the text's topics and an impression of its key concerns. Notably, Richard Selfe and Cynthia Selfe (2013) advocated using word clouds as a heuristic to define technical communication's boundaries, artifacts, and identities. Given word clouds' ability to offer high-level, word/phrase-based summaries of data, they are well suited for our purpose to identify TPC keywords. Drawing upon Selfe and Selfe's heuristic, we used a five-step process to create word clouds.

Step 1. Identify rhetorical context

At this stage, we considered the purpose, audience, and content of our word clouds. These considerations guided the subsequent steps, allowing us to focus the word clouds for their intended use context. Our purpose in creating word clouds is to find enduring and emerging keywords in TPC research, education, and practice. The keywords that emerge from the word clouds will be examined in short essays, which are intended for all TPC scholars, educators, students, and practitioners. These essays may be especially valuable to newcomers to TPC by orienting them to the focus of the field and by distilling complex key concepts.

Step 2. Identify source data appropriate for the rhetorical context

Journal publications are an important indicator of the changing focuses and concerns of a disciplinary field and, as such, represent promising source data. Given our rhetorical contexts, we included in our corpus journals that have a considerable publication history and influence and that, collectively, emphasize all aspects of the field—from original research to pedagogical studies to industry practices. With these considerations, five journals were included; by alphabetic order, they are *IEEE Transactions on Professional Communication, Journal of Business and Technical Communication, Journal of Technical Writing and Communication, Technical Communication,* and *Technical Communication Quarterly.* Our journal choices coincided with those of recent field-mapping publications (Boettger & Friess, 2020; Carradini, 2021), which we learned after finishing our selection process.

Given our purpose, we needed to trace a historical trajectory of the field but also focus on the more recent and emerging developments. Given these considerations, we decided to include ten years of publications (2009–2019) from the five journals. The full texts of all these publications, however, would be a data set too unwieldy for word cloud generators and subsequent analyses. Drawing upon Selfe and Selfe (2013), we narrowed the scope of our analysis to the abstracts of the published articles—summaries meant to capture the essence of the articles. With this decision, we then exported all available abstracts from the identified journals and publication range from the Scopus database.[1]

1. At the time of our study, *Technical Communication* did not have its 2019 publications available in our subscribed databases, so for this journal, data from 2009–2018 were included.

We recognize that, besides journals, other possible corpora exist, notably, technical communication job advertisements, which can illuminate the core competencies required of industry technical communication practitioners (see, e.g., Brumberger & Lauer, 2015). We decided not to use these source texts for several reasons. First, our project does not have the narrower purpose of preparing students for the workplace but the broader purpose of orienting newcomers to the history, disciplinary concerns, and identity of TPC. Job advertisements are less capable of reaching this broader purpose. Second, while practitioners are part of our intended audience, we do not envision them using this work to assist their day-to-day, on-the-job practices. Rather, we envision them encountering this work in an academic context as students, precisely the spaces where academic approaches can illuminate the changes, tensions, and issues underlying pragmatic industry practices. Finally, there already exists a practitioner-oriented glossary book (*The Language of Technical Communication*, edited by Ray Gallon, 2016), which offers extended definitions of specific terms such as *eBook*, *HTML5*, and *XML processors*, in addition to many of the terms our authors cover, such as *accessibility*, *user experience*, and *project management* (see Figure 4).

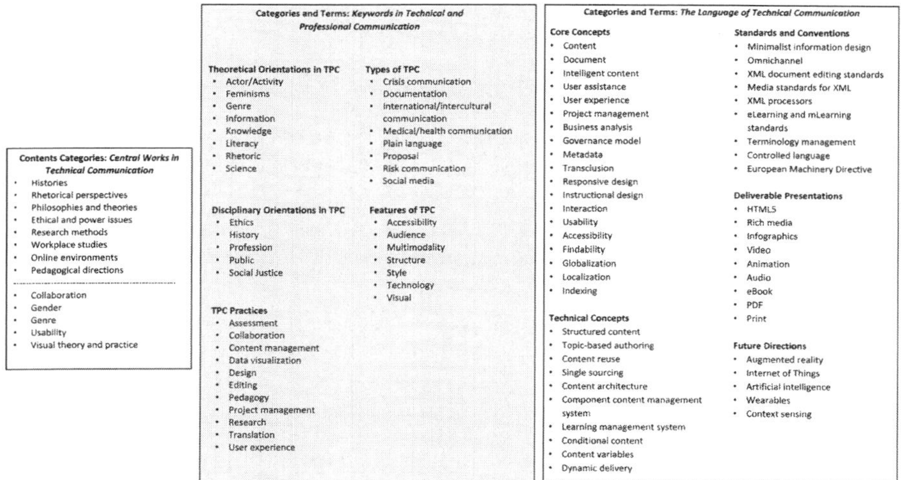

Figure 4. Keywords as connections between theory and practice, academy and industry. The left box lists categories used by Johnson-Eilola and Selber to organize the academic essays in Central Works in Technical Communication, *with the categories from the "Alternative Contents" listed below the dashed line. The right box includes the terms and section categories in Gallon's* The Language of Technical Communication, *an elaborated glossary for TPC practitioners. The middle box lists our keywords as they are sorted in the "Thematically Organized Contents." We envision these terms as bridging the concerns of TPC as an academic discipline and as a field of practice.*

It is worth noting that Gallon's practitioner-focused terms cross-articulate with the keywords of our collection in different ways. For example, Gallon's authors engage concepts our authors cover (such as *multimodality*) through specific instances or sub-terms (such as *animation*, *audio*, and *video*). At the same time, our collection provides nuanced descriptions of concepts that Gallon's collection takes for granted (for example, *audience* and *genre*). These different granularities and premises reflect the different purposes of the two projects. They also demonstrate the continuity of theory and practice as well as the potential for our nuanced investigations of keywords to enrich how practitioners think about practice. Indeed, as Figure 4 attempts to demonstrate, the keyword essays of our collection are situated to inform conversations within and between the academic and pragmatic traditions of TPC.

Step 3. Decide on a word cloud generator and identify rules for structuring terms

Given our purpose, we did not need to turn our source texts into aesthetically pleasing visual representations, which is a focus of some word cloud generators (for example, WordArt). Rather, we needed a set of terms where the frequently used ones are highlighted visually and each term's numerical frequency is given. Also, we wanted to extract not only singular words but also phrases that may function as keywords. With these needs in mind, we compared multiple word cloud generators before choosing ToCloud. ToCloud is a free online generator available at tocloud.com. It outputs terms in a simple list format and can use font sizes and/or colors to visually denote frequency as well as specify numerical frequency in parentheses following the terms. Users can choose to list the terms alphabetically or by frequency; they can extract not only singular words but also phrases. ToCloud automatically filters out words such as *a*, *the*, and *that*; a user can specify additional words to be filtered out. Figure 5 shows a sample ToCloud mapping result.

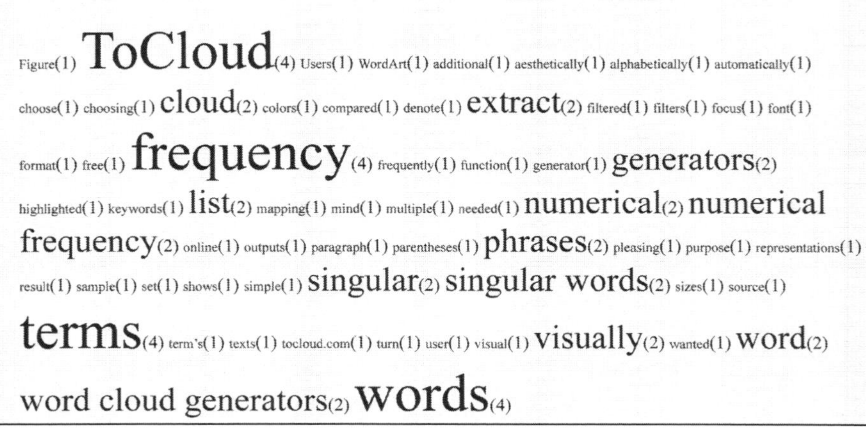

Figure 5. An example of a ToCloud word cloud.

Following Selfe and Selfe (2013), we manually manipulated our source text before submitting it to ToCloud. We removed proper nouns, such as *Sage Publications,* and boilerplate terms, such as the standardized abstract headings *purpose* and *literature review,* that may be mistaken for high-frequency keywords.

Steps 4 and 5. Adjust granularity and interpret/compare word clouds

Even though we narrowed our source text to abstracts, the dataset is still quite large, containing more than 153,000 words, many of which appear less than five times. To make the resultant cloud more manageable, we applied ToCloud's threshold functionality and mapped only those terms with a minimum of 20 appearances. With this adjustment, the resulting cloud contained a total of 948 unique terms.

As Selfe and Selfe (2013) reminded us, word clouding cannot be treated as a wholly computerized process; active reading and authorial interpretation is needed to make sense of the results. Given our purpose, in our interpretation, we tried to identify terms that are frequently used and thus quantitatively significant. Equally importantly, we tried to identify terms that, based on our knowledge of the field, are qualitatively significant and the focal points of disciplinary efforts and debates. In addition, we needed to identify terms that share a common root (e.g., *user*, *usability*, and *usability testing*) to ensure we do not miss key interests shared by these otherwise different terms.

To facilitate our reading and interpretation, we generated two versions of the cloud. One lists the terms by frequency so we can more easily compare use frequency; the other lists the terms alphabetically so we can more easily identify related terms that share a common root. Comparisons between the two clouds allowed us to balance our needs. Notably, we found that the most frequently used terms do not necessarily serve our purpose. In our word clouds, the most frequently used term is *communication*, which appeared a total of 1,361 times. Although communication *is* essential to our discipline, this term is too broad for useful description—indeed, this entire project is conceived to identify keywords that unpack technical and professional communication. Similarly frequent and broad terms include *writing* (430 times), *data* (286 times), and *English* (148 times). On the other hand, terms that appear less often can represent emerging concepts important to the field's development, for example, *social justice* (24 times). Through such constant comparisons, we arrived at a short list of 123 terms, which we then cross-examined in other phases of our methods.

Phase 1.2. Use Excel to Analyze Journal Keywords

Similar to abstracts, keywords—both those submitted by article authors and those indexed by journals—are signposts of the foci and concerns of journal publications. Because these source texts already exist in a "keyword" format, Microsoft Excel offered a more expedient way to analyze them. As with Phase 1.1, we exported from Scopus all author keywords and indexed keywords from our identified journals and publication range. Using Excel formatting tools, including "Text to Columns," we created flat lists of all key terms (one term per cell) and then counted the frequency of each unique term in those lists using the function COUNTIF(A:A, A*n*).

Within author-submitted keywords were a total of 2,551 unique terms. The vast majority of them, over 2,000, appeared only once, and less than 50 terms appeared eight or more times. Journal-indexed keywords have a similar trend: Among a total of 2,383 unique terms, over 1,800 appeared only once, and just over 70 terms appeared eight or more times. Between the two lists, the most frequent term is again the broad *communication*, which appeared 166 times, followed by similarly broad terms such as *technical communication*, *teaching*, and *technical writing*. After about ten such terms, we started to see specific terms such as *usability* (55 times) and *rhetoric* (46 times).

We next compared these results with Phase 1.1 results for a combined analysis, again balancing the needs for quantitative and qualitative significance. At this point, we also took practical factors into consideration. Other keyword collections we reviewed typically contain essays on 30 to 50 terms to allow a substantial coverage but also sufficient elaboration on each term, and thus we aimed for a number in that range. The result of this phase was a total of 43 unique terms, including slash-bound terms. For example, although *international technical communication* and *intercultural technical communication* both appeared in the corpus, these terms have affinities and contrasts that—in our view—would be best examined in a single essay; hence, we combined them into *international/intercultural communication*.

Phase 2. Survey

To supplement and check the interpretive perspectives the two of us brought to the process described above, we created a survey and distributed it via the Association of Teachers of Technical Writing listserv and the Council for Programs in Technical and Scientific Communication listserv. The survey invited participants

1. to submit what they believe are terms significant for the field and in need of explanation to newcomers,
2. to evaluate the terms we identified in Phase 1, and
3. to suggest names of contributors well suited to write about the keywords.

Toward the end of the survey, participants were invited to share their disciplinary backgrounds. Depending on participants' answers on which role they primarily identify with in the field of TPC (faculty, graduate student, industry practitioner, or other), they were then taken to different background questions.

A total of 43 participants completed the survey, though not all participants answered all questions. Most participants reported having either six to ten years or 20+ years of experience in the field, with the majority being technical communication faculty at four-year universities in ranks ranging from instructor to full professor. Many of these participants direct technical communication programs, including certificates, major and minor programs, service course programs, master's programs, and Ph.D. programs. Six participants identified themselves as graduate students or industry practitioners, and two identified themselves as performing more than one role, for example, faculty member and industry practitioner. Participants who are industry practitioners work in areas ranging from web content strategy and program management to social media.

We recognize that 43 participants is not a large sample and that, in particular, there were a limited number of participants who self-identified as industry practitioners. Despite the relatively small number of responses, we still obtained rich data informative for our purpose. In addition, we were not so much relying on the survey to generate data as using it to help refine and support our interpretation of Phase 1 data. Specifically, we used participant responses to identify key terms we might have missed in Phase 1 and to gather additional perspectives on the terms we already identified.

To limit biasing participant responses, we asked open-ended questions first before inviting participants to evaluate our Phase 1 keywords. In the open-ended questions, participants were asked to identify the five most important topics in technical communication today as well as the five terms they think newcomers to the field struggle with the most.

The key terms suggested by survey participants were analyzed using the same methods outlined in Phase 1.2. A total of 147 unique terms were suggested. Of these, 101 terms appeared only once, and 21 terms appeared more than four times. The most frequently mentioned term is *rhetoric* (26 times), followed by *usability* (19 times), though if *usability* is combined with related terms (*user experience* and *user-centered*), it becomes the most frequently mentioned concept. The terms that participants identified as important and/or difficult included themes not covered by our Phase 1 results; they are, notably, accessibility, audience, particularly challenging genres (*grants* and *documentation*), social media practices, specific standards and markup languages (e.g., DITA and XML), and structured authoring.

Participant evaluations of the 43 terms we generated in Phase 1 confirmed the importance of those terms. All 43 terms received at least some votes of "very important," though many also received votes of "not important." Table 1 summarizes these evaluations. In addition to rating the terms, participants could offer qualitative comments, though no consistent patterns emerged from these comments. For example, as mentioned earlier, some championed the inclusion of terms such as *feminism*, while others questioned their centrality to the field. Some applauded the coverage of the 43 terms, while others wondered if some of the terms are already common knowledge for the field. Some participants also questioned if some terms, such as *service*, are too general. Such comments aided us in refining the list of terms; for example, we ultimately decided to cut *service* from the list of essays.

Final Analysis and Interpretation

In our final analysis, we cross-examined and synthesized Phase 1 and Phase 2 results, again balancing quantitative and qualitative considerations. With Phase 1 terms that participants questioned as being too general, we reconsidered their relevance. Many of those with the lowest importance ratings were cut, some terms were conceptually broadened (e.g., *digital technology* became *technology*), and others were combined under a single term that could encompass several themes (e.g., *user experience* now covers multiple user-related terms). In some cases, we returned to the journal data to find modifiers to limit the terms or create notes for future essay writers to specify the terms in their writing. For example, with the term *design*, we recorded that it is associated with terms such as *document design*, *participatory design*, and *web design*.

Table 1. Ratings of Phase 1 Keywords by Survey Participants

Term	Very Important		Somewhat Important		Not Important		Importance Score*	Rank**
	%	*N*	%	*N*	%	*N*		
Actor/activity	38.9%	14	44.4%	16	16.7%	6	0.56	9
Content analysis	41.7%	15	50.0%	18	8.3%	3	0.58	7
Content management	72.2%	26	25.0%	9	2.8%	1	0.63	3
Crisis communication	45.7%	16	48.6%	17	5.7%	2	0.59	7
Data visualization	83.8%	31	16.2%	6	0.0%	0	0.65	1
Design	81.6%	31	18.4%	7	0.0%	0	0.64	2
Digital technology	64.9%	24	32.4%	12	2.7%	1	0.62	4
Discourse analysis	27.8%	10	61.1%	22	11.1%	4	0.54	10
Distance education	22.2%	8	44.4%	16	33.3%	12	0.45	13
Entrepreneurship	16.7%	6	47.2%	17	36.1%	13	0.41	14
Environment	41.7%	15	47.2%	17	11.1%	4	0.57	8
Ethics	88.9%	32	11.1%	4	0.0%	0	0.65	1
Feminism	44.4%	16	41.7%	15	13.9%	5	0.58	7
Genre	58.3%	21	36.1%	13	5.6%	2	0.61	5
Globalization	64.9%	24	35.1%	13	0.0%	0	0.62	4
Information	63.9%	23	30.6%	11	5.6%	2	0.62	4
International / intercultural communication	88.9%	32	11.1%	4	0.0%	0	0.65	1
Knowledge	41.7%	15	44.4%	16	13.9%	5	0.57	8
Literacy	58.3%	21	33.3%	12	8.3%	3	0.61	5

Term	Very Important		Somewhat Important		Not Important		Importance Score*	Rank**
	%	*N*	%	*N*	%	*N*		
Localization	62.2%	23	32.4%	12	5.4%	2	0.62	4
Medical/health communication	63.9%	23	33.3%	12	2.8%	1	0.62	4
Multimodality	59.5%	22	32.4%	12	8.1%	3	0.61	5
Narrative/ storytelling	33.3%	12	55.6%	20	11.1%	4	0.56	9
Organizational culture	33.3%	12	63.9%	23	2.8%	1	0.57	8
Plain language	55.6%	20	38.9%	14	5.6%	2	0.60	6
Professionalization	33.3%	12	55.6%	20	11.1%	4	0.56	9
Programmatic research	29.7%	11	43.2%	16	27.0%	10	0.51	12
Project management	58.3%	21	38.9%	14	2.8%	1	0.61	5
Public engagement	63.9%	23	27.8%	10	8.3%	3	0.62	4
Research methods	66.7%	24	30.6%	11	2.8%	1	0.62	4
Rhetoric	83.3%	30	8.3%	3	8.3%	3	0.65	1
Risk communication	61.1%	22	33.3%	12	5.6%	2	0.61	5
Science	58.3%	21	33.3%	12	8.3%	3	0.61	5
Service	28.6%	10	51.4%	18	20.0%	7	0.53	11
Social justice	47.2%	17	30.6%	11	22.2%	8	0.57	8
Style	41.7%	15	47.2%	17	11.1%	4	0.57	8
Technical editing	61.1%	22	36.1%	13	2.8%	1	0.62	4
Technical translation	52.8%	19	44.4%	16	2.8%	1	0.60	6
Usability	89.2%	33	10.8%	4	0.0%	0	0.65	1
User experience (UX)	84.2%	32	15.8%	6	0.0%	0	0.65	1
User interface (UI)	70.3%	26	29.7%	11	0.0%	0	0.63	3
Virtual collaboration	35.1%	13	54.1%	20	10.8%	4	0.56	9
Visual rhetoric	80.6%	29	13.9%	5	5.6%	2	0.64	2

**The "importance score" is a weighted average of each term's rating that assigns 2 points for every "Very Important" rating, 1 point for every "Somewhat Important" rating, and -1 for every "Not Important" rating.*

***Several terms had the same score; thus, there are only 14 ranks to account for tied scores.*

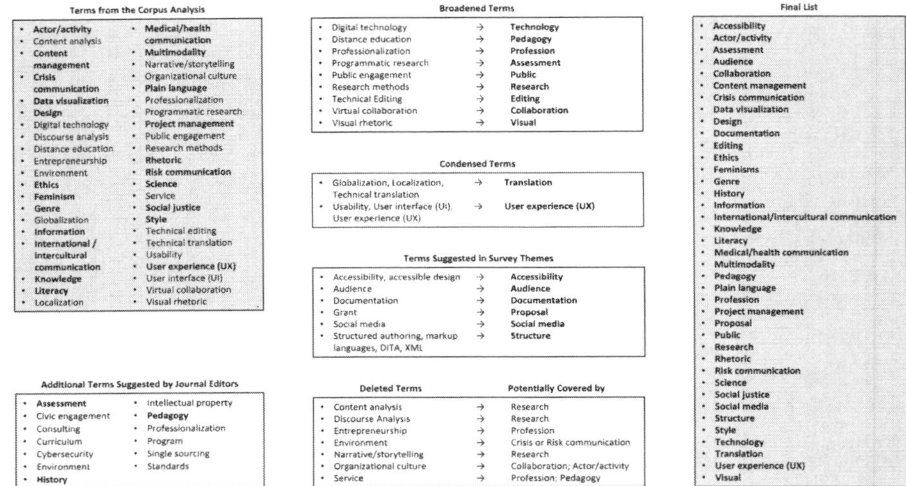

Figure 6. Summary of the development of our roster of keywords. Terms in bold were included in the final list.

With additional terms that were suggested by multiple participants, we examined the journal data to assess their quantitative frequency. We also solicited feedback from several trusted colleagues who edit or have edited major journals, which resulted in additional keywords. Through several iterations of this process, we arrived at our final list of the 39 keywords examined in this volume. Figure 6 summarizes our changing roster of keywords.

To frame the collection, we also solicited two additional essays. We invited Johndan Johnson-Eilola and Stuart A. Selber to write a keyword foreword ("Technical and Professional Communication"), which provides reflections on both this collection and the current state of the field. We also asked Kristen Moore, Lauren Cagle and Nicole Lowman to write an afterword ("Diversity, Equity, and Inclusion through Citational Practice") to document their process and findings for the inclusive citation audit they conducted on earlier drafts of the contributed essays. This audit is described in more detail after the following description of how we selected our contributors.

Our Contributors

As mentioned previously, our survey invited participants to suggest contributors well suited to write essays on our identified keywords. Among our final list of 39 keywords, all but eight keywords received at least one and often multiple contributor suggestions. Some of the suggested contributors are clearly situated in the field of TPC; that is, they work in TPC programs and publish in TPC journals. Other suggested contributors are situated more closely in related fields such as cultural studies or composition and rhetoric. Given our purpose for this project, in selecting contributors, we focused on those clearly situated in the field of TPC. In addition, when multiple survey participants suggested the same contributor for a keyword, we also favored that contributor.

For each keyword, we also performed literature searches (using that particular keyword) in technical communication publications to identify potential contributors who have published on that keyword. The publications we searched included the journals used in our keyword data mining but also other TPC journals and book publications. We focused on authors whose work exhibits extensive knowledge of a keyword, for example, people who have published multiple articles on the topic or who have written or edited books on the topic.

Through these processes, we were able to decide on potential contributors for all 39 keywords. In some cases, we identified two contributors for a single keyword because they frequently co-author on the topic. We then contacted potential contributors and invited them to participate in our project. In almost all cases, they agreed to participate. In a few cases where they couldn't (usually because of time constraints), we repeated the above process to identify another contributor.

As the above description shows, the process we used to identify contributors is not a science. There is no denying the potential biases that we, as well as our survey participants, brought into the process. Most importantly, by focusing on (consciously or unconsciously) experts who have published extensively on a topic, our selection favors those more established in the field and is biased against emerging scholars. Potentially, then, the resulting keyword essays may be more backward-looking than forward-looking. We tried to address this limitation in our guidelines to contributors (more about this below). In addition, our process may be biased against marginalized scholars who are systemically underrepresented in citations, publications, and the field's collective memory. However, eight of our 42 keyword essay contributors self-identify as multiply marginalized or underrepresented (MMU) scholars on the "MMU Scholar List" maintained by Cana Itchuaqiyaq (2022). Overall, our contributors represent a wide range of disciplinary perspectives, from the most seasoned teacher-scholar-practitioners to mid-career scholars expanding the boundaries of the field. We have been delighted and honored to work with this group of experts who collectively have cultivated centuries of expertise in TPC.

Guidelines for Contributors

To create a certain level of consistency between our keyword chapters—without prescribing a rigid structure or pattern—we established several guidelines for our contributors. These guidelines are taken from and modified based on the contributor guidelines for *Keywords for Children's Literature*, generously shared by one of the editors, Phil Nel.

Contributors were asked to start their essays with a paragraph summarizing the history of their keywords in English and (if appropriate) in other languages. For English etymology, we recommended that the contributors use *The Oxford English Dictionary*. Alternatively, for keywords that may not benefit from an etymological opening, we suggested an initial paragraph that summarizes the changing contemporary and disciplinary contexts where a keyword has been used. As the essays took shape, some introductions were revised to emphasize a term's significance for TPC praxis before discussing its history.

Beyond the first paragraph, we asked contributors to structure their essays around the significant debates surrounding their keywords. We use the word "debate" loosely. It can include particular problems that emerge in the use of the keyword in the research, teaching, and practice of TPC. It can refer to the ways that multiple theoretical perspectives have been used to interpret the keyword in TPC. It can include the critical projects/perspectives that the keyword enables or hinders in TPC.

For the ending of their essays, contributors were asked to create a forward-looking paragraph, discussing how their keywords could be used in the future and/or whether they need to be rethought in our current environment. Our foreword authors were similarly asked to take both a historical view of the importance of the keywords in this volume and to look forward to how these keywords could evolve in the future.

Finally, we also shared with our contributors the entire list of our keywords and asked that they identify cross-reference potentials. That is, when their chapters mention another keyword included in the volume, they should bold that other term upon its first use to signal cross-references. These cross-references, we hope, will allow readers to form multi-dimensional understandings of the field. Indeed, as the 442 lines representing them in Figure 7 suggest, the links between keywords form a complex network of connections.

Inclusive Citation Audit

As this project got underway, Kristen Moore (author of the essay on *public* in this volume) offered to conduct an inclusive citation audit for the volume, which she completed with Lauren Cagle and Nicole Lowman. An inclusive citation audit helps to ensure that a collection actively cites underrepresented and marginalized scholars, recognizes their scholarly contributions, and includes them in a critical reflection of the field. Doing so is important in all our work but essential in a keywords collection where authors claim to identify key topics, discussions, and debates. After Kristen proposed the idea, we immediately agreed that an inclusive citation audit, conducted by scholars other than us, could help to modulate subjectivity and bias across the collection.

The audit was performed on early drafts of the chapters. Moore, Cagle, and Lowman provided contributors with chapter-specific feedback, suggesting possible angles to relate their writing to issues of inclusion and social justice as well as citations of work by underrepresented scholars that can inform any discussions of the term. More details about this audit can be found in the afterword of the volume.

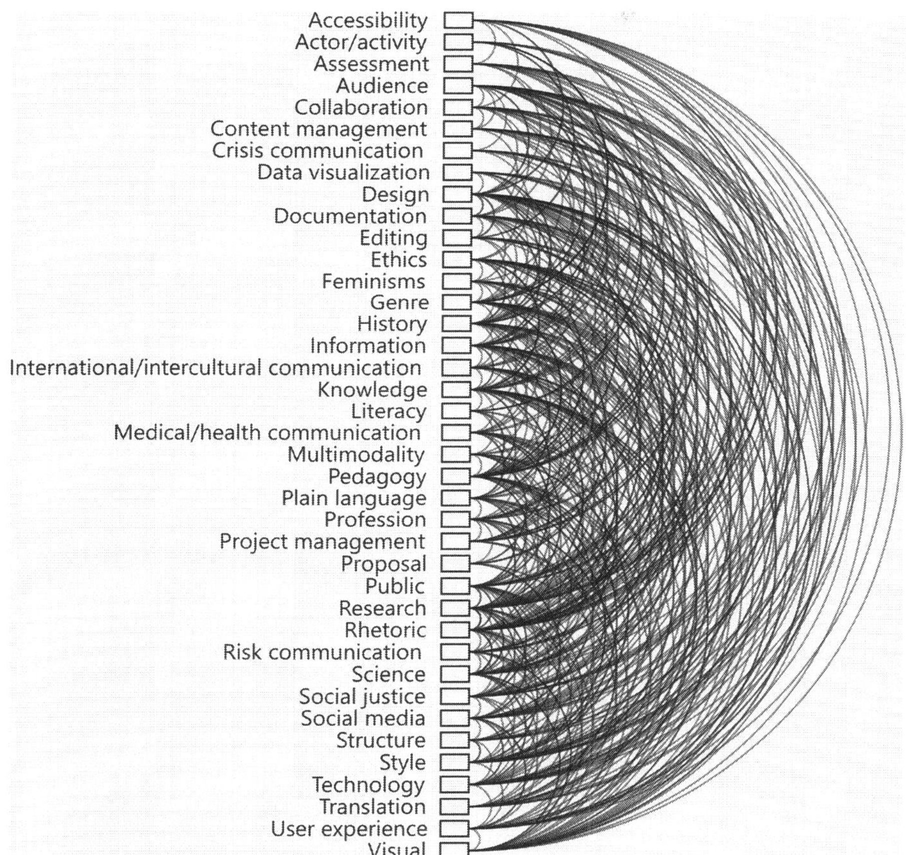

Figure 7. Chart mapping cross-references between keywords. Each arced line represents a cross-reference.

We also want to point out that, prior to this citation audit, many of our contributors already explicitly engaged with issues of diversity, equity, power, and inclusion in their early drafts. These include not only chapters that, given their keywords, have an explicit focus on social justice and non-hegemonic theoretical frames (such as chapters on *social justice* or *feminisms*) but many other chapters, for example, those on *profession*, *history*, and *literacy*, to name just a few.

Ways to Use This Collection

We envision two ways that readers may use this collection. First, we believe it can function as a useful scholarly source. By tracing the genealogy of terms central to TPC and revealing their evolving, divergent, and contested meanings, these essays can help researchers to critically engage related studies. This use may take the forms of researchers acknowledging conflicting viewpoints, adopting different theoretical lenses, or strengthening their original arguments. We also hope, as mentioned earlier, that researchers will actively interact with this collection, whether by updating the essays contained here, by critiquing or responding to them, or by creating new essays. Doing so will allow all of us to continue questioning and enriching our understanding of the field's keywords and their interconnections.

Another important way we envision this collection being used is as a pedagogical tool. Essays in this collection, with their focused intent, short length, ambitious goals, and rich examples, are well suited for orienting students into the field. The entire collection, or selected essays, might be assigned to students to provide background information related to course topics. Alternatively, students may be charged to revise or expand select essays using information gained in a course or through additional research. More ambitious assignments might ask students to write new keyword essays, either individually or collaboratively, using terms that are central in the context of a given course.

For instructors interested in such an assignment, Burgett and Hendler (2014) suggested a two-step process. In step one, students create a repository of use examples of a particular keyword. Depending on the nature of a course, students can use as source materials their assigned readings, additional readings, or multimedia materials such as "images and sound, conversations overheard on the street, or exchanges on a bus" (Burgett & Hendler, 2014, p. 10). In step two, students write about the usages they curated. This two-step process can help students develop a range of relevant skills, from close reading and observation to multimodal data gathering and organization to synthesis and collaboration.

The above represents just a small list of possibilities for using the essays collected here. We look forward to learning from our readers about how they use, respond to, and interact with this collection.

Conclusion

It has been a pleasure and privilege—and a long journey—to work on this project. We hope our readers find this collection helpful as they enter the field of TPC, navigate its terminological complexities, expand understandings of established concepts, and develop the new terms that will move the field forward.

References

Boettger, R. & Friess, E. (2020). Content and authorship patterns in technical communication journals (1996–2017): A quantitative content analysis. *Technical Communication, 67*(3), 4–24.

Brumberger, E. & Lauer, C. (2015). The evolution of technical communication: An analysis of industry job postings. *Technical Communication, 62*(4), 224–243.

Burgett, B. & Hendler, G. (2014). *Keywords for American cultural studies.* New York University Press.

Burke, K. (1966). *Language as symbolic action: Essays on life, literature, and method.* University of California Press. https://doi.org/10.1525/9780520340664.

Carlson, H. V., Mayo, R. H., Philler, T. A. & Schmidt, D. J. (1983). *An annotated bibliography on technical writing, editing, graphics, and publishing: 1966–1980.* Society for Technical Communication.

Carradini, S. (2021). The Ship of Theseus: Change over time in topics of technical communication research abstracts. In J. Schreiber & L. Melonçon (Eds.), *Assembling critical components: A framework for sustaining technical and professional communication* (pp. 39–68). The WAC Clearinghouse; University Press of Colorado. https://doi.org/10.37514/TPC-B.2022.1381.2.02.

Dryer, D. (2019). Divided by primes: Competing meanings among writing studies' keywords. *College English, 81*(3), 214–255.

Gallon, R. (2016). *The language of technical communication.* XML Press.

Heilker, P. & Vandenberg, P. (1996). *Keywords in composition studies.* Boynton/Cook Publishers.

Heilker, P. & Vandenberg, P. (2015). *Keywords in writing studies.* Utah State University Press. https://doi.org/10.7330/9780874219746.

Itchuaqiyaq, C. U. (2022, March 15). *MMU scholar list.* Cana Uluak Itchuaqiyaq. https://www.itchuaqiyaq.com/mmu-scholar-list.

Johnson-Eilola, J. & Selber, S. A. (2004). *Central works in technical communication.* Oxford University Press.

Nel, P. & Paul, L. (2011). *Keywords for children's literature.* New York University Press.

Philler, T. A., Hersch, R. K. & Carlson, H. V. (1966). *An annotated bibliography on technical writing, editing, graphics, and publishing: 1950–1965.* Society of Technical Writers & Publishers; Carnegie Library of Pittsburgh.

Rude, C. (2015). Technical communication. In P. Heilker & P. Vandenberg (Eds.), *Keywords in writing studies* (pp. 165–168). Utah State University Press. https://doi.org/10.7330/9780874219746.c033.

Selfe, R. J. & Selfe, C. L. (2013). What are the boundaries, artifacts, and identities of technical communication? In J. Johnson-Eilola & S. A. Selber (Eds.), *Solving problems in technical communication* (pp. 19–49). The University of Chicago Press.

Steinbock, D., Pea, R. & Reeves, B. (2007). Wearable tag clouds: Visualizations to facilitate new collaborations. In C. A. Chinn, G. Erkens & S. Puntambekar (Eds.), *Proceedings of the 8th International Conference on Computer Supported Collaborative Learning* (pp. 672–674). International Society of the Learning Sciences. https://doi.org/10.3115/1599600.1599725.

Vargas, D., La Fountain-Stokes, L. & Mirabal, N. (2017). *Keywords for Latina/o studies.* New York University Press. https://doi.org/10.2307/j.ctt1pwtbpj.

Williams, R. (1983). *Keywords: A vocabulary of culture and society* (revised edition). Oxford University Press. (Original work published 1976)

1. Accessibility

Sushil K. Oswal
University of Washington

The terms *access* and *accessibility* have been in circulation for at least a century, and their usage generally connotes availability and physical accessibility for a certain population (Guy, 1983). The word *access* in English originated from Latin, meaning "accession" (Hoad, 1996). In contemporary policy discourse in English, however, access is defined as making information and communication technologies (ICTs) widely available to all citizens (Wise, 1997). Echoing this meaning of access, the *Oxford English Dictionary* defines accessibility as "the quality or condition of being accessible (in various senses)" (Oxford University Press, n.d.). It further defines accessible as "capable of being conveniently used or accessed by people with disabilities; of or designating goods, services, or facilities designed to meet the needs of the disabled." To discuss accessibility, understanding how access, accessibility, and accessible ***design*** have become common terms (with fluid definitions) in technical communication today is important. This essay will unpack these terms by considering both historical definitions and contemporary perspectives.

Not only do the terms access and accessibility have different meanings, but researchers also differ in how they relate the terms and establish their connection to disability. These differing views represent the perspectives of technical standards organizations, digital rhetoricians, disability activists, and disability studies-centered design scholars. The International Organization for Standardization (2014) defines accessibility as the "extent to which products, systems, services, environments and facilities can be used by people from a population with the widest range of characteristics and capabilities to achieve a specified goal in a specified context of use" (n.p.). James Porter, a digital ***rhetoric*** scholar, makes a distinction between access and accessibility. "Access," Porter explains, "is the more general term related to whether a person has the necessary hardware, software, and network connectivity in order to use the Internet—and to whether certain groups of persons have a disadvantaged level of access due to their race, ethnicity, socioeconomic status, gender, age, or other factors" (2009, p. 216). Accessibility, on the other hand, "refers to the level of connectedness of one particular group of persons—those with disabilities" (Porter, 2009, p. 216). Porter also adds that "the reason to write/design for accessibility is not only to allow people with disabilities to consume ***information***, but to help them produce it" (p. 216). Activists in the disability field do not always make Porter's distinction. For example, the University of Leeds' Centre for Disability Studies employs *access* as the search term for all accessibility-related entries on its website. Speaking from a disability

DOI: https://doi.org/10.37514/TPC-B.2023.1923.2.01

studies-centered design perspective, Sushil Oswal (2013) describes accessibility more broadly as "the ability to use, enjoy, perform, work on, avail of, and participate in a resource, technology, activity, opportunity, or product at an equal or comparable level with others" (n.p.).

As these varying definitions from different disciplines suggest, accessibility is shaped by a number of factors—which can change from context to context and are both spatial and temporal in nature: physical distance from resources and opportunities, the availability of ***technology*** and means to overcome that distance, and the infrastructural and legal resources to overcome barriers to workplace entry. For example, for full integration into society, a disabled person not only needs a job to support themself and their family but also laws to protect them from discrimination by employers and providers of services, consistent access to adaptive technology and special training for holding on to a job, accessible opportunities to participate in recreational activities, and of course, availability of inclusively designed consumer goods for living a comfortable life (Wilson & Lewiecki-Wilson, 2001).

Since technical communication is preoccupied with design and communication of information, and since information is imbedded everywhere in human environments, the field's scope extends into the accessibility of both the brick-and-mortar and digital spaces (Whitehouse, 1999). Accessibility in the former can consist of signage, directories, and spatial maps—digital and otherwise—whereas accessibility in the latter refers to a range of ICTs, including "computer hardware and software, digital broadcast technologies, telecommunications technologies such as mobile phones, as well as electronic information resources such as the world wide web" (Selwyn, 2004, pp. 346–347). The World Wide Web Consortium's (W3C) web content accessibility guidelines (2.1) break down accessibility into four elements in terms of the interactivity of the web for the disabled user: perceivable, operable, understandable, and robust (Web Accessibility Initiative, 2019). Though W3C's intent for these guidelines is to assist developers in designing accessible websites from the bottom up, more often, these are used for checking the accessibility of already built websites for the purpose of retrofitting them with accessibility (Wentz et al., 2011).

In technical communication practice, teaching, and ***research***, accessibility has been advocated by the community members dedicated to the needs of disabled users. It is not often included in ***user experience*** design discussions, although it should be (Oswal, 2019; Zdenek, 2019). This exclusion might result from when and how accessibility is included in the design process—it is often an afterthought, or comes up as a result of a quality check at the tail end of the design cycle. In either of these situations, accessibility gets retrofitted to an already developed product and rarely results in an equitable user experience. Another problematic reason could be that the designers forgot to include disabled consumers among their imagined users. Such omissions are more common in professional practice than one would expect after all of the accessibility activism of the last three decades (Charlton, 1998; Finkelstein, 1993).

The growth of the World Wide Web in the last three decades has not only resulted in an information explosion, but it has also introduced new questions about access to any informational content for disabled users. Considering the central place of the web in technical communication work, the accessibility in this area can be broken down into several subcategories: web interfaces for assistive technology, such as screen readers and voice browsers; accessible data input, navigation, and content; intuitive page layout and design; and accessible web authoring and development tools. Another important area of concern is the design of human-computer interactions. While the World Wide Web code in itself is not inaccessible, the interactions it enables can erect access barriers, unless these interactions have been conceptualized with disabled users in mind. For example, screen readers can process both text and links on a webpage without a problem. Web code also permits alternative text descriptions for images, which can be read by a screen reader. However, when a designer attributes an interactive element to an image, such as a link, that interaction becomes inaccessible to the screen reader. Designers and developers often forget that screen readers are text readers and lack the ability to read and interpret images.

In conceptualizing different aspects of access and accessibility, it is important to pay attention to how the relating terms are operationalized. Thus far, design fields, including technical communication, have often operationalized definitions of *information*, *place*, *cyberspace*, and *accessibility* that exclude disabled users, or have left them open-ended and matter of situational interpretations in different social and technical domains (Janelle & Hodge, 2013, p. 3). The debate surrounding the definitions of accessibility is murky, and the disabled users are often left out of this discussion. Instead of a focus on how different users access and interact with spaces—virtual or not—researchers are more interested in studying the changes in these technologies.

It is also important to note that accessibility is different from universal design (UD). On the surface, the design practice based on UD suggests access for all, hence the nomenclature "universal design." However, when put to practice loosely for divergent purposes, it can easily be reduced to a checklist for legal compliance, lead to tokenism, and water down the original intent of UD principles (Connell et al., 1997; Mace, 1985; Oswal & Melonçon, 2017; Sandhu, 2011). Take, for example, the accessibility for wheelchair users: The ramp designs and locations are seldom conceptualized according to the convenience of their users, and are rarely integrated into the original design of buildings in a way that doesn't stigmatize, or separate, this user population. Even the signs for these problematically located ramps are often hidden, or are hard to read from the position of the wheelchair rider. A good example of the pervasive tokenism toward blind and visually impaired users in contemporary architectural design is the use of braille and large print even though spatial access is affected far more by layout, acoustics, and ambient lighting. Most buildings have only one design feature that relates to this group—braille signs, which are often mounted upside down, might display inaccurate information, or are placed so far from the intuitive locations that blind users might fail to find the sign by touch. Tactile maps are rare, even in university and public buildings serving thousands of people and constructed at the expense of tens of millions of dollars. Seldom are indoor and outdoor public spaces designed for users with a range of common mental, ***visual***, and hearing disabilities, and they often give unending grief to these users due to their confusing layouts, odd features (four steps up and then three steps down, requiring unnecessary exertion), and unexpected location of specific amenities such as restrooms, elevators, and information desks.

In the context of learning spaces, curricular, and pedagogies, the universal design debate has another accessibility dimension. This debate has its origins in the universal design for learning (UDL) movement, which built on the universal design principles for built environments (Gronseth & Dalton, 2019; Rose, 2000). While the UD principles were directly rooted in the accessibility of built environments for disabled users, UDL was developed to meet the legal mandate to provide secondary education to all children (Individuals with Disabilities Education Act, 1997). Consequently, the developers of UDL focused on the learning environment rather than the individual needs of disabled students. They did not see accessibility as a part of design, as is obvious from the following claim: "accessibility is a function of compliance with regulations or criteria that establish a minimum level of design necessary to accommodate people with disabilities" (Salmen, 2011, p. 6.1). While there are exceptions, many researchers in this group strongly differentiate between universal design and accessibility because their focus is on the technology of universal design rather than its users. For example, during the COVID-19 pandemic, UDL has been inserted in many discussions about remote teaching in higher education without a regard for accessibility (Dickinson & Gronseth, 2020).

UD argues for simple and intuitive use requiring no special technological ***knowledge***, language proficiency, or mental concentration. Though it makes claims of equitable use that doesn't isolate, stigmatize, or disadvantage a particular group, it often sacrifices the accessibility needs of users with severe disabilities to accommodate all other constituencies on this omnibus version of universal design. It seems to accommodate everyone, but due to the watered-down affordances of such design, more often it only succeeds in serving the needs of users with less severe disabilities. Universal design has so many other ambitions—"improved design standards, better information, and new products and lower costs" (Greer, 1987, p. 58)—that distract it from the purposes of accessibility and accessible design for disabled users. The universal design advocates critique design approaches that compensate disabled people's functional limitations (Connell & Sandford, 1999; Salmen & Ostroff, 1997; Weisman, 1999). Despite their assertions about not stigmatizing disability and accessibility, universal designers reflect similar attitudes by pushing disability under the rug (Steinfeld, 1994). These universal designers forget that many disabled people see their disability not only as a bodily or mental limitation, but also a mark of identity and pride (Brown, 2003; Charlton, 1998; Fleischer & Zames, 2011; Johnson, 1987).

As the emerging literature on sensory architecture has begun to inform us, blindness is not necessarily an absence (Pallasmaa, 2012). Architecture as seen from the combination of other senses—sound, touch, smell, and taste—can be luxurious. But in spite of all the developments in phenomenological ***sciences*** about the multisensory aspects of human perception, neither the designers of the physical, nor of web ***structures***, have a standard practice of engaging disabled users in early phases of project development (Oswal, 2014; Pallasmaa, 2012). The participatory design movement has been with us for half a century (Ehn, 1989, 2017), but designers and developers of built environments, technologies, and websites have seldom made a concerted effort to involve disabled users as co-designers and knowledge partners (Chandrashekar et al., 2006; Krantz, 2013; Lewthwaite et al., 2018; Oswal, 2014; Sahib et al., 2013). Architects could learn a great deal through participatory design with disabled users drawing on their experiential and embodied knowledge about spaces.

Stressing the fact that prevalent designs fail users with severe sensory disabilities such as blindness and deafness sounds redundant. However, without attention to their particular accessibility needs, no design can be assumed inclusive, accessible, and complete. On the other hand, meaningful accessible designs that don't depend on the ocular and aural experiences alone can open new paths for blind and deaf users to enjoy fuller embodied experiences both in virtual and physical spaces. The opportunity to access fulfilling experiences of this nature can result in blind and deaf users creating a centerspace for themselves as designers and creators to share their multisensory perspectives to build interiors, public spaces, and digital sites with the design community, thus altering the current one-way traffic between designers and users into an enriching exchange of ideas (Butler, 2016; Doiphode, 2019; Oswal, 2019).

The technical and professional communication field can not only expand its footprint into accessible web design practice by preparing students in this area, it can also command a leadership role through laboratory and field design ***collaborations*** with disabled users, designers, and industry practitioners to standardize methods for accessible web development, conceptualize accessible digital interfaces within physical spaces with architects and interior designers, and partner with urban planners to imagine disabled-friendly open spaces employing ubiquitous technologies.

Acknowledgments

My sincere thanks go to my son, Hitender Oswal, for his assistance in researching for the readings for this project. I also want to thank Han Yu for her feedback on the multiple drafts of this keyword essay.

References

Brown, S. E. (2003). *Movie stars and sensuous scars: Essays on the journey from disability shame to disability pride*. People with Disabilities Press.

Butler, J. (2016). Where access meets multimodality: The case of ASL music videos. *Kairos: A Journal of Rhetoric, Technology, and Pedagogy, 21*(1). https://kairos.technorhetoric.net/21.1/topoi/butler/index.html.

Chandrashekar, S., Stockman, T., Fels, D. & Benedyk, R. (2006, October). Using think aloud protocol with blind users: A case for inclusive usability evaluation methods. In *Proceedings of the 8th international ACM SIGACCESS conference on Computers and accessibility* (pp. 251–252). https://doi.org/10.1145/1168987.1169040.

Charlton, J. I. (1998). *Nothing about us without us: Disability oppression and empowerment*. University of California Press. https://doi.org/10.1525/9780520925441.

Connell, B. R., Jones, M., Mace, R., Mueller, J., Mullick, A., Ostroff, E., Sanford, E., Story, M. & Vanderheiden, G. (1997). *Principles of universal design*. Raleigh: North Carolina State University, Center for universal design. Retrieved April 12, 2023, from http://www.design.ncsu.edu/cud/about_ud/udprinciples.htm.

Connell, B. & Sandford, J. (1999). Research implications of universal design. In E. Steinfeld & G. Danford (Eds.), *Enabling environments* (pp. 35–57). Kluwer Academic. https://doi.org/10.1007/978-1-4615-4841-6_3.

Dickinson, K. J. & Gronseth, S. L. (2020). Application of Universal Design for Learning (UDL) principles to surgical education during the COVID-19 pandemic. *Journal of Surgical Education, 77*(5), 1008–1012. https://doi.org/10.1016/j.jsurg.2020.06.005.

Doiphode, K. (2019). *More than meets the eye: A foray into designing architecture for the multiple senses*. (Unpublished master's thesis). Unitec Institute of Technology, Auckland, New Zealand. Retrieved from https://hdl.handle.net/10652/4813.

Ehn, P. (1989). *Work-oriented design of computer artifacts* (2nd ed.). Lawrence Erlbaum Associates.

Ehn, P. (2017). Scandinavian design: On participation and skill. In D. Schuler & A. Namioka (Eds.), *Participatory design: Principles and practices* (pp. 41–77). CRC Press. https://doi.org/10.1201/9780203744338-4.

Finkelstein, V. (1993). The commonality of disability. In J. Swain, V. Finkelstein, S. French & M. Oliver (Eds.), *Disabling barriers—Enabling environments* (pp. 9–16). Sage.

Fleischer, D. Z. & Zames, F. (2011). *The disability rights movement: From charity to confrontation*. Temple University Press. http://www.jstor.org/stable/j.ctt14bt7kv.20.

Greer, N. (1987). The state of art design for accessibility. *Architecture, 76*(1), 58–60.

Gronseth, S. L. & Dalton, E. M. (2019). *Universal access through inclusive instructional design*. Routledge. https://doi.org/10.4324/9780429435515.

Guy, C. M. (1983). The assessment of access to local shopping opportunities: A comparison of accessibility measures. *Environment and Planning B: Planning and Design, 10*(2), 219–237. https://doi.org/10.1068/b100219.

Hoad, T. (1996). Access. In T. Hoad (Ed.), *The concise Oxford dictionary of English etymology*. Oxford University Press. https://doi.org/10.1093/acref/9780192830982.001.0001.

Individuals with Disabilities Education Act (IDEA). (1997). http://uscode.house.gov/view.xhtml?path=/prelim@title20/chapter33&edition=prelim.

International Organization for Standardization. (2014). *Guide for addressing accessibility in standards* (ISO/IEC Guide 71:2014). https://www.iso.org/standard/57385.html.

Janelle, D. G. & Hodge, D. C. (Eds.). (2013). *Information, place, and cyberspace: Issues in accessibility*. Springer Science & Business Media.

Johnson, M. (1987). Emotion and pride: The search for a Disability Culture. *Disability Rag*, *8*(1), 4–10.

Krantz, G. (2013). Leveling the participatory field: The Mind's Eye Program at the Guggenheim Museum. *Disability Studies Quarterly*, *33*(3). https://doi.org/10.18061/dsq.v33i3.3738.

Lewthwaite, S., Sloan, D. & Horton, S. (2018). A web for all: A manifesto for critical disability studies in accessibility and user experience design. In K. Ellis, R. Garland-Thomson, M. Kent & R. Robertson (Eds.), *Manifestos for the future of critical disability studies: Volume 1* (pp. 130–141). Routledge. https://doi.org/10.4324/9781351053341-12.

Mace, R. (1985). Universal design, barrier-free environments for everyone. *Designers West*, *33*(1), 147–152.

Oswal, S. K. (2013). Ableism. *Kairos: A Journal of Rhetoric, Technology, and Pedagogy*, *18*(1). https://kairos.technorhetoric.net/18.1/coverweb/yergeau-et-al/pages/ableism/index.html.

Oswal, S. K. (2014). Participatory design: Barriers and possibilities. *Communication Design Quarterly Review*, *2*(3), 14–19. https://doi.org/10.4324/9781351053341-12.

Oswal, S. K. (2019, October). Breaking the exclusionary boundary between user experience and access: Steps toward making UX inclusive of users with disabilities. In *Proceedings of the 37th ACM International Conference on the Design of Communication* (pp. 1–8). https://doi.org/10.1145/3328020.3353957.

Oswal, S. K. & Melonçon, L. (2017). Saying no to the checklist: Shifting from an Ideology of normalcy to an ideology of inclusion in online writing instruction. *WPA: Writing Program Administration-Journal of the Council of Writing Program Administrators*, *40*(3), 63–77.

Oxford University Press. (n.d.). Accessibility. In *Oxford English Dictionary*. Retrieved January 2, 2021, from www.oed.com.

Oxford University Press. (n.d.). Accessible. In *Oxford English Dictionary*. Retrieved January 2, 2021, from www.oed.com.

Pallasmaa, J. (2012). *The eyes of the skin: Architecture and the senses*. John Wiley & Sons.

Porter, J. E. (2009). Recovering delivery for digital rhetoric. *Computers and Composition*, *26*(4), 207–224. https://doi.org/10.1016/j.compcom.2009.09.004.

Rose, D. (2000). Universal design for learning. *Journal of Special Education Technology*, *15*(3), 45–49. https://doi.org/10.1177/016264340001500307.

Sahib, N. G., Stockman, T., Tombros, A. & Metatla, O. (2013, September). Participatory design with blind users: A scenario-based approach. In *IFIP Conference on Human-Computer Interaction* (pp. 685–701). Springer. https://doi.org/10.1007/978-3-642-40483-2_48.

Salmen, J. P. S. (2011). U.S. accessibility codes and standards: Challenges for universal design. In W. F. E. Preiser & K. H. Smith (Eds.), *Universal design handbook* (pp. 6.1–6.7). McGraw-Hill.

Salmen, J. & Ostroff, E. (1997). Universal design and accessible design. In D. Watson (Ed.), *Time-saver standards for architectural design data: The reference of architectural fundamentals* (pp. 1–8). McGraw Hill.

Sandhu, J. (2011). The rhinoceros syndrome: A contrarian view of universal design. In W. Preiser & K. Smith (Eds.), *Universal design handbook* (pp. 44.3–44.12). McGraw-Hill.

Selwyn, N. (2004). Reconsidering political and popular understandings of the digital divide. *New Media & Society*, *6*(3), 341–362. https://doi.org/10.1177/1461444804042519.

Steinfeld, E. (1994, June 19). *The concept of universal design* [Paper presentation]. Sixth Ibero-American conference on accessibility, Center for Independent Living, Rio de Janeiro.

Union of the Physically Impaired Against Segregation. (1976). *Fundamental principles of disability*.

University of Leeds. (n.d.). *The disability archive*. Retrieved January 2, 2021, from https://disability-studies.leeds.ac.uk/library/.

Web Accessibility Initiative. (2019). *Web content accessibility guidelines 2.1*. World Wide Web Consortium. https://www.w3.org/WAI/WCAG21/quickref/.

Weisman, L. (1999). *Creating justice, sustaining life: The role of universal design in the 21st century* [Keynote address]. Adaptive Environments Center, 20th anniversary celebration, The Computer Museum, Boston, Massachusetts.

Wentz, B., Jaeger, P. T. & Lazar, J. (2011). Retrofitting accessibility: The legal inequality of after-the-fact online access for persons with disabilities in the United States. *First Monday*, *16*(11). https://doi.org/10.5210/fm.v16i11.3666.

Whitehouse, R. (1999). The uniqueness of individual perception. In R. E. Jacobson (Ed.), *Information design* (pp. 103–129). The MIT Press.

Wilson, J. C. & Lewiecki-Wilson, C. (Eds.). (2001). *Embodied rhetorics: Disability in language and culture.* Southern Illinois University Press.

Wise, J. (1997). *Exploring technology and social space*. Sage.

Zdenek, S. (2019). Guest editor's introduction: Reimagining disability and accessibility in technical and professional communication. *Communication Design Quarterly Review*, *6*(4), 4–11. https://doi.org/10.1145/3309589.3309590.

2. Actor/Activity

Clay Spinuzzi
University of Texas at Austin

The *Oxford English Dictionary* provides definitions for *actor* and *activity* that are relevant to their current use in technical and professional communication. *Actor* is defined in part as "A person who performs or takes part in any action; a doer," while *activity* is defined in part as "Things that a person, animal, or group chooses to do" (Oxford University Press, n.d.). Both of these—agents and the things that they do—have been central to technical and professional communication (TPC) theory and ***research*** since the late 1980s and early 1990s, when TPC researchers began applying theories and methodologies from the social sciences to better understand technical and professional communication in practice. This turn to the social sciences entailed naming and describing social phenomena, among which are *actor* and *activity*.

The term *actor* has been used in several related senses to denote a social agent, which (as we'll see below) may or may not be an individual human being working with intentionality. Most generically, researchers have referred to individual writers and readers as "social agents" (Schryer 1993, 2000). But actor has been used in more specific ways grounded in particular theoretical stances. For instance, in sociocognitive approaches such as activity theory, situated cognition, and community of practice theory, the agent has been understood as an individual human being exercising individual agency within a specific sociocultural milieu. In posthumanist approaches such as actor-network theory, distributed cognition, the extended mind hypothesis, and new materialist theory, the agent can be human or nonhuman, and its agency is understood as networked or relational, i.e., emerging from the relationships among actors.

In activity theory (Kaptelinin & Nardi, 2006), an actor is specifically understood as a human being engaged in collective labor. Activity theory is essentially a sociocultural theory of human development within the context of cyclical, collective labor activity, and thus the term actor always refers to an individual human being who is engaged in that collective labor process. For instance, in their investigation of texts in a primary care clinic, Dawn Opel and William Hart-Davidson (2019, p.363) define the actors as human beings, including "providers in that same clinic, other providers such as specialists, pharmacists, home health aides, family members, and the patient herself." These actors are understood as separate from nonhumans such as tools, instruments, and infrastructure. Similarly, Kathleen Gygi and Mark Zachry (2010) studied how "a small group of industry professionals from a transnational corporation and academic researchers (the authors of this article) exchanged ideas about a project" (p.359). In this case, the actors were

DOI: https://doi.org/10.37514/TPC-B.2023.1923.2.02

identified as human beings, specifically human beings who interacted in order to develop the project's object ("a communication workshop for engineers," p.359). (For other examples, see Artemeva & Freedman, 2001; Bazerman et al., 2003; Haas, 1999; Hart-Davidson et al., 2008; McNely, 2009, 2019; Russell, 1997a; Sun, 2006.) Similarly, other sociocognitive approaches such as situated cognition and community of practice theory treat the agent as an individual human, although one who is thoroughly socialized (and not coincidentally, these approaches were lumped in with activity theory in the early to mid-2000s; see Artemeva, 2005; Tardy, 2003; Wegner, 2004).

In contrast, in actor-network theory (ANT), an actor is not necessarily a human being: Any human or nonhuman entity can be understood as exerting agency. ANT rejects classic sociological explanations that presume human agency and social structures and use them as ready-made explanations for observed phenomena (Latour, 1996, pp. 199–200), instead positing that human and nonhuman actors should be treated alike when considering how controversies are settled (Latour, 1987, p. 144). In this approach, actors are considered network effects rather than pre-existing entities (Law, 1994, pp. 33–34); they interdefine each other (Callon, 1986). Technical and professional communicators working in this vein have examined how actors emerge and exert agency. In Jason Swarts' (2010) study of recycled writing, for instance, he argues that when writers reuse writing, they ***rhetorically*** mobilize a range of actors that include people, policies, and ***style*** guides, aligning these actors to tap into the combined agency of the assemblage. (For other examples, see Dush, 2015; Fraiberg, 2017; Graham & Herndl, 2013; Potts, 2009, 2010; Potts & Jones, 2011; Read, 2016; Read & Swarts, 2015; and Jeff Rice, 2012.) Similarly, posthumanists or new materialists also use *actor* to refer to humans and nonhumans as they work in assemblages (Boyle, 2016; Gries, 2015; Mara & Hawk, 2010; Jenny Rice, 2012; see McNely et al., 2015 for an overview), as do those working with distributed cognition (e.g., Angeli, 2015; Spinuzzi, 2001; Swarts, 2006; Winsor, 2001).

Thus, in technical and professional communication, the term *actor* can be used in at least two senses: as an individual human working in a community to get something done (for instance, when writing a technical manual that tells an individual how to solve a bounded problem) or as a constructed bundle of agency emerging from the relationships of humans and nonhumans (for instance, when writing a handbook for an organization or workgroup, describing collective norms, tools, and infrastructure). These two senses are not necessarily exclusive.

The term *activity* has largely been used in technical and professional communication in reference to activity theory. This theory developed in the Marxist-Leninist milieu of the Soviet Union, and consequently understands organized human activity within the frame of labor. The term references the German "Tätigkeit (which has the synonyms work, job, function, business, trade, and doing) and distinguishes it from Aktivität" (Roth & Lee, 2007, p. 201), which is activity in a broader sense. Based on this distinction, activity theory's originators used the Russian term "*predmetnaya deyatel'nost'*, usually translated as 'object-oriented activity'" (Bakhurst, 2009, p. 202). In activity theory, an activity is a bounded, relatively durable instance of labor in which a subject (or actor) transforms a material object with the help of mediating instruments (Engeström, 1987). Activity theory entered technical and professional communication discussions in the mid-1990s when it was picked up by writing studies researchers such as Charles Bazerman, Carolyn Berkenkotter, Christina Haas, and David R. Russell by way of Yrjo Engeström (1987).

Since it described organized labor activity with definite boundaries, and since it encouraged focus on mediating instruments such as texts, this concept of activity was a strong fit for analyzing the qualitative case studies that began to fill technical and professional communication journals in the 1990s and 2000s. In such studies (Artemeva & Freedman, 2001; Bazerman et al., 2003; Berkenkotter & Huckin, 1995; Bracewell & Witte, 2003; Freedman & Smart, 1997; Haas & Witte, 2001; Kain & Wardle, 2005; Spafford et al., 2006; Walker, 2004; see Russell, 1997b for a review up to 1997), activity—often portrayed as an activity system with subjects or actors, mediating instruments or tools, an object or object(ive), rules, community, and division of labor—provided an analytical language suitable for dissecting context: bounding a case or a rhetorical situation via productive consensual orientation of a community to an object(ive). This notion of activity has given technical and professional communication practitioners a grounded framework for understanding and describing context in cases such as designing new content management systems (McCarthy et al., 2011), understanding user-generated documentation (Sherlock, 2009), identifying how texts support different functions in an organization (Jones, 2016), or developing engineering communication workshops (Gygi & Zachry, 2010).

With this background, we can understand some key debates around the terms as well as some key limitations.

For *actor*, the key debate is what counts as an actor. In earlier technical and professional communication research, the term typically represented an individual. In later research, the term came to additionally represent organizational roles; in some research, it also represents nonhuman or posthuman agents (e.g., Sackey et al., 2019). These different meanings of actor—as an individual agent vs. a networked agent defined through its relations—require different theoretical and methodological apparati as well as different understandings of how agency relates to intentionality. In technical and professional communication, we have come to generally recognize agency as distributed, but we have not yet come to agreement on how it is distributed or how it relates to intentionality. For instance, we may recognize that as individuals learn a ***genre***, they learn to participate in an ongoing activity. But in this case, do we consider the genre to be the residue of human agency, or should the genre itself be considered an agent (cf. McNely, 2019)?

The tension between the two senses of actor (as individual vs. networked agent), then, can cause occlusion or obstruction, especially as the term becomes more deeply sedimented in technical and professional communication. The senses are difficult to reconcile, and sometimes readers must check citations to determine which meaning is operant in a given source.

For *activity*, one key debate is how to bound an activity. When activity theory was introduced to technical and professional communication in the 1990s, the notion of activity provided a more structured, developmental, and objective-oriented alternative to vague terms such as "context," performance-oriented frameworks such as Burke's pentad, or concepts for describing social clusters such as "discourse communities." Specifically, it provided a way to bound qualitative case studies, one that goes beyond spatial, demographic, and organizational groupings to identify how people work together over time. (However, commentators have questioned how well this bounding works in practice, with some alleging that the activity system functions as both phenomenon and analysis; see Bracewell & Witte, 2003; Witte, 2005.) Yet activity theorists have steadily expanded the notion of activity, both spatially and temporally, resulting in case studies with larger bounds and arguably less precision (see Spinuzzi, 2011). In technical and professional communication, this expansion has sometimes resulted in "activity" being used vaguely and generically, essentially as a substitute for "context."

Another key debate is the question of the applicability of activity. As mentioned, the notion of activity is grounded in labor activity, which (in accordance with the Soviet outlook) was taken to be the very thing that makes us human (Engels, 1971; Leontyev, 2009) and thus was understood as universally applicable—that is, all human activity is rooted in labor activity. But this claim is not universally accepted: It is grounded in the Soviet outlook, which was modernist and instrumentalist. Thus, we should not be surprised that the concept of activity has sharp limits when applied to aspects of life beyond recurrent, bounded, collective efforts that are mediated by instruments. Specifically, associative and less structured forms of interaction are not well addressed by the term *activity*. For instance, although activity theory can clearly bound cases of collaborative work on a Wikipedia page (Slattery, 2009; Walsh, 2010), the Wikipedia *community* has less certain boundaries (Jones, 2008; Swarts, 2009; cf. Jemielniak, 2015); in such cases of social and peer production, the boundaries appear to fade away (Engeström, 2009). Similarly, phenomena that are not well defined by local object-oriented activity, such as ***public*** argumentation and structural racism, are not well modeled by activity theory. Finally, due to its instrumental labor focus, activity theory has trouble modeling and analyzing non-instrumental relations (see Miller, 2007 for a critique and Spinuzzi, 2008 for an extended discussion), and it "lacks a political edge" or critical analysis of politics suitable for cultural studies (Sun, 2020, p. 50).

The term *activity*, then, is becoming occluded due to tensions between its origination in an instrumentalist, work-oriented branch of Soviet psychology and its application to cases that do not necessarily fit this description, particularly in a field that must take non-instrumentalist relationships into account and that must analyze more associative, less structured phenomena. As technical and professional communication examines cultural and cross-cultural artifacts and practices (e.g., Fraiberg, 2017; Sackey et al., 2019; Sun, 2020; Walton, 2013) and ***social justice*** issues (Cox, 2019; Jones, 2017; Potts et al., 2019; Rose, 2016; Sackey, 2020), we can expect this term to be reexamined and rethought—or juxtaposed with different terms attached to theories that are better able to address such concerns.

References

Angeli, E. L. (2015). Three types of memory in emergency medical services communication. *Written Communication, 32*(1), 3–38. https://doi.org/10.1177/0741088314556598.

Artemeva, N. (2005). A time to speak, a time to act: A Rhetorical genre analysis of a novice engineer's calculated risk taking. *Journal of Business and Technical Communication, 19*(4), 389–421. https://doi.org/10.1177/1050651905278309.

Artemeva, N. & Freedman, A. (2001). "Just the boys playing on computers": An activity theory analysis of differences in the cultures of two engineering firms. *Journal of Business and Technical Communication, 15*(2), 164–194. https://doi.org/10.1177/105065190101500202.

Bakhurst, D. (2009). Reflections on activity theory. *Educational Review, 61*(2), 197–210. https://doi.org/10.1080/00131910902846916.

Bazerman, C., Little, J. & Chavkin, T. (2003). The Production of information for genred activity spaces: Informational motives and consequences of the environmental impact statement. *Written Communication, 20*(4), 455–477. https://doi.org/10.1177/0741088303260375.

Berkenkotter, C. & Huckin, T. N. (1995). *Genre knowledge in disciplinary communication: Cognition/culture/power*. Erlbaum.

Boyle, C. (2016). Pervasive citizenship through #SenseCommons. *Rhetoric Society Quarterly, 46*(3), 269–283. https://doi.org/10.1080/02773945.2016.1171695.

Bracewell, R. J. & Witte, S. P. (2003). Tasks, ensembles, and activity: Linkages between text production and situation of use in the workplace. *Written Communication, 20*(4), 511–559. https://doi.org/10.1177/0741088303260691.

Callon, M. (1986). Some elements of a sociology of translation: Domestication of the scallops and the fishermen of Saint Brieuc Bay. In J. Law (Ed.), *Power, action and belief: A new sociology of knowledge?* (pp. 67–83). Routledge.

Cox, M. B. (2019). Working closets: Mapping queer professional discourses and why professional communication studies need queer rhetorics. *Journal of Business and Technical Communication, 33*(1), 1–25. https://doi.org/10.1177/1050651918798691.

Dush, L. (2015). When writing becomes content. *College Composition and Communication, 67*(2), 173–196.

Engels, F. (1971). *Dialectics of nature*. International Publishers.

Engeström, Y. (1987). *Learning by expanding: An activity-theoretical approach to developmental research*. Orienta-Konsultit Oy. http://lchc.ucsd.edu/mca/Paper/Engestrom/expanding/toc.htm.

Engeström, Y. (2009). The future of activity theory: A rough draft. In A. Sannino, H. Daniels & K. Gutierrez (Eds.), *Learning and expanding with activity theory* (pp. 303–328). Cambridge. https://doi.org/10.1017/CBO9780511809989.020.

Fraiberg, S. (2017). Start-up nation: Studying transnational entrepreneurial practices in Israel's start-up ecosystem. *Journal of Business and Technical Communication*, *31*(3), 350–388. https://doi.org/10.1177/1050651917695541.

Freedman, A. & Smart, G. (1997). Navigating the current of economic policy: Written genres and the distribution of cognitive work at a financial institution. *Mind, Culture, and Activity*, *4*(4), 238–255. https://doi.org/10.1207/s15327884mca0404_3.

Graham, S. S. & Herndl, C. (2013). Multiple ontologies in pain management: Toward a postplural rhetoric of science. *Technical Communication Quarterly*, *22*(2), 103–125. https://doi.org/10.1080/10572252.2013.733674.

Gries, L. E. (2015). *Still life with rhetoric: A new materialist approach for visual rhetorics.* Utah State University Press. https://doi.org/10.7330/9780874219784.

Gygi, K. & Zachry, M. (2010). Productive tensions and the regulatory work of genres in the development of an engineering communication workshop in a transnational corporation. *Journal of Business and Technical Communication*, *24*(3), 358–381. https://doi.org/10.1177/1050651910363365.

Haas, C. (1999). On the relationship between old and new technologies. *Computers and Composition*, *16*(2), 209–228. https://doi.org/10.1016/S8755-4615(99)00003-1.

Haas, C. & Witte, S. (2001). Writing as embodied practice: The case of engineering standards. *Journal of Business and Technical Communication*, *15*(4), 413–457. https://doi.org/10.1177/105065190101500402.

Hart-Davidson, W., Bernhardt, G., McLeod, M., Rife, M. & Grabill, J. (2008). Coming to content management: Inventing infrastructure for organizational knowledge work. *Technical Communication Quarterly*, *17*(1), 10–34. https://doi.org/10.1080/10572250701588608.

Jemielniak, D. (2015). *Common knowledge? An ethnography of Wikipedia*. Stanford University Press.

Jones, J. (2008). Patterns of revision in online writing: A study of Wikipedia's featured articles. *Written Communication*, *25*, 262–289. https://doi.org/10.1177/0741088307312940.

Jones, N. N. (2016). Found things: Genre, Narrative, and identification in a networked activist organization. *Technical Communication Quarterly*, *25*(4), 298–318. https://doi.org/10.1080/10572252.2016.1228790.

Jones, N. N. (2017). Rhetorical narratives of Black entrepreneurs: The business of race, agency, and cultural empowerment. *Journal of Business and Technical Communication*, *31*(3), 319–349. https://doi.org/10.1177/1050651917695540.

Kain, D. & Wardle, E. (2005). Building context: Using activity theory to teach about genre in multi-major professional communication courses. *Technical Communication Quarterly*, *14*(2), 113–139. https://doi.org/10.1207/s15427625tcq1402_1.

Kaptelinin, V. & Nardi, B. A. (2006). *Acting with technology: Activity theory and interaction design*. MIT Press.

Latour, B. (1987). *Science in action: How to follow scientists and engineers through society.* Open University Press.

Latour, B. (1996). *Aramis, or the love of technology*. Harvard University Press.

Law, J. (1994). *Organizing modernity*. Blackwell.

Leontyev, A. N. (2009). *The development of mind.* Marxists Internet Archive.

Mara, A. & Hawk, B. (2010). Posthuman rhetorics and technical communication. *Technical Communication Quarterly*, *19*(1), 1–10. https://doi.org/10.1080/10572250903373031.

McCarthy, J. E., Grabill, J. T., Hart-Davidson, W. & McLeod, M. (2011). Content management in the workplace: Community, context, and a new way to organize writing. *Journal of Business and Technical Communication*, *25*(4), 367–395. https://doi.org/10.1177/1050651911410943.

McNely, B. (2019). Under pressure: Exploring agency–structure dynamics with a rhetorical approach to register. *Technical Communication Quarterly*, *28*(4), 317–331. https://doi.org/10.1080/10572252.2019.1621387.

McNely, B. J. (2009). Backchannel persistence and collaborative meaning-making. In B. Mehlenbacher, A. Protopsaltis, A. Williams & S. Slattery (Eds.), *SIGDOC '09: Proceedings of the 27th ACM International Conference on Design of Communication* (pp. 297–303). ACM. https://doi.org/10.1145/1621995.1622053.

McNely, B., Spinuzzi, C. & Teston, C. (2015). Contemporary research methodologies in technical communication. *Technical Communication Quarterly*, *24*(1), 1–13. https://doi.org/10.1080/10572252.2015.975958.

Miller, C. (2007). Review of *Tracing Genres through Organizations*. *Technical Communication Quarterly*, *16*(4), 476–480. https://doi.org/10.1080/10572250701551432.

Opel, D. S. & Hart-Davidson, W. (2019). The primary care clinic as writing space. *Written Communication*, *36*(3), 1–31. https://doi.org/10.1177/0741088319839968.

Oxford University Press. (n.d.). Actor. In *Oxford English Dictionary*. Retrieved March 5, 2021, from www.oed.com.

Oxford University Press. (n.d.). Activity. In *Oxford English Dictionary*. Retrieved March 5, 2021, from www.oed.com.

Potts, L. (2009). Using actor network theory to trace and improve multimodal communication design. *Technical Communication Quarterly*, *18*(3), 281–301. https://doi.org/10.1080/10572250902941812.

Potts, L. (2010). Consuming digital rights: Mapping the artifacts of entertainment. *Technical Communication*, *57*(3), 300–318. http://www.ingentaconnect.com/content/stc/tc/2010/00000057/00000003/art00005.

Potts, L. & Jones, D. (2011). Contextualizing experiences: Tracing the relationships between people and technologies in the social web. *Journal of Business and Technical Communication*, *25*(3), 338–358. https://doi.org/10.1177/1050651911400839.

Potts, L., Small, R. & Trice, M. (2019). Boycotting the knowledge makers: How Reddit demonstrates the rise of media blacklists and source rejection in online communities. *IEEE Transactions on Professional Communication*, *62*(4), 351–363. https://doi.org/10.1109/TPC.2019.2946942.

Read, S. (2016). The net work genre function. *Journal of Business and Technical Communication*, *30*(4), 419–450. https://doi.org/10.1177/1050651916651909.

Read, S. & Swarts, J. (2015). Visualizing and tracing: Articulated research methodologies for the study of networked, sociotechnical activity, otherwise known as knowledge work. *Technical Communication Quarterly*, *24*(1), 14–44. https://doi.org/10.1080/10572252.2015.975961.

Rice, Jeff. (2012). *Digital Detroit: Rhetoric and space in the age of the network*. Southern Illinois University.

Rice, Jenny. (2012). *Distant publics: Development rhetoric and the subject of crisis*. University of Pittsburgh Press. https://doi.org/10.2307/j.ctt5vkftk.

Rose, E. J. (2016). Design as advocacy. *Journal of Technical Writing and Communication*, *46*(4), 427–445. https://doi.org/10.1177/0047281616653494.

Roth, W. M. & Lee, Y. J. (2007). "Vygotsky's neglected Legacy": Cultural-historical activity theory. *Review of Educational Research*, *77*(2), 186–232. https://doi.org/10.3102/0034654306298273.

Russell, D. R. (1997a). Rethinking genre in school and society: An activity theory analysis. *Written Communication*, *14*(4), 504–554. https://doi.org/10.1177/0741088397014004004.

Russell, D. R. (1997b). Writing and genre in higher education and workplaces: A review of studies that use cultural-historical activity theory. *Mind, Culture, and Activity*, *4*(4), 224–237. https://doi.org/10.1207/s15327884mca0404_2.

Sackey, D. J. (2020). One-size-fits-none: A heuristic for proactive value sensitive environmental design. *Technical Communication Quarterly*, *29*(1), 33–48. https://doi.org/10.1080/10572252.2019.1634767.

Sackey, D. J., Boyle, C., Xiong, M., Rios, G., Arola, K. & Barnett, S. (2019). Perspectives on cultural and posthumanist rhetorics. *Rhetoric Review*, *38*(4), 375–401. https://doi.org/10.1080/07350198.2019.1654760.

Schryer, C. F. (1993). Records as genre. *Written Communication*, *10*(2), 200–234. https://doi.org/10.1177/0741088393010002003.

Schryer, C. F. (2000). Walking a fine line: Writing negative letters in an insurance company. *Journal of Business and Technical Communication*, *14*(4), 445–497. https://doi.org/10.1177/105065190001400402.

Sherlock, L. (2009). Genre, activity, and collaborative work and play in World of Warcraft: Places and problems of open systems in online gaming. *Journal of Business and Technical Communication*, *23*(3), 263–293. https://doi.org/10.1177/1050651909333150.

Slattery, S. (2009). "Edit this page": The socio-technological infrastructure of a Wikipedia article. In B. Mehlenbacher, A. Protopsaltis, A. Williams & S. Slattery (Eds.), *SIGDOC '09: Proceedings of the 27th ACM International Conference on Design of Communication* (pp. 289–295). ACM. https://doi.org/10.1145/1621995.1622052.

Spafford, M. M., Schryer, C. F., Mian, M. & Lingard, L. (2006). Look who's talking: Teaching and learning using the genre of medical case presentations. *Journal of Business and Technical Communication*, *20*(2), 121–158. https://doi.org/10.1177/1050651905284396.

Spinuzzi, C. (2001). Software development as mediated activity: Applying three analytical frameworks for studying compound mediation. In *Proceedings of the 19th annual International Conference on Computer Documentation* (pp. 58–67). ACM Press. https://doi.org/10.1145/501516.501528.

Spinuzzi, C. (2008). *Network: Theorizing knowledge work in telecommunications*. Cambridge University Press. https://doi.org/10.1017/CBO9780511509605.

Spinuzzi, C. (2011). Losing by Expanding: Corralling the runaway object. *Journal of Business and Technical Communication*, *25*(4), 449–486. https://doi.org/10.1177/1050651911411040.

Sun, H. (2006). The triumph of users: Achieving cultural usability goals with user localization. *Technical Communication Quarterly*, *15*(4), 457–481. https://doi.org/10.1207/s15427625tcq1504_3.

Sun, H. (2020). *Global social media design: Bridging differences across cultures*. Oxford University Press. https://doi.org/10.1093/oso/9780190845582.001.0001.

Swarts, J. (2006). Coherent fragments: The problem of mobility and genred information. *Written Communication*, *23*(2), 173–201. https://doi.org/10.1177/0741088306286393.

Swarts, J. (2009). The collaborative construction of "fact" on Wikipedia. In *Proceedings of the 27th ACM International Conference on Design of Communication—SIGDOC '09* (pp. 281–288). ACM. https://doi.org/10.1145/1621995.1622051.

Swarts, J. (2010). Recycled writing: Assembling actor networks from reusable content. *Journal of Business and Technical Communication, 24*(2), 127–163. https://doi.org/10.1177/1050651909353307.

Tardy, C. M. (2003). A genre system view of the funding of academic research. *Written Communication, 20*(1), 7–36. https://doi.org/10.1177/0741088303253569.

Walker, K. (2004). Activity systems and conflict resolution in an online professional communication course. *Business Communication Quarterly, 67*(2), 182–197. https://doi.org/10.1177/1080569904265422.

Walsh, L. (2010). Constructive interference: Wikis and service learning in the technical communication classroom. *Technical Communication Quarterly, 19*(2), 184–211. https://doi.org/10.1080/10572250903559381.

Walton, R. (2013). Stakeholder flux: Participation in technology-based international development projects. *Journal of Business and Technical Communication. 27*(4), 409–435. https://doi.org/10.1177/1050651913490940.

Wegner, D. (2004). The collaborative construction of a management report in a municipal community of practice: Text and context, genre and learning. *Journal of Business and Technical Communication, 18*(4), 411–451. https://doi.org/10.1177/1050651904266926.

Winsor, D. A. (2001). Learning to do knowledge work in systems of distributed cognition. *Journal of Business and Technical Communication, 15*(1), 5–28. https://doi.org/10.1177/105065190101500101.

Witte, S. P. (2005). Research in activity: An analysis of speed bumps as mediational means. *Written Communication, 22*(2), 127–165. https://doi.org/10.1177/0741088305274781.

3. Assessment

Norbert Elliot
New Jersey Institute of Technology

Since its appearance in 1956, the term *assessment* has been straightforward in definition and contested in use (Oxford University Press, n.d.). Harold Loukes of Oxford University first used the noun in his 1956 study of British education, *Secondary Modern*, in which the Quaker educationist was trying to understand how well a selection system was serving students. In England, Wales, and Northern Ireland, the Ministry of Education established secondary modern schools in 1944 for students between 11 and 15 years old who failed to earn high marks on the 11-plus examination. As a contemporary of Loukes saw it, these were schools whose "job it was to cope with all the nation's dull children" (Dent, 1958, p. 31). For Loukes (1956), a way out of this caste system was to find "a new means of assessment" (p. 112), one that would allow secondary modern schools to "find their own place" (1959, p. 139), as he later put it, especially in terms of the value for vocational education.

And so we discover, in the very first use of the term, an enduring tension between the definition of assessment (as "a systematic process to measure or evaluate the characteristics or performance of individuals, programs, or other entities") and complexities surrounding its use ("for the purpose of drawing inferences") (American Educational Research Association [AERA], American Psychological Association [APA], and National Council on Measurement in Education [NCME], 2014, p. 216). Administrative leadership can readily identify a systematic process that will yield findings about students and instructional programs. It is another matter to draw inferences from the findings and have them accepted by stakeholders embedded in complex ***rhetorical*** situations in which cultural and linguistic diversity are of paramount importance (Gonzales & Baca, 2017).

When we talk about assessment in technical and professional communication (TPC), we carry forward this 60–year-old genealogy of complexity. In their foreword to *Assessment in Technical Communication* (2010), the first and only edited collection on the topic, Margaret N. Hundleby and Jo Allen observe that assessment in our field has suffered from irregular attention, uncertainty about authentic strategies, and muddled identification of aims. Recently, Geoffrey Clegg and colleagues (2020) argued that the field of TPC is only now buttressing programmatic student learning outcomes—the objectives upon which an assessment is based—with field-wide data from undergraduate degree programs across the US. An enduring tension—rising in the gap between the straightforward definition of assessment and the complexity of inferences drawn from it—remains.

Today, TPC researchers acknowledge this tension, view it productively, and use it as a means to create innovative assessment programs (St.Amant &

DOI: https://doi.org/10.37514/TPC-B.2023.1923.2.03

Nahrwold, 2007), equip researchers for socially just work through transformative paradigms (Phelps, 2020), and introduce interstitial pedagogical practices that engage diversity (Lane, 2021). Researchers acknowledge that assessment is a situated rhetorical activity that exists on a continuum of rhetorically situated aims. As a situated rhetorical activity, the TPC assessment exists in a sociocultural and sociocognitive environment. Here, John M. Swales' (2017) sociocultural concept of discourse communities as centers of professional identity resonates well with Robert J. Mislevy's (2019) depiction of sociocognition as community discourse practices revealed in linguistic, cultural, and substantive language patterns. Once the deeply situated nature of language is acknowledged, it is then easier to get to the harder realization: that our inferences from assessment are also rhetorically situated and, as such, filled with values both apparent and tacit.

Following acknowledgement of situated language use, assessment stakeholders often adopt two productive strategies for TPC assessment. Each has come into consideration in the 21st century. While one has demonstrated the ability to inform critical research, the other is best considered as a needed reconceptualization.

The first strategy involves reconceptualizing evidence. In 2006, Michael T. Kane proposed that traditional evidence categories of validity (evidence used to support a given interpretation) and reliability (evidence used to support consistency) be understood in terms of interpretation and use. Arguments about interpretation and use, he proposed, allow us to draw inferences and make claims about a given assessment. Gone were totalizing statements ("a given assessment is valid"); present were precise claims supported with evidence ("a given assessment demonstrates evidence of construct validity"). As part of the process of validation, construct validity—evidence that the characteristic the assessment was designed to measure is sufficiently present—was central to a given validity argument. As Kane wrote, "It is the plausibility of the proposed interpretations and uses that is to be evaluated" (p. 23). By 2013, he shifted his terminology to emphasize the interpretation/use argument (IUA)—"the network of inferences and assumptions inherent in the proposed interpretation and use" (p. 2).

This shift was profound and signaled a new era for TPC assessment. As Julia M. Williams (2010) recognized in her explication of the *RosEvaluation* assessment system, first used at Rose-Hulman Institute of Technology in 1998, identification of outcomes shifted institutional focus from identifying resource inputs to defining goals for student learning. With outcomes established, in this case by Accreditation Board for Engineering and Technology (ABET, 2016), evidence of student learning could then be systematically collected (For current criteria, see ABET, 2022.). In turn, this ***information*** could be used to advance opportunity to learn (Moss et al., 2008). As Williams (2021) observed in reflecting on the over-two-decade-old *RosEvaluation* assessment system, one of its notable achievements has been dissemination of communication ***pedagogy*** among technical faculty members to inform the way they use the language of rhetoric in their technical courses. Through discussions of curricular objectives, faculty express willingness to reinforce and extend students' communication development in their classes. Because instruction and assessment are reliably extended across the curriculum, stakeholders see these activities as complementary.

Evidence-based approaches have been accompanied by attention to a category of evidence techniques for the TPC assessment. In 2015, Edward M. White and colleagues designed the first assessment system specifically designed for writing studies. Using the federal Classification of Instructional Programs (CIP) for rhetoric and composition/writing studies, White and his colleagues designed a system in which programs in technical and business writing (CIP 23.1303) and programs in rhetoric and composition (CIP 23.1304) could equally benefit by a unified evidential approach: Design for Assessment (DFA). DFA invited assessment designers to focus on traditional evidential forms of validity and reliability while adding other categories of evidence: consequence (unintended and intended positive and negative impact), theorization (development of ideas concerning a given construct), standpoint (situated perspectives), ***research*** (foundational ***knowledge***), ***documentation*** (evidence gathering), accountability (demonstrated responsibility), sustainability (mission-related resource allocation), process (actions related to program success), and communication (providing information to stakeholders). In a survey approach, these evidence centers have been used by Nancy Coppola and colleagues (2016) to examine TPC program outcomes (Ilyasova & Bridgeford, 2014) and plan evidence-based revision of them. Coppola concludes that evidence models including IUAs provide stakeholders with a principled way to undertake programmatic research. An alternative evidence model for continuous curricular improvement—dedicated to making visible "all of the interrelated work and perspectives" of TPC to ensure that instructional programs continue to "grow and address stakeholder needs in a sustainable way"—has been proposed by Joanna Schreiber and Lisa Melonçon (2019, p. 275). In this model, evidence was collected beyond the program objectives and interpreted by perspectives beyond those of the program administrators.

The second strategy for TPC assessment involves reconceptualized assessment aims. While we have seen research related to evidence-based approaches become significant, we are late to reexamine assessment aims and have yet to witness assessment strategies in our field that are centered on fairness. While no detailed ***history*** of assessment in the field of TPC has been written, Elliot (2010) proposed a conceptual history in which modernism (assessment as an artifact of scientific objectivity) receded as postmodernism (assessment as a contextualized activity) advanced. In general, these phases parallel pedagogical developments in TPC in which instruction dominated by an emphasis on ***style*** and correctness was replaced by social constructivist perspectives on writing (Rude, 2015). Accompanying the move from language objectivity to contextualism, educational measurement researchers have begun to attend to fairness as a category of evidence equal to validity and reliability (AERA, APA, NCME, 2014). To establish evidence of fairness, researchers collect and interpret information in these areas: fairness during the assessment process in areas such as ***accessibility*** for all learners through universal ***design***; measurement bias toward student subgroups in terms of gender assignment and identity, race and ethnicity, socioeconomic characteristics, and other relevant categories and their combination; and access to the constructs being measured through educational opportunity (AERA, APA & NCME, 2014). In TPC, calls for evidence of fairness have been accompanied by attention to ***social justice*** research—a collective and active effort to "reveal, reject, and replace" oppression (Walton et al., 2019, p. 50; see also Agboka & Dorpenyo, 2022, Inoue, 2015; Jones, 2016; Poe et al., 2018; Walton & Agboka, 2021). As Mya Poe (2023) and her colleagues have suggested, even the *Standards for Educational and Psychological Testing*, while necessary, may be insufficient to achieve social justice.

This shift towards reconceptualization of assessment aims must now be accompanied by a fairness approach for TPC assessment. In her guide to mapping institutional values to the technical writing curriculum in order to contextualize assessment, Allen (2010) reminded researchers to consider "the *heritage* [emphasis added] that inspires the institution's traditions," such as that of historically Black colleges and universities and the founders' motivations and vision for women's colleges (p. 41). Advancing this idea of contextualization, Michelle F. Eble (2020) has called for "de-colonial and critical race theory, feminist and queer, and other community participatory approaches" to instruction in technical communication" (p. 40).

In transferring theory into TPC assessment practice, however, researchers have not yet realized the gains associated with evidence of fairness. Here we realize the truth of Miriam F. Williams, our field's first Black Association of Teachers of Technical Writing Fellow, that there is little research that addresses "the unique ways that historically marginalized racial and ethnic groups within the U.S. created or responded to technical communication" (Williams, 2013, p. 86). Put straightforwardly, the consequences of our TPC assessments are unknown in terms of their intersectional impact (Crenshaw, 1991). If assessment is to be a meaningful keyword in our field, then stakeholders will have to use theoretical concepts of diversity such as Black Feminist Theory to generate sources of evidence related to fairness (Collins, 2000). Following Loukes, we need a new means of assessment—an innovation focusing on fairness and consequences as sources of evidence—if we are to advance opportunity to learn and achieve universal justice for all our students.

References

Accreditation Board for Engineering and Technology (ABET). (2016). *Criteria for accrediting engineering programs, 2017–2018*. https://www.abet.org/accreditation/accreditation-criteria/criteria-for-accrediting-engineering-programs-2017-2018/.

Accreditation Board for Engineering and Technology (ABET). (2022). *Criteria for accrediting engineering programs, 2023–2024*. https://www.abet.org/wp-content/uploads/2023/01/23-24-ETAC-Criteria_FINAL.pdf.

Agboka, G. Y. & Dorpenyo, I. K. (2022). The role of technical communicators in confronting injustice—*Everywhere* [Special issue]. *IEEE Transactions on Professional Communication*, *65*(1), 409–411. https://doi.org/10.1109/TPC.2021.3133151.

Allen, J. (2010). Mapping institutional values and the technical communication curriculum: A strategy for grounding assessment. In H. Hundleby & J. Allen (Eds.), *Assessment in technical and professional communication* (pp. 39–56). Baywood.

American Educational Research Association, American Psychological Association, and National Council on Measurement in Education. (2014). *Standards for educational and psychological testing*. American Educational Research Association.

Clegg, G., Lauer, J., Phelps, J. & Melonçon. L. (2020). Programmatic outcomes in undergraduate technical and professional communication programs. *Technical Communication Quarterly*, *30*(1), 19–33. https://doi.org/10.1080/10572252.2020.1774662.

Collins, P. H. (2000). *Black feminist thought: Knowledge, consciousness, and the politics of empowerment* (2nd ed). Routledge.

Coppola, N. W., Elliot, N., Newsham, F. & Klobucar, A. (2016). Programmatic research in technical communication: An interpretive framework for writing program assessment. *Programmatic Perspectives*, *8*(2), 5–45. https://cptsc.org/wp-content/uploads/2018/04/vol8-2.pdf.

Crenshaw, K. (1991). Mapping the margins: Intersectionality, identity politics, and violence against women of color. *Stanford Law Review*, *43*(6), 1241–1299. https://doi.org/10.2307/1229039.

Dent, H. C. (1958). *Secondary modern schools: An interim report*. Routledge and Kegan Paul.

Eble, M. F. (2020). Response: Turning toward social justice approaches to technical and professional communication. In N. Elliot & A. Horning (Eds.), *Talking back: Senior scholars and their colleagues deliberate the past, present, and future of writing studies* (pp. 38–42). Utah State University Press.

Elliot, N. (2010). Assessing technical communication: A conceptual history. In H. Hundleby & J. Allen (Eds.), *Assessment in technical and professional communication* (pp. 17–35). Baywood.

Gonzales, L. & Baca, I. (2017). Developing culturally and linguistically diverse online technical communication programs: Emerging frameworks at University of Texas at El Paso. *Technical Communication Quarterly*, *26*(3), 273–286. https://doi.org/10.1080/10572252.2017.1339488.

Hundleby, M. & Allen, J. (Eds.). (2010). *Assessment in technical and professional communication*. Baywood.

Ilyasova, K. A. & Bridgeford, T. (2014). Establishing an outcomes statement for technical communication. In T. Bridgeford, K. S. Kitalong & B. Williamson (Eds.), *Sharing our intellectual traces: Narrative reflections from administrators of professional, technical, and scientific communication programs* (pp. 53–80). Baywood.

Inoue, A. B. (2015). *Antiracist writing assessment ecologies: Teaching and assessing writing for a socially just future*. The WAC Clearinghouse; Parlor Press. https://doi.org/10.37514/PER-B.2015.0698.

Jones, N. N. (2016). The technical communicator as advocate: Integrating a social justice approach in technical communication. *Journal of Technical Writing and Communication*, *46*(3), 342–361. https://doi.org/10.1177/0047281616639472.

Kane, M. T. (2006). Validation. In R. L. Brennan (Ed.), *Educational measurement* (pp. 17–64). American Council on Education/Praeger.

Kane, M. T. (2013). Validating the interpretation and uses of test scores. *Journal of Educational Measurement*, 50(1), 1–73. https://doi.org/10.1111/jedm.12000.

Lane, L. (2021). Interstitial design processes: How design thinking and social design processes bridge theory and practice in TPC pedagogy. In M. Klein (Ed.), *Effective teaching of technical communication: Theory, practice, and application* (pp. 29–43). The WAC Clearinghouse; University Press of Colorado. https://doi.org/10.37514/TPC-B.2020.1121.2.02.

Loukes, H. (1956). *Secondary modern*. Harrap.

Loukes, H. (1959). The pedigree of the modern school. *British Journal of Educational Studies*, *7*(2), 125–130. https://doi.org/10.1080/00071005.1959.9973019.

Moss, P. A., Pullin, D. C., Gee, J. P., Haertel, E. H. & Young, L. J. (Eds.). (2008). *Assessment, equity, and opportunity to learn*. Cambridge University Press. https://doi.org/10.1017/CBO9780511802157.

Mislevy, R. J. (2018). *Sociocognitive foundations of educational measurement*. Routledge. https://doi.org/10.4324/9781315871691.

Oxford University Press. (n.d.). Assessment. In *Oxford English Dictionary*. Retrieved June 23, 2023, from www.oed.com.

Phelps, J. P. (2020). The transformative paradigm: Equipping technical communication researchers for socially just work. *Technical Communication Quarterly*, *30*(2), 204–215, https://doi.org/10.1080/10572252.2020.1803412.

Poe, M., Inoue, A. B. & Elliot, N. (2018). *Writing assessment, social justice, and the advancement of opportunity*. The WAC Clearinghouse; University Press of Colorado. https://doi.org/10.37514/PER-B.2018.0155.

Poe, M., Oliveri, M.E. & Elliot, N. (2023). The *Standards* will never be enough: A racial justice extension. *Applied Measurement in Education, 36*(3), 193–215. https://doi.org/10.1080/08957347.2023.2214656.

Rude, C. (2015). Technical communication. In P. Heikler & P. Vandenberg (Eds.), *Keywords in writing studies* (pp. 165–168). Utah State University Press. https://doi.org/10.7330/9780874219746.c033.

Schreiber, J. & Melonçon, L. (2019). Creating a continuous improvement model for sustaining programs in technical and professional communication. *Journal of Technical Writing and Communication*, *49*(3), 252–278. https://doi.org/10.1177/0047281618759916.

St.Amant, K. & Nahrwold, C. (2007). Acknowledging complexity: Rethinking program review and assessment in technical communication. *Technical Communication*, *54*(4), 409–411.

Swales, J. M. (2017). The concept of a discourse community: Some recent personal history. *Composition Forum*, *37*. https://compositionforum.com/issue/37/swales-retrospective.php.

Walton, R. & Agboka. G. Y. (2021). *Equipping technical communicators for social justice work: Theories, methodologies, and pedagogies*. Utah State University Press.

Walton, R., Moore, K. R. & Jones, N. N. (2019). *Technical communication after the social justice turn: Building coalitions for action*. Routledge. https://doi.org/10.4324/9780429198748.

White, E. M., Elliot, N. & Peckham, I. (2015). *Very like a whale: The assessment of writing programs.* Utah State University Press.

Williams, J. M. (2010). Evaluating what students know: Using the RosE portfolio system for institutional and program outcomes assessment tutorial. *IEEE Transactions on Professional Communication*, *53*(1), 46–57. https://doi.org/10.1109/TPC.2009.2038737.

Williams, J. W. (2021). Assessment beyond accreditation: Improving the communication skills of engineering and computer science students. In. D. Kelly-Riley & N. Elliot (Eds.), *Improving outcomes: Disciplinary writing, local assessment, and the aim of fairness* (pp. 187–197). Modern Language Association.
Williams, M. F. (2013). A survey of emerging research: Debunking the fallacy of color-blind technical communication. *Programmatic Perspectives*, 5(1), 86–93. https://cptsc.org/wp-content/uploads/2018/04/Vol_5.1.pdf.

4. Audience

Ann M. Blakeslee
Eastern Michigan University

Audience has always been at the core of technical communication, both as a defining concept and as a cornerstone of the field's identity.[1] Two of the most commonly taught principles are "know your audience" and "write for your audience," which students begin hearing in their very first courses—and continue hearing throughout their studies and careers. Historically, the notion of audience and its importance are rooted in classical ***rhetoric***, dating back to at least the fifth century BC. Aristotle's (1926) definition of rhetoric as the "faculty of observing in any given case the available means of persuasion" establishes the importance of those whom a rhetor seeks to persuade. *The Oxford English Dictionary* defined audience initially in relation to judicial hearings and courts of law (Oxford University Press, n.d.). These definitions date back to the 12th century and are rooted in oral traditions. *Hearing*, *being given a hearing*, *being heard*, *attention to what is being spoken*, *performance*, *listeners*, and similar terms and statements are prevalent across the *Oxford English Dictionary* definitions of audience. Less prevalent are the terms *reading*, *readership*, *publication*, and *writer*, which appear in the 18th century, after printed texts had become more commonplace.

Technical communication gained prominence as a ***professional*** field after World War II. Early scholars often considered audience as they defined technical communication or described what technical communicators do. Charles Stratton (1979), for example, said a technical writer in "a particular art, science, discipline, or trade . . . helps audiences approach subjects" (p. 10). Another early scholar, W. Earl Britton (1975), implies an audience, albeit a passive one, when he says, "The primary, though certainly not sole, characteristic of technical and scientific writing lies in the effort of the author to convey one meaning and only one meaning in what he says" (p. 11). A few years later, David Dobrin (1983), in "What's Technical About Technical Writing?," suggested as a new definition: "Technical writing is writing that accommodates technology to the user" (p. 242). Dobrin explained that he focused on "user" rather than "reader," "because technology is meant to be used" (p. 243). As the field has matured, one constant has been the value placed on understanding and writing effectively for audiences

1. Ideas in this chapter, especially those expressed at the end about future directions for audience research and about the fluid roles of writers and readers, were influenced by research and conversations carried out in collaboration with Rachel Spilka, formerly of the University of Wisconsin-Milwaukee. These conversations occurred between 2010 and 2015, and these ideas are connected to concepts that Dr. Spilka and the author developed together.

DOI: https://doi.org/10.37514/TPC-B.2023.1923.2.04

in the workplace, and, in the classroom, on teaching students how to write for audiences. Additionally, audience became what distinguished technical communicators: It is often their ***knowledge*** and skill in addressing audiences that is recognized as "adding value" in the workplace; technical communicators are those best positioned to function as audience or user advocates.

As fields, both technical communication and rhetoric and composition have long and conflicted histories of stressing the importance of audience. Audience figures prominently in Chaïm Perelman and Lucie Olbrechts-Tyteca's (1969) *The New Rhetoric*. They define audience, consonant with Aristotle, "as *the ensemble of those whom the speaker wishes to influence by his argumentation*" (p. 19). They also put forward an idea that has been carried forward in numerous considerations of audience—that knowing an audience with certainty is impossible.

In their germinal 1984 article, Lisa Ede and Andrea Lunsford presented two notions that continue to guide both our scholarship and ***pedagogy***: "audience addressed" and "audience invoked." While the former refers to the "concrete reality of the writer's audience" (p. 156), the latter depicts the audience of "written discourse" as "a construction of the writer" (p. 160). With regard to the latter, they said,

> The central task of the writer, then, is not to analyze an audience and adapt discourse to meet its needs. Rather, the writer uses the semantic and syntactic resources of language to provide cues for the reader—cues which help to define the role or roles the writer wishes the reader to adopt in responding to the text. (p. 160)

There also is a distinction for Ede and Lunsford—and others—between speakers and writers, with speakers, as Perelman and Olbrechts-Tyteca also acknowledge, having the ability to know their audiences with greater certainty. Ede and Lunsford also acknowledge the work of Herbert Simons, who presents "a continuum of audiences based on opportunities for interaction" (Simons, 1976, as cited in Ede & Lunsford, 1984, p. 162). The importance of interactions with audience members has grown over time as ***research*** and theory have placed great emphasis on the roles and responses of readers.

In their efforts to explain audience, many scholars, early on, developed ***visual*** renderings or models, which typically depicted a stable and usually one-directional movement of ***information*** from writers to readers (e.g., Corbett, 1982; Kinneavy, 1971). James Porter (1992) describes such models as misleading (p. xi). He also calls attention to the uneven distribution of power they tended to depict: "Such a conception isolates rhetor from audience, thereby creating a political division that privileges the rhetor with access to knowledge (and hence, truth and power) and that places the audience in a non-participatory subordinate role" (Porter, 1992, p. xi).

A few of the early models were more sophisticated and ahead of their time. J.C. Mathes and Dwight Stevenson (1976), for example, portrayed different "players" (not just writers) interacting in intertextual, interactive contexts while planning, designing, evaluating, and finalizing documents. Their "Interactive Audience Chart" was criticized for being too complex; however, they were innovators in portraying the "*range* of possibilities" (Porter, 1992) and in acknowledging the importance of relationships and interactions between writers and audience members. This is something numerous scholars eventually have also addressed (e.g., Albers, 2003; Beaufort, 2008; Blakeslee, 1993, 2001; Johnson, 1997, 1998; Kitalong, 2004; Long, 1980, 1990; Mirel, 1992, 1998, 2002, 2004; Mirel & Spilka, 2002; Rafoth, 1989; Rosenbaum & Walters, 1986; Roth, 1987; Spilka, 1988a, 1988b, 1990).

As scholars in both composition and technical communication began paying greater attention to concepts like ***genre***, document ***design***, and discourse communities, conceptions of audience evolved (Berkenkotter & Huckin, 1995; Faber & Johnson-Eilola, 2002, 2003; Flower, 1979; Flower & Hayes, 1981; Johnson-Eilola, 1996; Mirel, 2002; Porter, 1992; Schriver, 1997; Spinuzzi, 2003; Swales, 1990). Various theoretical turns have also influenced the field's approaches to audience. For example, reader response theory prompted the field to focus more on how readers respond to writing. Poststructuralism shifted the field's focus to even more fluid and dynamic conceptions of audience, foregrounding interactions between readers and writers and the contributions of readers to developing and evolving texts (Porter, 1992). Over time, the field has shifted from a view of writing as mostly ***style***- and writer-focused to more complex views that attend to how writing actually occurs and to how readers respond to writing. These views also consider how writers might anticipate readers' responses as they plan and design documents, and how writers and readers, through different kinds of interactions and relationships, may even co-construct texts. These shifts have been supported by the field's greater attention to ***collaboration*** and social constructionism, social epistemic perspectives, and, more recently, usability and ***user experience*** research.

In addition to Porter's scholarship, work by scholars such as Ann Blakeslee (1993, 2001), Robert Johnson (1997, 1998), Barbara Mirel (1992, 1998, 2002, 2004), Karla Kitalong (2004), and Rachel Spilka (1988a, 1988b, 1990), among others, also supported these more collaborative and participatory conceptions of audience—with power distributed across an array of "players" as opposed to being situated exclusively in the writer. Kitalong (2004) addresses traditional audience analysis categories, contending that with the proliferation of ***technology*** comes a "proliferation of users, who are now more fully diversified than ever before in terms of the traditional audience-analysis categories of educational background, profession, age, gender, race, and economic status" (p. 171). Blakeslee's (2010) workplace case studies of digital writing suggest a more contextualized approach to analyzing audiences rooted in problem solving. Her findings counter Porter's argument for a universal digital audience. Other workplace researchers also make important contributions to the field's understanding of audience (e.g., Beaufort, 2008; Johnson, 1997; Spilka, 1988a, 1988b, 1990; Spinuzzi, 2003; Winsor, 2003). Although not all of these researchers focused directly on audience, their findings shed light on its complexity. Dorothy Winsor (2003), in *Writing Power*, provides a perspective on the complexity of audiences in the workplace, and Clay Spinuzzi (2003) critiques our field's paternalistic assumption that readers in the workplace are helpless without our support. His critique, however, does not offer suggestions for resolving this. In fact, much that we understand about audience in technical communication comes from works, like these, that address it as an aspect of some other focus. Because of this, they fall short of helping the field develop and test successful, evidence-based conceptions of and approaches to considering audience.

Other important sources for understanding the field's approach to and perspectives on audience are its textbooks. Generally, technical communication textbooks reveal that we continue to rely, even after decades of stressing the importance of audience analysis, on universally applicable and abstract principles in considering audience. Most textbooks still depict audience analysis as a linear, one-way process of identifying, analyzing, and then writing to or accommodating the audience. This process is also still generally portrayed as being controlled by the writer. For example, students often are told to describe their readers using demographic categories, and textbooks also often emphasize using tools such as profiles or late usability tests. However, without access to firsthand information about their audiences, students may mis-categorize and/or simply guess, make up, or overlook important aspects of their readers' experiences and identities (e.g., their ableness, languages, backgrounds, cultures). Our increased and much-needed attention in our field to critical topics like disability, ***social justice***, and anti-racism point to the importance of much more detailed and nuanced considerations of audience that eschew profiling, generalizing, and categorizing in ways that perpetuate the violence and oppression of perspectives like ableism, racism, white supremacy, and xenophobia (Browning & Cagle, 2017; Cedillo, 2018; Colton & Walton, 2015; Condon & Young, 2016; Haas, 2012; Melonçon, 2013; Mutnik, 2015; Oswal, 2013; Palmeri, 2006; Yu, 2012; Zdenek, 2020). Training writers in how best to analyze an audience in a way that is limited to activities of identifying and categorizing them precludes a strong research-backed and inclusive focus on types of analysis that can and must go deeper. In short, students often are taught to cobble together information about audiences from varied sources and to work from more generalized instead of particular, more specific, accurate, and representative conceptions. Few textbooks, for example, advise writers to interact with or research readers directly, which more recent scholarship suggests has value (e.g., Blakeslee, 2001; 2010).

In general, scholarship on audience in technical communication—and rhetoric and composition—has focused mainly on early invention activities—identifying, thinking about, and analyzing audience, generally viewed as a collective. Later stages, including accommodating and influencing audience(s), are still less well understood. There is benefit—and a need—to call into question the status quo around audience and to strive, through empirical research and re-theorizing, to arrive at more expansive and encompassing conceptions. Porter addressed this need in his 1992 work, and it still exists—and is even more urgent, particularly as we interrogate our professional practices for those aspects of them that ultimately are biased, exclusionary, and unjust. Revitalizing and expanding both scholars' and practitioners' understandings of the rhetorical dynamics and complexity of audiences, especially in contemporary contexts, is vital. We must explore and advocate for ways to understand and honor the multiple identities, backgrounds, and lived experiences of our readers.

This is also true in relation to recent and ongoing transformations in how individuals write and work. For example, advancements in technology and in ways we communicate have increasingly blurred the roles of and relationships between writers and readers. In ***social media*** (see, for example, Breuch, 2017, and Potts, 2009, 2014) and other realms, we see how the audience can become writers at any time, and how the principal roles of some writers can be to read and respond to audience input. This points to conceptions of audience that are increasingly relational, discursive, and participatory. Technical communicators need to understand that regardless of the extent of their experience and familiarity with an audience, they must research, continuously, both recurring and new audiences (and this may well necessitate engaging, firsthand, with those audiences). This will assist them with deciding how best to negotiate the ever-changing rhetorical and social contexts of each writing task. Rather than "writing *to* or *for* an audience," we should be thinking instead about "writing *with* an audience" or "writing *as part of* an audience."[2] Audience, moving forward, must be addressed in the context of 21st century writing, technology, workplace contexts, social consciousness, and cultural responsiveness. Rather than privileging writers in relation to readers and end users, and as is often the case only certain readers and end users, technical communication scholars can strive to develop new theories and practices that align more closely with current trends in digital ***literacy***, participatory rhetoric, anti-racist pedagogy, social justice, disability studies, and user engagement.

References

Albers, M. (2003). Multidimensional audience analysis for dynamic information. *Journal of Technical Writing and Communication*, *33*, 263–279. https://doi.org/10.2190/6KJN-95QV-JMD3-E5EE.

Aristotle. (1926). *The art of rhetoric* (J. H. Freese, Trans.). Harvard University Press. https://doi.org/10.4159/DLCL.aristotle-art_rhetoric.1926.

Beaufort, A. (2008). Writing in the professions. In C. Bazerman (Ed.), *Handbook of research on writing: History, society, school, individual, text* (pp. 221–235). Lawrence Erlbaum.

Berkenkotter, C. & Huckin, T. N. (1995). *Genre knowledge in disciplinary communication: Cognition/culture/power.* Lawrence Erlbaum Associates. https://doi.org/10.2307/358302.

2. This idea emerged from the 2010 to 2015 work and conversations of Rachel Spilka and the author.

Blakeslee, A. M. (1993). Readers and authors: Fictionalized constructs or dynamic collaborations? *Technical Communication Quarterly*, *2*, 23–35. https://doi.org/10.1080/10572259309364521.
Blakeslee, A. M. (2001). *Interacting with audiences: Social influences on the production of scientific writing*. Lawrence Erlbaum Associates.
Blakeslee, A. M. (2010). Addressing audiences in a digital age. In R. Spilka (Ed.), *Digital literacy for technical communication: 21st century theory and practice* (pp. 199–229). Routledge.
Breuch, L. K. (2017). Audience involvement: The role of social media in web usability. In *IEEE International Professional Communication Conference (ProComm)* (pp. 1–6). IEEE. https://doi.org/10.1109/IPCC.2017.8013950.
Britton, W. E. (1975). What is technical writing? A redefinition. In D. H. Cunningham & H. A. Estrin (Eds.), *The teaching of technical writing*. National Council of Teachers of English.
Browning, E. R. & Cagle, L. E. (2017). Teaching a "critical accessibility case study": Developing disability studies curricula for the technical communication classroom. *Journal of Technical Writing and Communication*, *47*, 440–463. https://doi.org/10.1177/0047281616646750.
Cedillo, C. (2018). What does it mean to move?: Race, disability and critical embodiment pedagogy. *Composition Forum*, *39*. http://compositionforum.com/issue/39/to-move.php.
Colton, J. S. & Walton, R. (2015). Disability as insight into social justice pedagogy in technical communication. *The Journal of Interactive Technology & Pedagogy*, *8*. https://jitp.commons.gc.cuny.edu/disability-as-insight-into-social-justice-pedagogy-in-technical-communication/.
Condon, F. & Young, V. A. (Eds.). (2017). *Performing antiracist pedagogy in rhetoric, writing, and communication*. University of Colorado Press. https://doi.org/10.37514/ATD-B.2016.0933.
Corbett, E. P. J. (1982). *The little rhetoric & handbook* (2nd ed.). Scott Foresman.
Dobrin, D. N. (1983). What's technical about technical writing? In P. V. Anderson, R. J. Brockman & C. R. Miller (Eds.), *New essays in technical and scientific communication: Research, theory, and practice*. Routledge.
Ede, L. & Lunsford, A. (1984). Audience addressed/audience invoked: The role of audience in composition theory and pedagogy. *College Composition and Communication*, *35*, 155–171. https://doi.org/10.2307/358093.
Faber, B. & Johnson-Eilola, J. (2002). Migrations: Strategic thinking about the future(s) of technical communication. In B. Mirel & R. Spilka (Eds.), *Reshaping technical communication: New directions and challenges for the 21st century* (pp. 135–163). Lawrence Erlbaum.
Faber, B. & Johnson-Eilola, J. (2003). Universities, corporate universities, and the new professionals: Professionalism and the knowledge economy. In T. Kynell-Hunt & G. J. Savage (Eds.), *Power and legitimacy in technical communication. Volume 1: The historical and contemporary struggle for professional status* (pp. 209–234). Baywood Publishing.
Flower, L. (1979). Writer-based prose: A cognitive basis for problems in writing. *College English*, *41*, 19–38. https://doi.org/10.2307/376357.
Flower, L. & Hayes, J. R. (1981). A cognitive process theory of writing. *College Composition and Communication*, *32*, 365–387. https://doi.org/10.2307/356600.

Haas, A. M. (2012). Race, rhetoric, and technology: A case study of decolonial technical communication theory, methodology, and pedagogy. *Journal of Business and Technical Communication*, *26*, 277–310. https://doi.org/10.1177/1050651912439539.

Johnson, R. R. (1998). *User-centered technology: A rhetorical theory for computers and other mundane artifacts*. State University of New York Press.

Johnson, R. R. (1997). Audience involved: Toward a participatory model of writing. *Computers and Composition*, *14*, 361–376. https://doi.org/10.1016/S8755-4615(97)90006-2.

Johnson-Eilola, J. (1996). Relocating the value of work: Technical communication in a post-industrial age. *Technical Communication Quarterly*, *5*, 245–270. https://doi.org/10.1207/s15427625tcq0503_1.

Kinneavy, J. L. (1971). *A theory of discourse: The aims of discourse*. Norton.

Kitalong, K. S. (2004). Who are the users? Media representations as audience-analysis teaching tools. In T. Bridgeford, K. S. Kitalong & D. Selfe (Eds.), *Innovative approaches to teaching technical communication* (pp. 168–182). Utah State University Press. https://doi.org/10.2307/j.ctt46nzds.13.

Long, R. (1980). Writer-audience relationships: Analysis or invention? *College Composition and Communication*, *31*, 221–226. https://doi.org/10.2307/356377.

Long, R. C. (1990). The writer's audience: Fact or fiction? In G. Kirsch & D. H. Roen (Eds.), *A sense of audience in written communication. Written communication annual: An international survey of research and theory* (Vol. 5, pp. 73–84). Sage.

Mathes, J. C. & Stevenson, D. W. (1976). *Designing technical reports: Writing for audiences in organizations*. Bobbs-Merrill.

Melonçon, L. (Ed.). (2013). *Rhetorical accessibility: At the intersection of technical communication and disability studies*. Routledge.

Mirel, B. (1992). Analyzing audiences for software manuals: A survey of instructional needs for "real world tasks." *Technical Communication Quarterly*, *1*, 13–38. https://doi.org/10.1080/10572259209359489.

Mirel, B. (1998). "Applied constructivism" for user documentation: Alternatives to conventional task orientation. *Journal of Business and Technical Communication*, *12*, 7–49. https://doi.org/10.1177/1050651998012001002.

Mirel, B. (2002). Advancing a vision of usability. In B. Mirel & R. Spilka (Eds.), *Reshaping technical communication: New directions and challenges for the 21st century* (pp. 165–187). Lawrence Erlbaum. https://doi.org/10.4324/9781410603739.

Mirel, B. (2004). *Interaction design for complex problem solving: Developing useful and usable software*. Elsevier. https://doi.org/10.1016/B978-155860831-3/50000-X.

Mirel, B. & Spilka, R. (2002). Appendix: Proposed research agenda for technical communication. In B. Mirel & R. Spilka (Eds.), *Reshaping technical communication: New directions and challenges for the 21st century* (pp. 197–201). Lawrence Erlbaum. https://doi.org/10.4324/9781410603739.

Mutnik, D. (2015). The rhetorics of race and racism: Teaching writing in an age of colorblindness. *Literacy in Composition Studies*, *3*, 71–81. https://doi.org/10.21623/1.3.1.5.

Oswal, S. K. (2013). Exploring accessibility as a potential area of research for technical communication: A modest proposal. *Communication Design Quarterly*, *1*, 50–60. https://doi.org/10.1145/2524248.2524261.

Oxford University Press. (n.d.). Audience. In *Oxford English Dictionary*. Retrieved August 1, 2020, from www.oed.com/.

Palmeri, J. (2006). Disability studies, cultural analysis, and the critical practice of technical communication pedagogy. *Technical Communication Quarterly*, *15*, 49–65. https://doi.org/10.1207/s15427625tcq1501_5.

Perelman, C. & Olbrechts-Tyteca, L. (1969). *The new rhetoric: A treatise on argumentation* (J. Wilenson & P. Weaver, Trans.). University of Notre Dame Press.

Porter, J. E. (1992). *Audience and rhetoric: An archeological composition of the discourse community.* Prentice Hall.

Potts, L. (2009). Designing for disaster: Social software use in times of crisis. *International Journal of Sociotechnology and Knowledge Development*, *1*(2), 33–46. http://www.igi-global.com/article/designing-disaster-social-software-use/4094.

Potts, L. (2014). *Social media in disaster response: How experience architects can build for participation.* Routledge. https://www.routledge.com/Social-Media-in-Disaster-Response-How-Experience-Architects-Can-Build/Potts/p/book/9780415817417.

Rafoth, B. A. (1989). Audience and information. *Research in the Teaching of English*, *23*, 273–290.

Rosenbaum, S. & Walters, R. D. (1986). Audience diversity: A major challenge in computer documentation. *IEEE Transactions on Professional Communication*, *29*, 48–55. https://doi.org/10.1109/TPC.1986.6448989.

Roth, R. G. (1987). The evolving audience: Alternatives to audience accommodation. *College Composition and Communication*, *38*, 47–55. https://doi.org/10.2307/357586.

Schriver, K. A. (1997). *Dynamics in document design: Creating text for readers.* Wiley.

Simons, H. W. (1976). *Persuasion: Understanding, practice, and analysis.* Addison-Wesley.

Spilka, R. (1988a). *Adapting discourse to multiple audiences: Invention strategies of seven corporate engineers* [Unpublished doctoral dissertation]. Carnegie Mellon University.

Spilka, R. (1988b). Studying writer-reader interactions in the workplace. *The Technical Writing Teacher*, *15*, 208–221.

Spilka, R. (1990). Orality and literacy in the workplace: Process and text-based strategies for multiple audience adaptation. *Journal of Business and Technical Communication*, *4*, 44–67. https://doi.org/10.1177/105065199000400103.

Spinuzzi, C. (2003). *Tracing genres through organizations: A sociocultural approach to information.* MIT Press. https://doi.org/10.7551/mitpress/6875.001.0001.

Stratton, C. R. (1979). Technical writing: What it is and what it isn't. *Journal of Technical Writing and Communication*, *9*, 9–16. https://doi.org/10.2190/N664-6J52-WGB1-RYLX.

Swales, J. M. (1990). *Genre analysis: English in academic and research settings.* Cambridge University Press.

Winsor, D. A. (2003). *Writing power: Communication in an engineering center.* State University of New York Press.

Yu, H. (2012). Intercultural competence in technical communication: A working definition and review of assessment methods. *Journal of Technical Writing and Communication*, *42*, 159–181. https://doi.org/10.1080/10572252.2012.643443.

Zdenek, S. (2020). Transforming access and inclusion in composition studies and technical communication. *College English*, *82*, 536–544.

5. Collaboration

Pam Estes Brewer
Mercer University

Collaboration is an important tool for technical communicators and has been since the early days of the discipline; it will become even more significant in the future as global markets expand. Thus, the topic of collaboration is featured in the articles, books, conferences, and workplaces of the discipline—always in the background and often in the foreground. *One could say that collaboration is part of the foundation of the practice and study of technical communication.* However, practitioners and educators must consider collaboration in ways that they did not several decades or even several years ago. This essay briefly addresses the ***history*** of collaboration in the discipline, shares some common definitions of the term, and then examines several of the most important perspectives on collaboration today, including types of collaboration, ***technology's*** impact on collaboration, and ***information*** development with collaboration.

The history of collaboration in technical communication can be seen in its evolving definitions. The origins of the word are from the Latin *collaborare*, meaning to labor together. Examples in the *Oxford English Dictionary* show *collaboration* used interchangeably with *cooperation* (Oxford University Press, n.d.). Some early technical communication scholarship on collaboration focuses on collaboration as social construction; this scholarship claims that "***knowledge***, reality, and even facts are community generated . . . with knowledge being composed by collaboration" (Hedden, 1992, p. 27).

As collaborators began to increase their use of technology, scholarship on collaboration began to incorporate computer-based collaboration and suggested that it provided a more egalitarian setting (Selfe, 1992). We began to see technology changing the very nature of collaboration, enabling collaborators to question and negotiate meaning more freely (Selfe, 1992). In educational settings, students who may have been marginalized and less inclined to speak in face-to-face collaboration could find a voice in computer-based collaboration (Trimbur, 1983). Increasingly today, educators and employers seek ways to prepare students and employees to take advantage of remote collaboration (Brewer, 2015; Brewer et al., 2015; Mitchell, 2012; Mitchell et al., 2010; Wojahn et al., 2010). They also study the effects of trust and psychological safety on performance (Robinson et al., 2016).

In the workplace, collaboration is an intellectual endeavor that produces intellectual property. Purely theoretical definitions of the term are of limited use. In industry settings, "collaboration may more resemble cooperation in that the team's responsibilities include ensuring coverage, avoiding duplication, creating links, and ensuring consistency of organization" (Hewett et al., 2010, p. 4).

DOI: https://doi.org/10.37514/TPC-B.2023.1923.2.05

Workplace definitions of collaboration are based on products and productivity; collaboration is transactional in nature. It is viewed as a tool that helps people conduct business. In her editorial on the professionalization of technical communication, Nancy Coppola (2011) notes that managers expect collaboration among subject-matter experts and coworkers. In fact, the boundaries between terms like *group*, *team*, *collaboration*, and *cooperation* become blurry in current workplace use. Nevertheless, we can find useful definitions of collaboration that are specific to technical communication and writing. Beth L. Hewett and colleagues (2010) write that collaboration "involves strategic and generative interactivity among individuals seeking to achieve a common goal, such as problem solving, knowledge sharing, and advancing discovery" (p. 9). Rebecca Burnett et al. (2013) define collaboration as "an intentional, sustained interaction toward a common goal" (p. 454). Peter S. England and Pam Estes Brewer (2018) write that "true collaboration results in outputs better than what could have been achieved by a single person" (p. 161).

Because the definitions of collaboration are nuanced, there is some debate about what constitutes *effective* collaboration within technical communication and *when* it should take place in the information-development process. There is no shortage of stories about collaboration gone wrong (e.g., Brewer, 2015; Mamishev & Williams, 2010). Barriers to effective collaboration in virtual writing include training and technology, an organization's ability to create a culture for effective collaboration, personality characteristics, and team composition (Carney, 2010). *To be effective, collaboration in technical communication today requires common goals, a focus on the whole rather than the individual, effective use of technology, and sustained communication.* In addition, the types of collaboration must meet the needs of the project.

Technical communicators can choose the best type(s) of collaboration for the context in which they are working. Most types of collaboration can be described based on two characteristics: power structure and synchronicity. Power structure refers to whether the relationships among collaborators are largely horizontal (where all collaborators have a relatively equal voice) or vertical (where one or several of the collaborators have more authority over the collaboration). Synchronicity refers to whether or not collaborators are present together in time.

Alternatively, Hewett et al. (2010) provide a useful schema with three types of collaboration: serial, parallel, and collective. These types are identified in the context of virtual collaborative writing, but they represent well the types of collaboration common in technical communication. Briefly, serial collaborators work one after the other, while parallel collaborators work on different pieces of a project at the same time. Collective collaborators use both serial and parallel collaboration while working on the project as a whole (Hewett et al., 2010). For example, the composition program at Texas Tech University used decentralized grading groups (Carney, 2010) wherein graduate instructors collaborated collectively to improve assessment skills and in parallel to grade student projects. Similar to the collective model of collaboration identified by Hewett et al. (2010) is the interlaced model of collaboration advocated by Robinson et al. (2016): "[Interlaced collaborative writing] is a distributed practice, predicated on psychological safety that promotes iterative CCK [co-construction of knowledge] by allowing for both parallel and synchronous discussion and production of texts with intense periods of simultaneous production."

In comparison, Scott L. Jones (2007) identifies three primary classifications of collaboration based on a survey of 1,790 members of the Society for Technical Communication: contextual collaboration (using templates, ***genres***, and existing ***documentation***); hierarchical collaboration ("carefully, and often rigidly, structured, driven by highly specific goals, and carried out by people playing clearly defined and delimited roles" [Ede & Lunsford, 1990, p. 133]); and group collaboration ("involv[ing] a collection of people who largely plan, draft, and revise together" [Jones, 2005, p. 454]). Note that Jones (2007) adds a category called "contextual collaboration" wherein communicators collaborate with artifacts produced by others rather than with people directly. In this type of collaboration, communicators work with existing artifacts, such as documentation. Jones' classifications move from what he calls less overt to more overt communication, with contextual collaboration representing the least overt.

Quickly evolving technologies have enabled more and more diverse forms of collaboration than ever before, and the speed of change shows no signs of slowing. Software tools that support collaboration include information communication technologies (ICTs) like web conferencing and email; ***content***, learning, and ***project management*** systems; virtual worlds (e.g., Bosch-Sijtsema & Sivunen, 2013; Brewer et al., 2015); development software; and some ***social media***. Suites of tools support collaboration by enabling conversation, storage, scheduling, and more. With these tools, collaboration today is inter/intraorganizational, inter/intradisciplinary, ***inter/intranational***, and inter/intra-market sector in ways that it was not prior to advances in technology. Technical communicators and their organizations can be attentive to these opportunities or ignore them at the risk of surrendering the benefits to competitors.

In order to develop these collaborative opportunities, one must fit the technologies to the task just as one fits type of collaboration to the task. One of the best ways to do so is to consider the affordances of technologies. The technologies themselves may change, but the affordances that collaborators need remain relatively stable. Hewett et al. (2010) developed a list of four affordances of technology that can be helpful in evaluating technology choices for collaboration:

- Presence awareness is "the degree to which individuals in virtual settings know that others are present or available to communicate."
- Synchronicity is "the length of time it takes for individuals to interact using virtual collaborative technology."

- Hybridity is "the use of tools that combine different elements of communication, such as speech and written language."
- Interactivity is "the extent to which individuals can maintain a dynamic flow of communication across virtual space and interactions made when a tool seems to diminish spatial distance." (p. 12)

For example, when a technical communicator wishes to have a dialog with a colleague, they can choose technology that offers rich or lean communication. Rich media support multiple cues (similar to face-to-face communication). The closer a medium is to face-to-face communication, the richer it is. For example, video conferencing is a rich medium as it offers audio and visual cues in real time; it offers high levels of presence awareness, synchronicity, hybridity, and interactivity. The more ambiguous or complex a task, the more richness is needed for the communication. Lean media support fewer cues—for example, email is a lean medium, as it offers only text with some delay. Lean media can be very effective for communicating concrete information because they decrease unnecessary cues. Technical communicators might use such a list of affordances to guide them in choosing the technology to support the collaboration for a given project.

Technology has not only changed the way that technical communicators collaborate to create content, it has changed their roles and required them to collaborate in higher order tasks, such as information architecture (Jones, 2005), in order to manage the technologies, collaborations, and products. Within the field of technical communication, the primary goal of collaboration is information development. Because information development has become far more complex than it was several decades ago, the collaboration that supports it has also become more complex. As predicted by Brad Mehlenbacher (2013), "Future technical communicators will serve as knowledgeable team members, learning, researching, organizing, and synthesizing the many support materials that are required to mediate between communication design, humans, and complex technological processes and products" (p. 205). Instead of collaboration taking place face-to-face, it often takes place remotely. Instead of products being released as stand-alone versions on individual platforms, they are often released in small updates and for multiple platforms. And an increased collaboration between producers and their users significantly affects ***design***. For reasons like these, information products are most often developed via collaboration of many people, and projects require new roles for technical communicators as information coordinators.

Collaboration in technical communication today is complex, facilitated by many choices in both structure and technology. Effective collaboration requires thought and planning, whether that collaboration takes place face to face or at a distance. As a world market, we will need the many types and tools of collaboration to meet such challenges as protecting the environment (Nidumolu et al., 2014); creating networks among ***science***, education, and business (Basov &

Minina, 2018); and addressing global health crises. In fact, the future of technical communication likely depends on effective collaboration to enable technical communicators to function as a part of the development and innovation process (Giammona, 2004) and to create professional presence in a global market.

References

Basov, N. & Minina, V. (2018). Personal communication ties and organizational collaborations in networks of science, education, and business. Journal of Business and Technical Communication, 32(3), 373–404. https://doi.org/10.1177/1050651918762027.

Bosch-Sijtsema, P. M. & Sivunen, A. (2013). Professional virtual worlds supporting computer-mediated communication, collaboration, and learning in geographically distributed contexts. IEEE Transactions on Professional Communication, 56(2), 160–175. https://doi.org/10.1109/TPC.2012.2237256.

Brewer, P. E. (2015). International virtual teams: Engineering global success. IEEE Press; Wiley. https://doi.org/10.1002/9781118886465.

Brewer, P. E., Mitchell, A., Sanders, R., Wallace, P. & Wood, D. D. (2015). Teaching and learning in cross-disciplinary virtual teams. IEEE Transactions on Professional Communication, 58(2), 208–229. https://doi.org/10.1109/TPC.2015.2429973.

Burnett, R. E., Cooper, L. A. & Welhausen, C. A. (2013). What do technical communicators need to know about collaboration? In J. Johnson-Eilola & S. A. Selber (Eds.), Solving problems in technical communication (pp. 454–478). The University of Chicago Press.

Carney, W. (2010). Removing barriers to collaborating in virtual writing projects. In B. L. Hewett & C. Robidoux (Eds.), Virtual collaborative writing in the workplace: Computer-mediated communication technologies and processes (pp. 128–143). Information Science Reference. https://doi.org/10.4018/978-1-60566-994-6.ch007.

Coppola, N. W. (2011). Professionalization of technical communication: Zeitgeist for our age. Technical Communication, 58(4), 277–284.

Ede, L. & Lunsford, A. (1990). Singular texts/plural authors. Southern Illinois University Press.

England, P. S. & Brewer, P. E. (2018). What do instructors need to know about teaching collaboration? In T. Bridgeford (Ed.), Teaching professsional and technical communication (pp. 159–175). Utah State University Press. https://doi.org/10.7330/9781607326809.c010.

Giammona, B. A. (2004). The future of technical communication: How innovation, technology, information management, and other forces are shaping the future of the profession. *Technical Communication, 51*(3), 349–366.

Hedden, C. (1992). Hypertext and collaboration: Observations on Edward Barrett's philosophy. Technical Communication Quarterly, 1(4), 27–41. https://doi.org/10.1080/10572259209359511.

Hewett, B. L., Robidoux, C. & Remley, D. (2010). Principles for exploring virtual collaborative writing. In B. L. Hewett & C. Robidoux (Eds.), Virtual collaborative writing in the workplace: Computer-mediated communication technologies and processes (pp. 1–27). Information Science Reference. https://doi.org/10.4018/978-1-60566-994-6.

Jones, S. L. (2005). From writers to information coordinators: Technology and the changing face of collaboration. Journal of Business and Technical Communication, 19(4), 449–467. https://doi.org/10.1177/1050651905278318.

Jones, S. L. (2007). How we collaborate: Reported frequency of technical communicators' collaborative writing activites. Technical Communication, 54(3), 283–294.

Mamishev, A. V. & Williams, S. D. (2010). Technical writing for teams. Wiley. https://doi.org/10.1002/9780470602706.

Mehlenbacher, B. (2013). What is the future of technical communication? In J. Johnson-Eilola & S. A. Selber (Eds.), Solving problems in technical communication (pp. 187–208). University of Chicago Press.

Mitchell, A. (2012). Interventions for effectively leading in a virtual setting. Business Horizons, 55, 431–439. https://doi.org/10.1016/j.bushor.2012.03.007.

Mitchell, A., Chen, C. & Medlin, D. (2010). Teaching and learning with Skype. In C. Wankel (Ed.), Cutting-edge social media approaches to business education: Teaching with LinkedIn, Facebook, Twitter, Second Life, and blogs (pp. 39–56). Information Age Publishing.

Nidumolu, R., Ellison, J., Whalen, J. & Billman, E. (2014, April). The collaborative imperative. Harvard Business Review, 92(4), 76–84.

Oxford University Press. (n.d.). Collaboration. In *Oxford English Dictionary*. Retrieved May 31, 2021, from https://www.oed.com/.

Robinson, J., Dusenberry, L. & Lawrence, H. M. (2016). Collaborative strategies for distributed teams: Innovation through interlaced collaborative writing [Paper presentation]. IEEE International Professional Communication Conference, Austin, TX, United States. https://doi.org/10.1109/IPCC.2016.7740489.

Selfe, C. L. (1992). Computer-based conversations and the changing nature of collaboration. In J. Forman (Ed.), New visions of collaborative writing (pp. 147–169). Boynton/Cook.

Trimbur, J. (1983). Consensus and difference in collaborative learning. College English, 5, 602–616. https://doi.org/10.2307/377955.

Wojahn, P. G., Blicharz, K., A. & Taylor, S. K. (2010). Engaging in virtual collaborative writing: Issues, obstacles, and strategies. In B. L. Hewett & C. Robidoux (Eds.), Virtual collaborative writing in the workplace: Computer-mediated communication technologies and processes (pp. 65–88). Information Science Reference. https://doi.org/10.4018/978-1-60566-994-6.ch004.

6. Content Management

Tatiana Batova
University of Virginia

Rebekka Andersen
University of California, Davis

Content management (CM) refers to the methodologies, processes, standards, and technologies that allow communicators to create and manage ***information*** as modular units for the purpose of reuse and multi-channel publishing. Having first emerged as an interdisciplinary area of practice in the mid-1990s, content management has redefined what it means to be a communicator and conduct technical communication (TC) work. For technical communicators, the practice has introduced new approaches to writing, new processes for managing and publishing content, new roles such as information architects and content strategists, and new competencies such as structured authoring and business analysis.

CM allows for writing content once and reusing it by different people at different times and in different contexts to create any number of information products. These products can be published through various delivery channels (e.g., websites, mobile applications, and ebooks) and accessed from various devices. In CM literature, the term *content* is typically described as "any text, image, video, decoration, or user-consumable elements" that help people understand "an organization's products or services, stories, and brand" (Abel, 2014, p. 12). It is what we produce (Abel, 2014) but also "how we produce and update" (Hart, 2013, p. 30). In technical terms, content is the meaning that is held within and transported by a container (Abel, 2014)—a set of standard markup tags that contain the content and allow for automated processing of content. For example, a single content unit might be a "medication description" that can be simultaneously published to a PDF of an informed consent, an online Q&A, and a medication insert. Content units can also be building blocks, allowing customers to generate a user guide on demand based on the product features that are relevant for them and for the device they are using.

In the field of TC, CM has often been used to refer to both web content management (WCM) and component content management (CCM). Whereas WCM has focused on approaches and technologies for creating, presenting, and maintaining content on websites (Clark, 2008) and for managing the web ***user experience*** (Gollner, 2015), CCM has focused on approaches for creating and managing content as small units of information rather than as entire documents. However, these distinctions in approaches are increasingly blurring because organizations now must produce content that can be rendered in different outputs for

DOI: https://doi.org/10.37514/TPC-B.2023.1923.2.06

different delivery channels, a process that necessarily relies on principles of reuse, granularity, and ***structure***.

Reusable content has to have the potential to become various types of information (Gollner, 2013). Content thus must be freed from the confines of presentation so that it can be manipulated in multiple ways; markup tags that describe the content enable this manipulation. Content as potential information possesses the following qualities: it is dynamic (able to stay fresh and be subject to ongoing revision), customizable (able to change based on ***audiences'*** needs and preferences), linked and distributed (able to be reused), granular (able to communicate meaning at a micro-level), and interactive (able to provide users the support they need when they need it; Hart-Davidson, 2005, p. 29).

Granular content is the smallest unit of usable information (Sapienza, 2007), e.g., a warning statement or the procedure for accomplishing a particular task. In contrast, content at the document level is that of complete information products, e.g., user guides, training modules, technical bulletins. It is important to note that the relationship between the two levels is dynamic: what we consider a complete information product can in some cases also be the smallest usable unit, e.g., a mission statement. While several terms have been used to describe granular content, the term *topic* grew to be the most commonly and extensively defined. Topic derives from Darwin Information Typing Architecture (DITA), the open content standard that defines a common structure for content. In DITA, the term *topic* describes the content type and structure allowed for that content type.

Structured content enables reuse and multi-channel publishing, key goals of CM, through its use of "semantic rules that allow machine processing to meet specific business requirements" (Day, 2014, p. 62). Its mobile affordances give it the potential to automatically adjust to specific user requests and device capabilities such as screen size and orientation. Such content has been described as "adaptive" (Cooper, 2014; McGrane, 2012), "future-ready" (Wachter-Boettcher, 2012), "intelligent" (Gollner, 2010, 2014; Rockley & Cooper, 2012), "nimble" (Lovinger, 2010), "portable" (Bailie, 2009), and "smart" (Bock et al., 2010).

CM has a rich ***history*** in TC and has been a prominent practice since the mid-1990s. At that time, the need to keep pace with shorter product development cycles, to improve content quality and consistency, to expand product ***documentation*** into additional languages—and to do it all with smaller budgets—led some early adopter TC work groups to replace the desktop publishing approach to technical information with the CM approach. Early on, CM was most commonly centered on product documentation because the main purpose of the approach was to efficiently and effectively reuse information between similar products or versions of the same product.

Towards and into the early and mid-2000s, definitions and descriptions of CM as a new approach to technical publishing began to appear in the literature. These definitions and descriptions primarily focused on the separation of form and content and the shift from the craftsperson (one author crafting a complete text) to the industrial (assembly-line texts created from parts written by multiple authors) approach to writing. During this time, the term single-sourcing was most commonly applied to describe CM (see, e.g., Albers, 2003; Ament, 2003; Rockley, 2001). Single-sourcing refers to a method for writing small content units once, storing them in a single information source, and reusing them in multiple contexts for multiple purposes (Ament, 2003; O'Keefe, 2009; Rockley et al., 2010). Whereas trade publications led the way in defining single-sourcing and its best practices, scholarly publications offered more critical perspectives, such as questioning the readily-accepted fact that separation of content and form is good (Clark, 2008) or theorizing single-sourcing as a ***rhetorical*** act (see, e.g., Albers, 2000; Hart-Davidson, 2005; Sapienza, 2007).

From the mid-2000s into the early 2010s, concerns shifted towards the many problematic and sometimes failed implementations of CM (see, e.g., Andersen, 2011; Bailie, 2007; Schumate, 2011), particularly content management systems (CMSs), which are packages of integrated technologies (XML authoring tools, schemas or document type definitions, database platforms, and publishing engines) used to collect, manage, and publish large quantities of content components (Andersen & Batova, 2015). Given these concerns, authors of scholarly publications sought to better understand CMS adoption challenges and contributed ***research***-based heuristics and theoretical frameworks for studying CMS adoption (see, e.g., Andersen, 2014; Batova & Clark, 2015; Dayton, 2006) as well as theoretical frameworks for understanding content reuse and ***knowledge*** work in CM contexts (Hart-Davidson, 2009; Swarts, 2010, 2011).

During this period, ***translation*** and localization practices also received increased attention, because CM promised significant return on investment (ROI) in these areas. Trade publications typically focused on the "why" (making a business case for CM) and "how" of multilingual CM (e.g., indexing DITA topics for translation, adapting XML for localization purposes, publishing multilingual content with a CMS, and integrating translation memory with a CMS; e.g., Cowan, 2010; Freeman, 2006; Hackos, 2008, 2010; Swisher, 2014).

Potential issues of using CM for translation and localization were also points of discussion. These issues, among others, included micro levels of segmentation leading to ungrammatical translation for highly inflected languages, lack of training for translators who are traditionally freelancers, and problematic implications for job satisfaction and motivation (Batova, 2018b; Byrne, 2013; Gattis, 2008; Swisher, 2011). The issues surrounding translation and localization continued into the 2010s, with academic authors calling for more collaborative, user-focused, highly contextualized strategies for translation and localization quality assurance (Batova, 2014, 2018a, 2019; Batova & Clark, 2015).

The rate at which industry was adopting CM in the 2010s incited many academic authors to research and develop approaches to teaching CM and the competencies and skills needed to perform CM work. Authors published teaching cases (e.g., Duin & Tham, 2018; Evia et al., 2015; Robidoux, 2008) and reviews of the CCM teaching landscape (e.g., Batova & Andersen, 2017; McDaniel & Steward, 2011); they contributed to edited collections focused on competency and curriculum development (Bridgeford, 2020; Getto et al., 2019) and created practical strategies for teaching structured content (Evia, 2018).

What is more, during the 2010s, maturing technologies, such as CMSs, high-speed networks, artificial intelligence, and XML-based languages and standards, combined with the explosion of smart devices and conversational interfaces, created the need for "intelligent content" (see, e.g., Gollner, 2010, 2014; Rockley & Cooper, 2012). Intelligent content is "content that can be managed efficiently and dynamically delivered to an unlimited range of targets using high-precision automation" (Gollner, 2011). In other words, it is content that is well-structured and semantically rich, as well as both human- and machine-readable. This content could now be "designed and engineered to interact with chatbots, voice assistants, and intelligent machines and to populate PDFs, online help, mobile, video, and other content delivery channels" (Evia & Andersen, 2020, p. 216). The process of creating, managing, and publishing content that could achieve these goals became immensely more complex, requiring an organization-wide content strategy and engineering approach, particularly as CM outgrew the realms of TC departments.

In the early 2020s, given this complexity, terms such as *content strategy*, *content engineering*, and *content operations* have gained prominence as content management no longer sufficiently describes the various disciplines of content (see Evia & Andersen, 2020).

Content strategy moves beyond the management paradigm of CM to include the entire content lifecycle, or the phases of development through which content moves. While definitions of content strategy, just as with CM, come primarily from industry sources and vary based on consultants who produce these definitions, the common themes in the descriptions of content strategy are that it is a systematic plan that defines the vision for how content will be created, managed, and delivered and that grows out of business goals and needs as well as customer goals and needs (see, e.g., Bailie & Urbina, 2013; O'Keefe & Pringle, 2012; Rockley & Gollner, 2011).

Not surprisingly, the relevance of CM has grown for all areas of content production in organizations (e.g., marketing, training, product support, technical documentation), as it offers a way for teams to share and reuse content and to publish content to a multitude of devices and platforms (Leibtag, 2014; McGrane, 2012; Wachter-Boettcher, 2012), including web portals where customers access pre- and post-sales content. Key to enabling this larger organizational adoption of CM is an integrated content strategy that serves as a unifying vision and action plan for producing, governing, and publishing content across the organization (Rockley & Cooper, 2012).

Whereas the discipline of content strategy focuses on the strategic vision and plan for content (the "what"), the discipline of content engineering focuses on the technical aspects of publishing workflows (the "how"). Content engineering is concerned with defining "the content structure, metadata, content reuse planning, taxonomy and other content relationships" (Saunders, 2015, p. 17). It focuses on how content is created, manipulated, and processed to achieve business goals; content engineers do not write the content but rather create the tools and processes that allow content to be created more efficiently and with less variability (Baker, 2013). The emergence of the disciplines of content has allowed for a more precise and narrow definition of CM, now more commonly described as the discipline focused on managing content after it has been created (Saunders, 2015).

Most recently, the term *content operations* has gained traction for its focus on how the disciplines of content relate and interact (see Barker, 2016; Jones, 2019; Saunders, 2015). Content operations has been defined as effective management of content that happens behind the scenes and that encompasses people, process, and technology (Jones, 2019); it accounts for everything between content strategy and content management.

As this brief history shows, the disciplines of content will become increasingly important knowledge and skill areas for technical communicators who want to contribute to content activities in meaningful ways.

References

Abel, S. (2014). Content. In S. Abel & R. A. Bailie (Eds.), *The language of content strategy* (pp. 12–13). XML Press.

Albers, M. (2000). The technical editor and document databases: What the future may hold. Technical Communication Quarterly, 9(2), 191–206. https://doi.org/10.1080/10572250009364693.

Albers, M. J. (2003). Single sourcing and the technical communication career path. *Technical Communication*, *50*(3), 335–344.

Ament, K. (2003). *Single sourcing: Building modular documentation*. William Andrew Publishing. https://doi.org/10.1016/B978-081551491-6.50003-8.

Andersen, R. (2011). Component content management: Shaping the discourse through innovation diffusion research and reciprocity. *Technical Communication Quarterly*, *20*, 384–411. https://doi.org/10.1080/10572252.2011.590178.

Andersen, R. (2014). Planning for the shaping force of cultural dynamics in a component content-management system implementation. *IEEE Transactions on Professional Communication*, *57*(3), 216–234. https://doi.org/10.1109/TPC.2014.2342336.

Andersen, R. & Batova, T. (2015). The current state of component content management: An integrative literature review. *IEEE Transactions on Professional Communication*, *58*(3), 247–270. https://doi.org/10.1109/TPC.2016.2516619.

Bailie, R. A. (2007). Top ten mistakes in content management. *Intercom*, *54*, 18–21.

Bailie, R. A. (2009). *Anticipating the impact of content convergence*. Multilingual Magazine. https://multilingual.com/articles/anticipating-the-impact-of-content-convergence/.

Bailie, R. A. & Urbina, N. (2013). *Content strategy for decision makers: Connecting the dots between business, brand and benefits*. XML Press.

Baker, M. (2013, November 13). Content engineering is not technical writing. Every Page is Page One. https://everypageispageone.com/2013/11/13/content-engineering-is-not-technical-writing/.

Barker, D. (2016, January 27). The need for content operations. *Gadgetopia*. https://gadgetopia.com/post/9307.

Batova, T. (2014). Component content management and quality of information products for global audiences: An integrative literature review. *IEEE Transactions on Professional Communication*, *57*(4), 325–339. https://doi.org/10.1109/TPC.2014.2373911.

Batova, T. (2018a). Negotiating multilingual quality in component content management environments: A case study. *IEEE Transactions on Professional Communication*, *61*(1), 77–100. https://doi.org/10.1109/TPC.2017.2747278.

Batova, T. (2018b). Work motivation in the rhetoric of component content management. *Journal of Business and Technical Communication*, *32*(3), 308–346. https://doi.org/10.1177/1050651918762030.

Batova, T. (2019). Lost in content management: Constructing quality as a global technical communication metric. *Technical Communication*, *66*(1), 30–52.

Batova, T. & Andersen, R. (2017). Introduction to the special issue: Content strategy—A unifying vision. *IEEE Transactions on Professional Communication*, *59*(1), 2–6. https://doi.org/10.1109/TPC.2016.2540727.

Batova, T. & Clark, D. (2015). The complexities of globalized content management. *Journal of Business and Technical Communication*, *29*, 221–235. https://doi.org/10.1177/1050651914562472.

Bock, G., Waldt, D. & Laplante, M. (2010). *Smart content in the enterprise: How next generation XML applications deliver new value to multiple stakeholders*. Gilbane Group.

Bridgeford, T. (Ed.). (2020). *Teaching content management in technical and professional communication*. Routledge. https://doi.org/10.4324/9780429059612.

Byrne, J. (2013). Of tomatoes, translators and the importance of context. Experimental Communicator. https://www.experimental.wtf/1382.

Clark, D. (2008). Content management and the separation of presentation and content. *Technical Communication Quarterly*, *17*(1), 35–60. https://doi.org/10.1080/10572250701588624.

Cooper, C. (2014). Adaptive content. In Abel, S. & Bailie, R. (Eds.), *The language of content strategy* (pp. 78–79). XML Press.

Cowan, C. (2010). *XML in technical communication* (2nd ed.). ISTC.

Day, D. (2014). Structured content. In Abel, S. & Bailie, R. (Eds.), *The language of content strategy* (pp. 62–63). XML Press.

Dayton, D. (2006). A hybrid analytical framework to guide studies of innovative IT adoption by work groups. *Technical Communication Quarterly*, *15*, 355–382. https://doi.org/10.1207/s15427625tcq1503_5

Duin, A. & Tham, J. (2018). Cultivating code literacy: Course redesign through advisory board engagement. *Communication Design Quarterly*, *6*(3), 44–58. https://doi.org/10.1145/3309578.3309583.

Evia, C. (2018). *Creating intelligent content with lightweight DITA*. Routledge. https://doi.org/10.4324/9781351187510.

Evia, C. & Andersen, R. (2020). Beyond management: Understanding the many forces that shape content today. In T. Bridgeford (Ed.), *Teaching content management in technical and professional communication* (pp. 213–231). Routledge. https://doi.org/10.4324/9780429059612-12.

Evia, C., Sharp, M. & Perez-Quinones, M. (2015). Teaching structured authoring and DITA through rhetorical and computational thinking. *IEEE Transactions on Professional Communication, 58*(3), 328–343. https://doi.org/10.1109/TPC.2016.2516639.

Freeman, B. (2006). Multilingual publishing with a content management system. *Intercom, 53*, 14–15.

Gattis, L. F. (2008). Applying cohesion and contrastive rhetoric research to content management practices. In G. Pullman & B. Gu (Eds.), *Content management: Bridging the gap between theory and practice* (pp. 201–216). Baywood. https://doi.org/10.2190/CMBC11.

Getto, G., Labrio, J & Ruszkiewics, J. (Eds.). (2019). *Content Strategy in Technical Communication*. New York, NT: Routledge. https://doi.org/10.4324/9780429201141.

Gibbon, C. (2014). Content model. In Abel, S. & Bailie, R. (Eds.), *The language of content strategy* (pp. 52–53). XML Press.

Gollner, J. (2010). *The emergence of intelligent content: The evolution of open content standards and their significance*. https://www.rockley.com/articles/The%20Emergence%20of%20Intelligent%20Content%20(JGollner%206%20Jan%202009).pdf.

Gollner, J. (2011). Intelligent content in the green desert. Content & Management. https://www.gollner.ca/2011/03/intelligent-content-in-the-green-desert.html.

Gollner, J. (2013). Products content strategy. *Content & Management*. http://www.gollner.ca/2013/10/product-content-strategy.html.

Gollner, J. (2014). A short primer on intelligent content. Content & Management. https://www.gollner.ca/2014/01/primer-on-intelligent-content.html.

Gollner, J. (2015). Would the real content management please stand up? Content & Management. http://www.gollner.ca/2015/03/real-content-management.html.

Hackos, J. (2008). *Best practice for indexing DITA topics for translation* [White paper]. OASIS. https://www.oasis-open.org/committees/download.php/27581/IndexingBestPracticesWhitePaper.pdf.

Hackos, J. (2010, October). DITA and localization: Improving your cost savings. *Center for Information Development Management, Best Practices Newsletter*.

Hart, G. (2013). Defining content: Going beyond the explicit to define the implicit information. *Intercom*, 26–30.

Hart-Davidson, W. (2005). Shaping texts that transform: Toward a rhetoric of objects, relationships, and views. In Lipson, C. & Day, M. (Eds), *Technical communication and the World Wide Web* (pp. 27–42). Lawrence Erlbaum Associates.

Hart-Davidson, W. (2009). Content management: Beyond single-sourcing. In R. Spilka (Ed.), *Digital literacy for technical communication: 21st Century theory and practice* (pp. 128–144). Routledge Publishing.

Jones, C. (2019). *The content advantage [Clout 2.0]*. New Riders.

Leibtag, A. (2014). *The digital crown: Winning at content on the web*. Elsevier Science.

Lovinger, R. (2010). *Nimble: A razorfish report on publishing in the digital age*. https://hypermedia2010.files.wordpress.com/2010/10/2010_razorfish-nimble-report.pdf.

McDaniel, R. & Steward, S. (2011). Technical communication pedagogy and the broadband divide: Academic and industrial perspectives. In A. P. Lamberti & A. R. Richards (Eds.), *Complex worlds: Digital culture, rhetoric, and professional communication* (pp. 195–212). Baywood Press. https://doi.org/10.2190/CWDC10.

McGrane, K. (2012). Adapting ourselves to adaptive content. https://karenmcgrane.wordpress.com/2012/09/04/adapting-ourselves-to-adaptive-content-video-slides-and-transcript-oh-my/.

O'Keefe, S. (2009). *Structured authoring and XML*. Scriptorium. http://www.scriptorium.com/structure.pdf.

O'Keefe, S. & Pringle, A. (2012). Transforming technical content into a business asset. *tcworld*. https://tcworld.info/e-magazine/technical-writing/transforming-technical-content-into-a-business-asset-337/.

Robidoux, C. (2008). Rhetorically structured content: Developing a collaborative single-sourcing curriculum. *Technical Communication Quarterly*, *17*, 110–135. https://doi.org/10.1080/10572250701595652.

Rockley, A. (2001). The impact of single sourcing and technology. *Technical Communication*, *48*(2), 189–193.

Rockley, A. & Cooper, C. (2012). *Managing enterprise content: A unified content strategy* (2nd ed.). New Riders.

Rockley, A. & Gollner, J. (2011). An intelligent content strategy for the enterprise. *Bulletin of the American Society for Information Science and Technology*, *37*(2), 33–39. https://doi.org/10.1002/bult.2011.1720370211.

Rockley, A., Manning, S. & Cooper, C. (2010). *DITA 101: Fundamentals for DITA for authors and managers* (2nd ed). Rockley Group.

Sapienza, F. (2007). A rhetorical approach to single-sourcing via intertextuality. *Technical Communication Quarterly*, *16*(1), 83–101. https://doi.org/10.1080/10572250709336578.

Saunders, C. (2015). *Content engineering for a multi-channel world: Enabling the next generation of customer experiences*. Simple A.

Schumate, C. (2011). Implementing the big 3: A project management view on lessons learned on implementing an XML/DITA/CCMS publishing environment. Best Practices: A Publication of the Center for Information-Development Management, 13(1), 1–7.

Swarts, J. (2010). Recycled writing: Assembling actor networks from reusable content. *Journal of Business and Technical Communication*, *24*(2), 127–163. https://doi.org/10.1177/1050651909353307.

Swarts, J. (2011). Technological literacy as network building. *Technical Communication Quarterly*, *20*, 274–302. https://doi.org/10.1080/10572252.2011.578239.

Swisher, V. (2011). The problem with translating DITA. *Content Rules*. http://www.contentrules.com/blog/the-problem-with-translating-dita/.

Swisher, V. (2014). *Global content strategy: A primer*. XML Press.

Wachter-Boettcher, S. (2012). Future ready content. *A List Apart*. http://alistapart.com/article/future-ready-content.

7. Crisis Communication

Elizabeth L. Angeli
Marquette University

Understanding crisis communication requires a clear definition of the word *crisis*. However, crises can be challenging to define because definitions risk slipping into tautologies. Quite often, crises are labeled in hindsight after events unfold that are characteristic of a crisis event. Adding to this complexity, the various definitions and uses of "crisis" are as varied as crisis situations themselves, as the definitions outlined by the *Oxford English Dictionary* (Oxford University Press, n.d.) suggest:

- *Pathology*. The point in the progress of a disease when an important development or change takes place which is decisive of recovery or death; the turning-point of a disease for better or worse; also applied to any marked or sudden variation occurring in the progress of a disease and to the phenomena accompanying it.
- *Astrology*. Said of a conjunction of the planets which determines the issue of a disease or critical point in the course of events.
- *Transferred* and *figurative*. A vitally important or decisive stage in the progress of anything; a turning-point; also, a state of affairs in which a decisive change for better or worse is imminent; now applied *esp.* to times of difficulty, insecurity, and suspense in politics or commerce.

However varied, these definitions share a few characteristics, including change, transition, and turning points. As such, any situation can become a crisis if the conditions are just right, and a crisis manifests when risk, fear, uncertainty, anticipation, and consequence converge, threaten upheaval, and overwhelm stakeholders involved. These stakeholders are not limited to human stakeholders but include organizations, ***technologies***, the environment, ecologies, and economies that are impacted by crisis events.

Perhaps best defined as "a risk manifested" (Heath & O'Hair, 2009, p. 9), crisis is deeply embedded with risk because without risk, there often is no crisis (Palenchar, 2010; Venette, 2008). Consider skydiving: It involves a great deal of risk due to uncertainty and consequences if the jump isn't successful. If the jump is successful, there is, typically, no crisis. If the jump is unsuccessful, crises may follow, including potential life-threatening, irreversible injuries or even death. Each of these outcomes has their own affiliated crisis, from high-risk medical decisions to financial consequences.

In technical and professional communication, scholars have tended more fully to ***risk communication*** instead of crisis communication. Our contributions

DOI: https://doi.org/10.37514/TPC-B.2023.1923.2.07

to crisis communication subsume "crisis" under "risk" or are labeled as "crisis rhetoric," and work under the latter key term falls under ***rhetorical*** theory rather than technical communication (Cherwitz & Zagacki, 1986; Dow, 1989; Hart & Tindall, 2009). Given the symbiotic relationship between risk and crisis, this meshing is understandable and perhaps best captured in Dorothy Winsor's work with the Challenger disaster, which is arguably the first instance of crisis work in technical and professional communication scholarship (1988, 1990). Although Winsor's publications are not labeled under "crisis," her work identifies how communication failures in high-risk situations, like a space flight launch, can lead to a crisis. In turn, she illustrates how technical and professional communicators with our rhetorical expertise can influence, and perhaps prevent, crises.

Crisis-related studies in technical and professional communication are grounded in rhetorical theories, draw on various ***research*** methods, and speak to a variety of crisis-related disciplines. Scholars have turned to ancient rhetorical theories, such as stasis theory (DeVasto et al., 2016) and topoi (Ding, 2018; Nielsen, 2017), assessment metrics (Applen, 2020), and visualization (Richards, 2015), as frameworks to make sense of the many facets of crises. At crisis communications' core, though, is message creation, dissemination, and implications for multiple ***audiences***, including the ***public***, students (Schlachte, 2019), researchers, and practitioners. For example, M. M. Brown's (2019) research on handwashing campaigns captures the affordances rhetoric and technical and professional communication bring to crisis research. Brown's purpose is not "to question hand hygiene's efficacy as a form of infection control," which is traditionally the purview of communication studies and public health research (Brown, 2019, p. 221). Instead, Brown's rhetorically driven research highlights the "broader implication[s]" of hand hygiene promotion in that it "moralizes the spread of infection" and raises questions about who profits from "the negative emotions often highlighted" in such campaigns (Brown, 2019, p. 221).

Due to its interdisciplinary nature, technical and professional communication crisis-related research interfaces with many related rhetorical fields of study, including public rhetorics, environmental rhetoric, rhetoric of health and medicine, ***medical/health communication***, medical humanities, digital rhetorics, ***intercultural communications*** and rhetorics, rhetoric of risk, and ***user experience*** studies. Within these fields, and although they are not tagged as such, much technical communication scholarship covers crisis-related topics. These topics usually are categorized under the keyword *risk* and include

- epidemics, pandemics, and healthcare (Angeli, 2012; Angeli & Norwood, 2017, 2019; Bloom-Pojar & DeVasto, 2019; Ding, 2014; Ding & Zhang, 2010; Keränen, 2019),
- intercultural and organizational communication (Dong, 2020; Hopton & Walton, 2019),

- emergency management (Amidon, 2014; Angeli, 2015, 2019; Richards, 2018; Seawright, 2017; Yu & Monas, 2020),
- engineering and hazardous environments (Amidon, 2020; Amidon et al., 2018; Sauer, 2003; Winsor, 1988, 1990),
- natural and international disasters (Baniya, 2019; DeVasto et al., 2016; Frost, 2013; Simmons, 2007),
- climate change and the environment (Cagle & Tillery, 2015; Ross, 2017; Walker, 2016), and
- social and mass media (Potts, 2014; Roundtree et al., 2011).

Although risk is a component of these topics, some of this scholarship speaks more to crisis communication than risk communication in part because technical and professional communication has not yet parsed through the symbiotic relationship of risk and crisis.

Adjacent to technical and professional communication, the field of communication studies has refined and applied "crisis" in myriad contexts (Coombs, 2009). This scholarship approaches crisis communication from a few angles: how people and institutions communicate about a crisis (Stephens et al., 2005), during a crisis (Heath, 2006), and in the backstage of a crisis (Cole & Fellows, 2008). Work in this field focuses on public-facing communication, exploring how communicators develop messaging about a crisis and analyze its effectiveness and impact (Borden & Zhang, 2019; Wang, 2016; Zhao, 2013; Zheng et al., 2018; Zhu et al., 2017). Ultimately, communication studies approaches crisis communication as a "strategic [process] designed to respond to various rhetorical problems in ways that can be evaluated by standards of empirical success, values, and ***ethics***" (Heath & O'Hair, 2009, pp. 17–18).

But when understood through a rhetorical lens, this "strategic process" is murky. At its foundation, crisis communication is rhetorical, rooted in a specific situation with various, targeted exigencies, audiences, and purposes. Technical and professional communicators use rhetorical strategies to study risk and related crises, and in turn, our approaches to crises explore implications of the language that is used—and not used—about an event. These implications often are best understood by looking at the ecologies of events surrounding a crisis, particularly how risk, fear, uncertainty, and authority impact such events. For example, Huiling Ding's (2014, 2018; Ding & Zhang, 2010) work on epidemics points to how communities and organizations navigate complex networks of power and media, and, in turn, her work highlights how these and other factors impact policy and messaging. Likewise, Esben Bjerggaard Nielsen's (2017) work on environmental crisis reimagines the topoi of time and place to be "discursive organizing tool[s]" that create a stronger identification with a "global and far-removed audience" (p. 102). As such, technical and professional communicators look beyond crisis communication as only a strategic process and pursue lines of inquiry that tease out nuances and tensions involved in communication. These lines of inquiry include:

- Who defines, creates, deploys, assesses, and upholds "standards of empirical success, values, and ethics" (Heath & O'Hair, 2009, pp. 17–18)?
- Who determines what is "worth" being feared and risked, and whether the threatened consequences are dire enough to call an event a crisis?
- How are all the terms surrounding "crisis" defined?
- When does a situation actually become a crisis?

In short, the answer to that last question is, "It depends on who you ask." Because risk and fear, in part, determine when a crisis manifests, the actual work of defining a crisis is subjective (Heath & O'Hair, 2009; Sandman, 2006; Stephens et al., 2005). In turn, whether events are called a crisis and responded to as such depends on who has power to define terms. Adding to subjectivities, mindsets of "it won't happen to me" or "that doesn't affect me" pervade much thinking and leadership, particularly in the United States, and prevent people in power from seeing crises as crises. The subjectivities associated with crises can be captured in many large-scale events, including the terrorist attacks of 9/11 and subsequent formation of the Transportation Security Administration (TSA), Hurricane Katrina, the Flint water crisis, and the COVID-19 pandemic. Before September 11, 2001, people generally did not fear terrorist attacks on airplanes in the United States, and airport security measures were limited compared to today's standards. That all changed within two months of 9/11. The fear and perceived risk of subsequent attacks was so high among United States leadership that in November 2001, the federal government created the TSA, and current airport screening procedures, such as taking off shoes, exist because mass fear and perceived risk motivated the United States government to prevent potential terrorist risks.

However, some crisis events and their aftermath go on for much longer than two months, and despite widespread, prolonged devastation and trauma, people in power do not define them as crises; responses are then delayed, ineffective, or non-existent. In these instances, hardest hit are communities of color, and Hurricane Katrina, the Flint water crisis, and, most recently, COVID-19 illustrate the relationship among power, privilege, perceived risk, and race (Atherton, 2021; Cole & Fellows, 2008; Dave, 2015; Henkel et al., 2006; Pauli, 2020). In these crises, racial inequities were often ignored, leaving communities to face trauma without resources. Like other crises, these events had and continue to have pervasive, life-threatening, large-scale impact on numerous stakeholders—environmental, economic, structural, technological—and on communities' and individuals' physical, emotional, mental, spiritual, and psychological well-being. Despite these consequences, systemic, timely changes were not implemented because the entities and individuals who could enact change and provide resources did not perceive the fear, threat, and risks to be at a tipping point. Consequently, inaction, delayed action, and ineffective action oppress and disempower racial minorities, leaving communities of color hit hardest by crises out of the very systems that are set up to address them. Identifying, plumbing, and responding to that power and those systems is where technical and professional communicators excel.

In technical and professional communication, our field has defined COVID-19 and racism as crises that demand responses, and these responses model how technical and professional communicators can engage in crisis communication practice, research, and teaching. In response to COVID-19, scholars have initiated public-facing outlets, such as "Communicating about COVID-19" (St.Amant, 2020), and when our larger organizations, like the Association for Teachers of Technical Writing, cancelled annual conferences, leadership engaged best practices of crisis communication to mitigate fear, to demonstrate organizational leadership, and to commit to members' safety. In response to racism, scholars have issued statements about zero tolerance policies and calls to action, drawing on feminist rhetorical theories that urge scholars to use "critical imagination" (N. Jones & M. Williams, personal communication, June 10, 2020), "an inquiring tool, a mechanism for seeing the noticed and the unnoticed, rethinking what is there and not there, and speculating about what could be there instead" (Royster & Kirsch, 2012, p. 20).

Because of our focus on language, action, and power, technical and professional communicators are well positioned to examine, understand, and respond to the complexities and layers involved in crisis events, which cross disciplinary boundaries. In kind, crisis communication work is interdisciplinary because the various rhetorical problems associated with crisis events are created through interrelated mechanisms, such as health, politics, environments, technologies, and economies. The complexity of these events demands contextualization and nuance, in turn, aligning with the interdisciplinary scope of technical and professional communication.

Crises are complex, often unpredictable events that involve much rhetorical work to anticipate, manage, and resolve. Despite the negative connotations of crises, they can also present opportunities, and this aspect is worth studying, especially how stakeholders leverage crisis-related artifacts, decisions, and consequences to create new policies, structures, or programs. Technical and professional communicators are well positioned to participate in this area of study given our expertise in the rhetorical nuances of communication.

References

Amidon, T. (2014). Firefighters' multimodal literacy practices [Doctoral dissertation, University of Rhode Island]. DigitalCommons@URI. http://digitalcommons.uri.edu/oa_diss/239.

Amidon, T. R. (2020). Brightness behind the eyes: Rendering firefighters' tacit literacies visible. *Kairos: A Journal of Rhetoric, Technology, and Pedagogy, 25*(1). http://kairos.technorhetoric.net/25.1/inventio/amidon/sizeup.html.

Amidon, T. R., Williams, E. A., Lipsey, T., Callahan, R., Nuckols, G. & Rice, S. (2018). Sensors and gizmos and data, oh my: Informating firefighters' personal protective equipment. *Communication Design Quarterly, 5*(4), 15–30. https://doi.org/10.1145/3188387.3188389.

Angeli, E. L. (2012). Metaphors in the rhetoric of pandemic flu: Electronic media coverage of H1N1 and swine flu. *Journal of Technical Writing and Communication, 42*(3), 203–222. https://doi.org/10.2190/TW.42.3.b.

Angeli, E. L. (2015). Three types of memory in emergency medical services communication. *Written Communication, 32*(1), 3–38. https://doi.org/10.1177/0741088314556598.

Angeli, E. L. (2019). *Rhetorical work in emergency medical services: communicating in the unpredictable workplace*. Routledge. https://doi.org/10.1177/0741088314556598.

Angeli, E. L. & Norwood, C. D. (2017). Responding to public health crises: Bridging collective mindfulness and user experience to create communication interventions. *Communication Design Quarterly Review, 5*(2). 29–39. https://doi.org/10.1145/3131201.3131204.

Angeli, E. L. & Norwood, C. D. (2019). The internal rhetorical work of a public health crisis response. *Rhetoric of Health & Medicine, 2*(2), 208–232. https://doi.org/10.5744/rhm.2019.1008.

Applen, J. D. (2020). Using Bayesian induction methods in risk assessment and communication. *Communication Design Quarterly, 8*(2), 6–15. https://doi.org/10.1145/3394264.3394265.

Atherton, R. (2021). "Missing/unspecified": Demographic data visualization during the COVID-19 pandemic. *Journal of Business and Technical Communication, 35*(1), 80–87. https://doi.org/10.1177/1050651920957982.

Baniya, S. (2019). Comparative study of networked communities, crisis communication, and technology: Rhetoric of disaster in Nepal earthquake and hurricane Maria. In Proceedings of the 37th ACM International Conference on the Design of Communication (pp. 1–2). ACM. https://doi.org/10.1145/3328020.3353913.

Bloom-Pojar, R. & DeVasto, D. (2019). Visualizing translation spaces for cross-cultural health communication. *Present Tense: A Journal of Rhetoric in Society, 7*(3). http://www.presenttensejournal.org/volume-7/visualizing-translation-spaces-for-cross-cultural/.

Borden, J. & Zhang, X. A. (2019). Linguistic crisis prediction: An integration of the linguistic category model in crisis communication. *Journal of Language and Social Psychology, 38*(5–6), 650–679. https://doi.org/10.1177/0261927X19860870.

Brown, M. M. (2019). Don't be the "fifth guy": Risk, responsibility, and the rhetoric of handwashing campaigns. *Journal of Medical Humanities, 40*(2), 211–224. https://doi.org/10.1007/s10912-017-9470-4.

Cagle, L. E. & Tillery, D. (2015). Climate change research across disciplines: The value and uses of multidisciplinary research reviews for technical communication. *Technical Communication Quarterly, 24*(2), 147–163. https://doi.org/10.1080/10572252.2015.1001296.

Cherwitz, R. A. & Zagacki, K. S. (1986). Consummatory versus justificatory crisis rhetoric. *Western Journal of Speech Communication: WJSC, 50*(4), 307–324. https://doi.org/10.1080/10570318609374240.

Cole, T. W. & Fellows, K. L. (2008). Risk communication failure: A case study of New Orleans and Hurricane Katrina. *Southern Communication Journal, 73*(3), 211–228. https://doi.org/10.1080/10417940802219702.

Coombs, W. T. (2009). Conceptualizing crisis communication. In R. L. Heath & D. O'Hair (Eds.), *Handbook of risk and crisis communication* (pp. 99–118). Routledge. https://doi.org/10.4324/9781003070726-6.

Dave, A. (2015). Categories as rhetorical barriers and the federal response to Hurricane Katrina. *Technical Communication Quarterly, 24*(3), 258–286. https://doi.org/10.1080/10572252.2015.1044121.

DeVasto, D., Graham, S. S. & Zamparutti, L. (2016). Stasis and matters of concern: The conviction of the L'Aquila Seven. *Journal of Business and Technical Communication, 30*(2), 131–164. https://doi.org/10.1177/1050651915620364.

Ding, H. (2014). *Rhetoric of a global epidemic: Transcultural communication about SARS*. SIU Press.

Ding, H. (2018). Cross-cultural whistle-blowing in an emerging outbreak: Revealing health risks through tactic communication and rhetorical hijacking. *Communication Design Quarterly, 6*(1), 10. https://doi.org/10.1145/3230970.3230975.

Ding, H. & Zhang, J. (2010). Social media and participatory risk communication during the H1N1 flu epidemic: A comparative study of the United States and China. *China Media Research, 6*(4), 80–91.

Dong, L. (2020). Rhetoric of public crises: Constructing communication networks in transcultural contexts. Georgia State University. https://doi.org/10.1109/ProComm48883.2020.00018.

Dow, B. J. (1989). The function of epideictic and deliberative strategies in presidential crisis rhetoric. *Western Journal of Speech Communication: WJSC, 53*(3), 294–310. https://doi.org/10.1080/10570318909374308.

Frost, E. A. (2013). Transcultural risk communication on Dauphin Island: An analysis of ironically located responses to the Deepwater Horizon disaster. *Technical Communication Quarterly, 22*(1), 50–66. https://doi.org/10.1080/10572252.2013.726483.

Hart, P. & Tindall, K. (2009). *Framing the global economic downturn: Crisis rhetoric and the politics of recessions*. ANU Press. https://library.oapen.org/handle/20.500.12657/33746.

Heath, R. L. (2006). Best practices in crisis communication: Evolution of practice through research. *Journal of Applied Communication Research, 34*(3), 245–248. https://doi.org/10.1080/00909880600771577.

Heath, R. L. & O'Hair, D. (Eds.). (2009). *Handbook of risk and crisis communication*. Routledge. https://doi.org/10.4324/9780203891629.

Henkel, K., Dovidio, J. & Gaertner, S. (2006). Institutional discrimination, individual racism, and Hurricane Katrina. *Analyses of Social Issues and Public Policy, 6*(1), 99–124. https://doi.org/10.1111/j.1530-2415.2006.00106.x.

Hopton, S. B. & Walton, R. (2019). Rigidity and flexibility: The dual nature of communicating care for Vietnamese Agent Orange victims. *Present Tense: A Journal of Rhetoric in Society, 7*(3). http://www.presenttensejournal.org/volume-7/rigidity-and-flexibility-the-dual-nature/.

Keränen, L. (2019). Biosecurity and communication. In B. C. Taylor & H. Bean (Eds.), *The handbook of communication and security* (pp. 223–246). Routledge. https://doi.org/10.4324/9781351180962-14.

Nielsen, E. B. (2017). Climate crisis made manifest. In D. G. Ross (Ed.), *Topic-driven environmental rhetoric* (pp. 87–105). Routledge. https://doi.org/10.4324/9781315442044-5.

Oxford University Press. (n.d.). Crisis. In *Oxford English Dictionary*. Retrieved March 16, 2021, from httpx://www.oed.com/.

Palenchar, M. (2010). Risk communication. In R. L. Heath (Ed.), *The SAGE handbook of public relations* (pp. 447–460). Sage.

Pauli, B. (2020). The Flint water crisis. *WIREs Water, 7*, 1–14. https://doi.org/10.1002/wat2.1420.

Potts, L. (2014). *Social media in disaster response: How experience architects can build for participation*. Routledge.

Richards, D. (2015). Testing the waters: Local users, sea level rise, and the productive usability of interactive geovisualizations. *Communication Design Quarterly, 3*(3), 20–24. https://doi.org/10.1145/2792989.2792992.

Richards, D. P. (2018). Not a cape, but a life preserver: The importance of designer localization in interactive sea level rise viewers. *Communication Design Quarterly, 6*(2), 13. https://doi.org/10.1145/2792989.2792992

Ross, D. G. (2017). *Topic-driven environmental rhetoric*. Taylor & Francis. https://doi.org/10.4324/9781315442044.

Roundtree, A. K., Dorsten, A. & Reif, J. J. (2011). Improving patient activation in crisis and chronic care through rhetorical approaches to new media technologies. *Poroi: An Interdisciplinary Journal of Rhetorical Analysis & Invention, 7*(1), 1–14. https://doi.org/10.13008/2151-2957.1081.

Royster, J. J. & Kirsch, G. E. (2012). *Feminist rhetorical practices: New horizons for rhetoric, composition, and literacy studies*. SIU Press.

Sandman, P. M. (2006). Crisis communication best practices: Some quibbles and additions. *Journal of Applied Communication Research, 34*(3), 257–262. https://doi.org/10.1080/00909880600771619.

Sauer, B. (2003). *The rhetoric of risk: Technical documentation in hazardous environments*. Lawrence Erlbaum Associates Publishers.

Schlachte, C. P. (2019). *Before the aftermath: A pedagogy for disaster responsiveness*. University of North Carolina at Greensboro.

Seawright, L. (2017). *Genre of power: Police report writers and readers in the justice system*. National Council of Teachers of English.

Simmons, W. M. (2007). *Participation and power: Civic discourse in environmental policy decisions*. State University of New York Press.

St.Amant, K. (2020). Communicating about COVID-19: Practices for today, planning for tomorrow. *Journal of Technical Writing and Communication, 50*(3), 211–223. https://doi.org/10.1177/0047281620923589.

Stephens, K. K., Malone, P. C. & Bailey, C. M. (2005). Communicating with stakeholders during a crisis. *Journal of Business Communication, 42*(4), 390–419. https://doi.org/10.1177/0021943605279057.

Venette, S. (2008). Risk as an inherent element in the study of crisis communication. *Southern Communication Journal, 73*(3), 197–210. https://doi.org/10.1080/10417940802219686.

Walker, K. C. (2016). Mapping the contours of translation: Visualized un/certainties in the ozone hole controversy. *Technical Communication Quarterly, 25*(2), 104–120. https://doi.org/10.1080/10572252.2016.1149620.

Wang, Y. (2016). Brand crisis communication through social media: A dialogue between brand competitors on Sina Weibo. *Corporate Communications: An International Journal, 21*(1), 56–72. https://doi.org/10.1108/CCIJ-10-2014-0065.

Winsor, D. A. (1988). Communication failures contributing to the Challenger accident: An example for technical communicators. *IEEE Transactions on Professional Communication, 31*(3), 101–107. https://doi.org/10.1109/47.7814.

Winsor, D. A. (1990). The construction of knowledge in organizations: Asking the right questions about the Challenger. *Journal of Business and Technical Communication, 4*(2), 7–20. https://doi.org/10.1177/105065199000400201.

Yu, H. & Monas, N. (2020). Recreating the scene: An investigation of police report writing. *Journal of Technical Writing and Communication, 50*(1), 35–55. https://doi.org/10.1177/0047281618812441.

Zhao, M. (2013). Beyond cops and robbers: The contextual challenge driving the multinational corporation public crisis in China and Russia. *Business Horizons, 56*(4), 491–501. https://doi.org/10.1016/j.bushor.2013.03.006.

Zheng, B., Liu, H. & Davison, R. M. (2018). Exploring the relationship between corporate reputation and the public's crisis communication on social media. *Public Relations Review, 44*(1), 56–64. https://doi.org/10.1016/j.pubrev.2017.12.006.

Zhu, L., Anagondahalli, D. & Zhang, A. (2017). Social media and culture in crisis communication: McDonald's and KFC crises management in China. *Public Relations Review, 43*(3), 487–492. https://doi.org/10.1016/j.pubrev.2017.03.006.

8. Data Visualization

Charles Kostelnick
Iowa State University

Data visualization is generally defined as the graphical representation of quantitative data related to virtually any subject or discipline—from business, engineering, and ***science*** to medicine, social science, mathematics, and statistics. Data displays appear in a wide range of communications: technical reports, journal articles, PowerPoint presentations, feasibility studies, annual reports, newsletters, and popular media, as well as in everyday documents like energy bills, investments statements, dashboards, and health and fitness records.

Data ***design*** can be classified into three basic plotting systems: rectilinear grids (bar charts, line graphs, scatterplots), circular configurations (pie charts, polar charts), and maps (Yau, 2013, pp. 104–107), though other classification systems have been proposed (Desnoyers, 2011). Robert Harris (1996) provided a detailed compendium of the hundreds of display ***genres*** used in the late 20th century, and since then, with the emergence of digital and interactive designs, many new and hybrid forms have begun to appear. However, traditional genres such as bar charts, line graphs, pie charts, maps, and scatterplots continue to be among the most popular forms of data design. Several of the genres commonly used today appear in online compendia (Eppler & Muntwiler, 2022; Ferdio ApS, 2021; Ribecca, 2023).

The ***history*** of data design has unfolded primarily in the past 300–400 years. Although rare earlier examples exist, graphical displays emerged in the 17th and 18th centuries as a means to chart weather and other scientific data and eventually to visualize engineering data. In the late 18th century, William Playfair (1801) applied graphical techniques to chart economic data, and during the so-called "golden age" of data design (Friendly, 2008; Funkhouser, 1937, p. 330) in the second half of the 19th century, visualizing data about population, health, and other human subjects developed rapidly, along with new forms, techniques, and genres. These developments coalesced with the creation of national atlases, most notably in the US and France, that visualized statistical data about nation states (see Kostelnick, 2004). Figure 8.1 shows a series of charts (mosaics) from the first *Statistical Atlas of the United States* (Walker & U.S. Census Office, 1874) that use rectilinear areas to show the relative populations of states and territories, arranged from smallest to largest.

With the emergence of modernism in the early 20th century, charts and graphs were simplified to create stronger ***visual*** impact and to appeal to larger ***public audiences*** (see Sloane & U.S. Bureau of the Census, 1914). At the same time, as data design began to establish a global presence, many additional discipline-specific

DOI: https://doi.org/10.37514/TPC-B.2023.1923.2.08

forms began to appear to show scientific, engineering, business, and medical data. In the later 20th century, digital tools enabled the rapid proliferation of data design, allowing anyone with the basic software to generate charts and graphs.

For most of data design's ***history***, charts and graphs have appeared in static paper form, whereby audiences interpret one fixed and immovable version. However, with the advent of interactive digital design, charts displaying quantitative data began to give users greater control by enabling them to choose which data to visualize, to modify the graphical display (genre, color), and to mouse over data points for additional details. This kind of display radically differs from static designs by giving users greater agency (Rawlins & Wilson, 2014).

The history of the early development of data design has been documented by H. Gray Funkhouser (1937), and the key graphical inventions and the pioneer designers who created them have been charted by Michael Friendly and Daniel Denis (2001–2018) in their website *Milestones in the History of Thematic Cartography, Statistical Graphics, and Data Visualization*. In addition, historical studies of genre, science, statistics, and mapping appear in *Visible Numbers: Essays on the History of Statistical Graphics* (Kimball & Kostelnick, 2016).

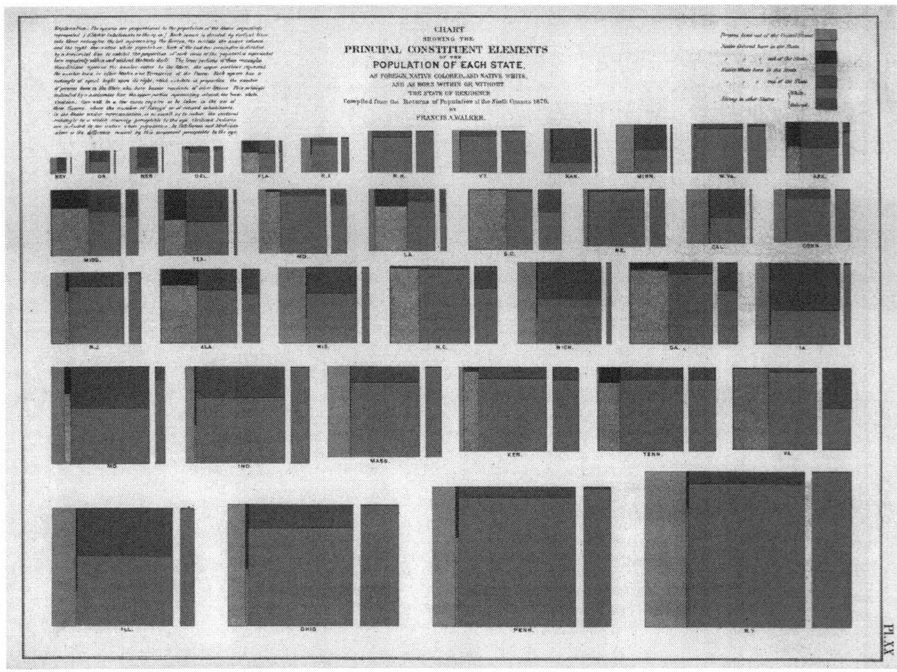

Figure 8.1. Mosaic charts from the Statistical Atlas of the United States *showing the composition of the population of states and territories (Walker & U.S. Census Office, 1874, Plate XX). Courtesy of the Library of Congress, Geography and Map Division.*

Data visualization has many contested areas, beginning with defining the kinds of images that actually fall within its realm, which can vary from one theorist or practitioner to another. Because data visualization can include virtually any image that represents "data" about some phenomenon, defining what fits within its boundaries can be a bit murky. Some visuals like pictures are usually not considered within the realm of data visualization because they typically don't represent quantitative data, though even pictures can be used for this purpose, especially when aligned in a series for comparison. Also typically outside the realm of data visualization are textual displays like diagrams, infographics, and organizational charts, though tables sometimes appear adjacent to graphical displays to provide precise numerical values. However, data visualization primarily refers to the graphical display of quantitative data.

Another area of debate concerns whether data visualization should be guided by perceptual principles or ***rhetorical*** factors. Many theorists and practitioners focus on perceptual efficiency: what forms audiences (or users) can most easily and efficiently process. For example, perceptual issues include the benefits and drawbacks of certain forms of graphical coding (lines, angles, areas, volume, gradients), which have been measured empirically (Cleveland & McGill, 1984; Heer & Bostock, 2010; see also Cochran et al., 1989; Macdonald-Ross, 1977). Other perceptual issues include "preattentive processing" (what users see in an instant), color recognition, and Gestalt principles (Few, 2012, pp. 66–68, 75–86). In the realm of digital media, perceptual issues include the effects of interactivity and animation (Fisher, 2010) on users' visual processing.

From a rhetorical perspective, these perceptual elements affect clarity, which is critical to any effective display; however, other aspects of audiences' interpretations also matter rhetorically: genres and their ability to meet audience expectations (Kostelnick, 2016a), ***ethical*** issues created by distortions or by lack of empathy (Dragga & Voss, 2001), emotional responses aroused by color or other graphical cues (Kostelnick, 2016b), agency afforded by interactivity (Rawlins & Wilson, 2014), and cultural and ideological influences on designers and on their audiences' interpretations (Barton & Barton, 1993a; Battle-Baptiste & Rusert, 2018; Brasseur, 2003; Li, 2020).

The perceptual and rhetorical approaches, however, need not be viewed as conflicting with each other. Indeed, they can complement each other as well, with the perceptual approach providing universal guidelines that guide functional communication and the rhetorical approach enabling designers to adapt their displays to specific audiences and situations.

Another distinction is often made between presentation charts, which are intended to inform or persuade audiences, and analytical charts, which are used as discovery tools to find hidden patterns in the data (see Fisher, 2010, pp. 338–339). Presentation charts are simpler and more explicit, while analytical charts are more complex, creative, and provisional. This dichotomy makes an important distinction about the data designer's processes and intentions: On the one hand, the designer carefully controls the data and the graphical elements in the display to ensure a good fit with audience and purpose, and on the other, the designer tries to model data with whatever tools and ***technology*** are available, with mixed and unpredictable results.

However, distinguishing between these two types of data design can be problematic, given the variations among audiences. A presentation chart that might be direct and compelling to one audience might be elementary or redundant to another; conversely, an analytical chart that's confusing and incomprehensible to one audience might be transparent and engaging to another. Nonetheless, the differences between these two modes of design are conceptually and operationally valid, with one emphasizing communication and the other emphasizing discovery.

Data designers often differ in their views about embellishing charts and graphs. Overly embellished charts, labeled "chartjunk" by Edward Tufte (1983, pp. 107–121), can impede clarity, and many data designers advise against excessive ornament. Modernist aesthetics also reinforced the minimalist approach to design, epitomized in the abstract pictographic system of Otto Neurath (1939). However, charts designed for popular media (e.g., those of Nigel Holmes, 1984) often contain illustrations and other embellishments to signal the subject of a given display and reveal its primary message. Moreover, with the advent of digital design, the uses of color and other non-data graphical elements have increased. So on the one hand, the purists prefer lean charts that reveal plentiful data as transparently as possible, while the artists see chart design as a way of engaging audiences, often those unmotivated to explore data. Both of these approaches have their place in data visualization, and the creativity of contemporary interactive design often bridges the two.

Ethical issues have long pervaded data design, though theorists and practitioners have defined them from several perspectives. Data design in popular media has been especially scrutinized because of deceptive practices, some of which Tufte (1983) demonstrates with his "Lie Factor," whereby perspective, volume, and area are misused (pp. 53–77; see also Brinton, 1914, pp. 20–27). Other examples of what are considered flawed (and potentially unethical) practices include starting the Y-axis scale above zero, stretching the plot frame vertically or horizontally to emphasize (or de-emphasize) trends or relationships, and using a double scale on the left and right sides of the plot frame. However, these methods might also be used ethically depending on the audience and situational context for a given display.

Data design can be evaluated according to general ethical principles (Kienzler, 1997), in relation to power and gender (D'Ignazio & Klein, 2020), as well as on its ability to project empathy and emotion (Dragga & Voss, 2001). However, the intentions of designers rarely matter, as designers are held accountable for displays that mislead their audiences. Still, ethical standards for data design vary, even for communications like annual reports, financial statements, and risk assessments, where charts and graphs might influence the audience's decision-making.

Data displays enable audiences to perceive the big patterns as well as explore the details, to see both the forest and the trees. Tufte (1990, pp. 37–51) described these two viewpoints as the "macro" and "micro" levels, and Ben Barton and Marthalee Barton (1993b) analyzed the differences between the "synoptic" and "analytic" ways of viewing. The macro-level (synoptic) view allows audiences to see the data patterns *at a glance*. As Playfair (1801) claimed early on in his pioneering work, charts and graphs enable us to perceive data "under one simple impression of vision," which makes them superior to tabular displays of data (p. x). Along similar lines, Jacques Bertin (1981) demonstrates the perceptual power and immediacy of our ability to "SEE" data graphically as opposed to interpreting data piecemeal (pp. 178–179).

Although the macro-level, big-picture view has always been touted as the most beneficial in visualizing data, ideally a chart should also allow users to explore data in detail and with precision at a more deliberate pace. Balancing these two levels of viewing creates challenges for designers, especially those working in print: gridlines, data labels, and more minute graphical coding can compromise macro-level processing, and space limitations can curb micro-level viewing. Interactive charts can address these problems by providing access to both levels through multiple viewing options. Because interactive charts afford user control over which data to display, they typically allow for both macro- and micro-level viewing.

Figure 8.2 shows an interactive animated scatterplot that uses both the macro- and micro-levels to visualize the relationship between life expectancy and income in countries around the world (Rosling et al., 2021). Users can see the big picture by viewing the animation of all countries over a span of over 200 years, or they can explore the details by stopping at a given year or by selecting a specific country from the menu on the right. Although creating interactive displays has heretofore been confined to a relatively small number of designers, software programs like Tableau (2023) are making interactive data visualization increasingly accessible.

The sprawling domain of data visualization is rapidly evolving and expanding as it shifts from print to digital forms, integrating the old and the new, generating hybrid forms, and sometimes reviving past forms. Currently, these developments are reshaping audiences' interactions with charts and graphs. In the future, digital design will likely generate novel and creative forms that will enable audiences to explore large data sets in stimulating and productive ways. At the same time, these new and inventive forms will challenge audiences perceptually and rhetorically, as audiences may have to experience a learning curve as they try to interpret them. However, audiences will be richly rewarded for their patience and interpretive flexibility. Digital design will also become increasingly dynamic in the future, as fluid data sets will be constantly replenished, giving audiences continuous visual access to the numbers. The data sets underlying these visualizations may be raw and unfiltered, but they will also be supple and timely.

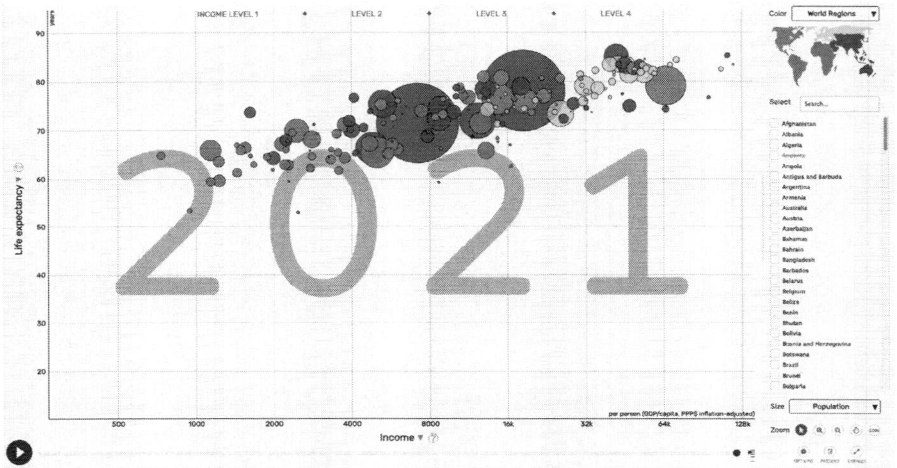

Figure 8.2. Gapminder animated scatterplot showing the relationship between life expectancy and income for people in countries around the world over the last two centuries (Rosling et al., 2021). Free material from www.gapminder.org. CC BY GAPMINDER.ORG.

Overall, then, by visualizing patterns, trends, and outliers, data design enables contemporary audiences to perceive quantitative aspects of the world around them in highly accessible forms. Data visualization is especially valuable in the field of technical communication, where facts and data must be communicated clearly and efficiently to both professional and lay audiences.

References

Barton, B. F. & Barton, M. S. (1993a). Ideology and the map: Toward a postmodern visual design practice. In N. Roundy Blyler & C. Thralls (Eds.), *Professional communication: The social perspective* (pp. 49–78). Sage.

Barton, B. F. & Barton, M. S. (1993b). Modes of power in technical and professional visuals. *Journal of Business and Technical Communication*, *7*(1), 138–162. https://doi.org/10.1177/1050651993007001007.

Battle-Baptiste, W. & Rusert, B. (Eds.). (2018). *W.E.B. Du Bois's data portraits: Visualizing Black America: The color line at the turn of the twentieth century*. The W.E.B. Du Bois Center at the University of Massachusetts Amherst; Princeton Architectural Press.

Bertin, J. (1981). *Graphics and graphic information processing*. (W. J. Berg & P. Scott, Trans.). Walter De Gruyter. https://doi.org/10.1515/9783110854688.

Brasseur, L. E. (2003). *Visualizing technical information: A cultural critique*. Baywood. https://doi.org/10.2190/VTI.

Brinton, W. C. (1914). *Graphic methods for presenting facts*. Engineering Magazine Co.

Cleveland, W. S. & McGill, R. (1984). Graphical perception: Theory, experimentation, and application to the development of graphical methods. *Journal of the American Statistical Association*, *79*(387), 531–554. https://doi.org/10.1080/01621459.1984.10478080.

Cochran, J. K., Albrecht, S. A. & Green, Y. A. (1989). Guidelines for evaluating graphical designs: A framework based on human perception skills. *Technical Communication*, *36*(1), 25–32.

Desnoyers, L. (2011). Toward a taxonomy of visuals in science communication. *Technical Communication*, *58*(2), 119–134.

D'Ignazio, C. & Klein, L. F. (2020). *Data Feminism*. MIT Press. https://doi.org/10.7551/mitpress/11805.001.0001.

Dragga, S. & Voss, D. (2001). Cruel pies: The inhumanity of technical illustrations. *Technical Communication*, *48*(3), 265–274.

Eppler, M. J. & Muntwiler, C. (2022). *Interactive knowledge maps*. Visual-literacy.org. https://www.visual-literacy.org/maps/.

Ferdio ApS. (2021). *Data Viz Project*. https://datavizproject.com/.

Few, S. (2012). *Show me the numbers: Designing tables and graphs to enlighten* (2nd ed.). Analytics Press.

Fisher, D. (2010). Animation for visualization: Opportunities and drawbacks. In J. Steele & N. Iliinsky (Eds.), *Beautiful visualization: Looking at data through the eyes of experts* (pp. 329–352). O'Reilly Media.

Friendly, M. (2008). The golden age of statistical graphics. *Statistical Science*, *23*(4), 502–535. https://doi.org/10.1214/08-STS268.

Friendly, M. & Denis, D. J. (2001–2018). *Milestones in the history of thematic cartography, statistical graphics, and data visualization.* http://www.datavis.ca/milestones/.

Funkhouser, H. G. (1937). Historical development of the graphical representation of statistical data. *Osiris*, *3*, 269–404. https://doi.org/10.1086/368480.

Harris, R. L. (1996). *Information graphics: A comprehensive illustrated reference.* Management Graphics.

Heer, J. & Bostock, M. (2010). Crowdsourcing graphical perception: Using Mechanical Turk to assess visualization design. In *CHI '10: Proceedings of the SIGCHI Conference on Human Factors in Computing Systems* (pp. 203–212). ACI Digital Library. https://doi.org/10.1145/1753326.1753357.

Holmes, N. (1984). *Designer's guide to creating charts & diagrams*. Watson-Guptill.

Kienzler, D. S. (1997). Visual ethics. *The Journal of Business Communication*, *34*(2), 171–187. https://doi.org/10.1177/002194369703400204.

Kimball, M.A. & Kostelnick, C. (Eds.). (2016). *Visible numbers: Essays on the history of statistical graphics*. Ashgate.

Kostelnick, C. (2004). Melting-pot ideology, modernist aesthetics, and the emergence of graphical conventions: The Statistical Atlases of the United States, 1874–1925. In C. A. Hill & M. Helmers (Eds.), *Defining visual rhetorics* (pp. 215–242). Erlbaum.

Kostelnick, C. (2016a). Mosaics, culture, and rhetorical resiliency: The convoluted genealogy of a data display genre. In M. A. Kimball & C. Kostelnick (Eds.), *Visible numbers: Essays on the history of statistical graphics* (pp. 177–206). Ashgate.

Kostelnick, C. (2016b). The re-emergence of emotional appeals in interactive data visualization. *Technical Communication*, *63*(2), 116–135.

Li, L. (2020). Visualizing Chinese immigrants in the *U.S. Statistical Atlases*: A case study in charting and mapping the other(s). *Technical Communication Quarterly*, *29*(1), 1–17. https://doi.org/10.1080/10572252.2019.1690695.

Macdonald-Ross, M. (1977). How numbers are shown: A review of research on the presentation of quantitative data in texts. *AV Communication Review*, *25*(4), 359–409. https://doi.org/10.1007/BF02769746.
Neurath, O. (1939). *Modern man in the making*. Alfred A. Knopf.
Playfair, W. (1801). *The commercial and political atlas, representing, by means of stained copper-plate charts, the progress of the commerce, revenues, expenditure, and debts of England, during the whole of the eighteenth century* (3rd ed.). T. Burton.
Rawlins, J. D. & Wilson, G. D. (2014). Agency and interactive data displays: Internet graphics as co-created rhetorical spaces. *Technical Communication Quarterly*, *23*(4), 303–322. https://doi.org/10.1080/10572252.2014.942468.
Ribecca, S. (2023). *The data visualisation catalogue*. https://datavizcatalogue.com/index.html.
Rosling, O., Rosling Rönnlund, A. & Rosling, H. (2021). *Wealth and health of nations* [Interactive bubble chart]. Gapminder. https://www.gapminder.org/tools/?from=world#$chart-type=bubbles.
Sloane, C. S. & U.S. Bureau of the Census (1914). *Statistical atlas of the United States.* U.S. Government Printing Office.
Tableau [Computer software]. (2023). https://www.tableau.com/.
Tufte, E. R. (1983). *The visual display of quantitative information*. Graphics Press.
Tufte, E. R. (1990). *Envisioning information*. Graphics Press.
Walker, F. A. & U.S. Census Office (1874). *Statistical atlas of the United States based on the results of the ninth census 1870*. Julius Bien; Library of Congress Geography and Map Division. https://www.loc.gov/item/05019329/.
Yau, N. (2013). *Data points: Visualization that means something*. John Wiley & Sons.

9. Design

Miles A. Kimball
Rensselaer Polytechnic Institute

Definitions of *design* are many—the *Oxford English Dictionary*, for example, provides 13 for the verb *to design* and another seven for the noun (Oxford University Press, n.d.). But for our purposes, let's start with these for the verb:

- "To point out or represent by some distinctive sign, mark or token" *(1a)*
- "To intend (a thing) to be or do something; to mean to serve some purpose or fulfill some plan" (9)
- "To plan and execute (a structure, work of art, etc.); to fashion with artistic skill; to furnish or adorn with a design" (14)

Together, these definitions create a terrain for us to understand what design is for the technical communicator: It is the skilled (even artistic) use of signs or marks to convey a message, one that may work on multiple levels. This composite definition offers the technical communicator a particular kind of open-ended methodology. Design in technical communication requires us to articulate our intentions and purposes; to develop and then demonstrate our plan to stakeholders through the use of prototypes; and to apply their feedback to subsequent sketches, delineations, and iterations.

Understanding this complex term requires looking briefly at its history. Before the 18th century, making new things was generally the result of craft ("cræft," in Anglo-Saxon; Langlands, 2018). To learn a craft, you would apprentice with a master craftsman, learning how to make useful things by rote, and eventually graduating to more complex products of your own design.

But as design historian Adrian Forty (1986) pointed out, designing things began to separate from making things in the 18th century. A key figure in this separation was Josiah Wedgwood (1730–1795), a wildly successful English pottery manufacturer whose products were mostly neoclassical-themed pots with *bas-relief* images of Greek or Roman figures. In the 1770s, Wedgwood printed a detailed catalog, accompanied by samples of shapes and glazings available—an early exemplar of technical specifications. This led to a conflict between his craftsmen, who typically included variation to show their skill in the *cræft* of pottery-making, and Wedgwood's purchasers, who wanted exact replicas of what they had ordered. Separating design from craft, Wedgwood hired London artists to create new designs and limited the craftsmen to mechanical application (literally, of *appliques*). Thus, the catalog—an illustrated technical document, complete with engravings of plate outlines, product numbers, and

DOI: https://doi.org/10.37514/TPC-B.2023.1923.2.09

evocative names—coordinated the expectations of buyers with the products of manufacturers.

This separation of design from production has allowed technical communication to flourish as a way to share ***information*** consistently for specific purposes, situations, and ***audiences***. A small number of designers could then provide designs to a manufactory level of production.

Yet this focus encourages novices in technical writing, in particular, to think of design as a series of questions tied to decoration that they approach with trepidation. Should this typeface be serif or sans serif? Should the menu's background color "match" other elements on the page, or contrast with them? How big should the leading be between lines of type? Certainly, these elements play a useful role in creating a design. But they approach design at a single level. These designers, like Wedgewood's factory hands, have gradually lost confidence in making such decisions. Even if we bring these symbolic/***visual*** issues to digital media, we are still dealing with the same issues—just through Cascading Style Sheets (CSS) and the quick succession of rather straightforward and even automated animations, such as the hover/on/off state of a typical link on a website, which is mostly managed by the browser, rather than by anything special done by the designer.

But does this mean technical document designers are limited to two-dimensional design and simple issues of ***style*** and format attendant on typographic issues of height and width? Not at all. In fact, as J. J. Gibson's well-known concept of ecological perception suggests, we do not experience any object as a silent, planar tableau. Instead, we see a document as a three-dimensional object that we explore naturally, as we do any other object we encounter. It involves a physicality that encourages us to seek new angles of view. We pick it up and feel its weight in our hands; we look at the front and the back, of course, but also the spine and the foredge. In this regard, technical documents are just like any other designed object that users interact with in multiple dimensions. And technical communicators are particularly good at integrating language and ***rhetoric*** with product design, if our employers let us do so.

Moreover, Richard Buchanan (1992) has argued that design extends to four levels of productive output, of which visual and symbolic communication is only the first:

1. The design of symbolic and visual communication (as described above)
2. The design of material objects
3. The design of activities and organized services
4. The design of complex systems or environments for living

Technical communicators have contributions to make in all four levels, in that they all rely on the use of symbolic and visual communication (Buchanan's level 1).

With the second level, the design of material objects, a well-designed chair is designed to look like something you can sit on. These communicative qualities of a material object are what Don Norman (2013) has called "affordances," or that quality of a thing that tells you how it is to be used or appreciated. This concept applies readily to technical communication, but extends beyond the two-dimensional space. As Buchanan argues, this is essentially an approach that involves using the elements of design rhetorically, in other words, to convince us of the object's affordances. And if we want to discourage sitting in a particular chair or using a chair inappropriately, such as standing on the chair seat, a warning label can readily be designed as part of the object. It's not by chance that technical communicators have developed a special expertise in risk management through the design of cautions, warnings, and dangerous situations.

At Buchanan's third level, design of activities and organized services, technical communicators might think about enterprise-level problems, rather than just documents or material objects. Consider, for example, a help-desk website. While the typeface might be important, the information the website makes available is likely more important. In other words, the activities and services of the help desk are a primary decision, while the appearance of the site itself (while not unimportant) is secondary.

Finally, at the fourth level, design extends to the largest of built structures, such as homes, office buildings, parks, and schools. A good example is the design of zoos, which tend to be organized either by the animals' place in the Linnean system (the great apes; the two-toed ungulates) or by ecosystem (the veldt; the polar regions; the altiplano). No matter what system the designers choose, it is bound to carry consequences, whether you are a lion or an antelope. The entire zoo rests on a visual/rhetorical design that conveys not only what things are, but what we must do about them.

And therein lies a problem—one that must be addressed before we start looking for solutions. Don Norman (2013) has long observed that most of us go off chasing solutions before we even begin to understand the problem at hand. The process he and others proposed was to do design ***research*** first, especially focusing upon iterative, participatory design: participatory, which asks members of the ***public*** to participate in the design team, and iterative, which involves developing a series of prototypes to address sequential issues. This approach brings focus to identifying what the problem is, so that design develops from superficiality to the foundations of human experience.

Our composite definition of design—the skilled (even artistic) use of signs or marks to convey a message, one that may work on multiple levels—combined with Buchanan's levels of design offers some implicit features that are important for technical communication:

- Design assumes intention on the part of the designer. The intention may be borrowed or operating by proxy, but nonetheless, design assumes adherence to a larger plan.
- Design includes typography and other meaningful signifiers, a visual medium depending upon distinctive signs, marks, and tokens. Consider,

for example, the ubiquitous media player controls ⏮ ⏸ ▶ ■ ⏭ whose meanings are profoundly dependent upon ***technology***.
- Design assumes a distant relationship between design and production, and therefore between designers and consumers/users.

In other words, design is a process or an approach that helps people to solve problems intentionally. It is not necessarily about art, shapes, colors, or any other symbolic forms of communication, though it often makes use of them. It takes the form of creating or changing something sensible in your range of perception. This latter distinction is important enough that it has acquired a more detailed and specific moniker: "design thinking."

Design thinking, as defined by Don Norman (2018), requires thinking systematically: "stepping back from the immediate issue and taking a broader look . . . [and] realizing that any problem is part of a larger whole, and that the solution is likely to require understanding the entire system." Shelley Goldman and Zaza Kabayadondo (2017) concur, defining design thinking in terms of its DIY (do-it-yourself) roots:

> Design thinking is a method of problem-solving that relies on a complex set of skills, processes, and mindsets that help people generate novel solutions to problems. . . . Once design thinking has been mastered anyone can go about redesigning the systems, infrastructures, and organizations that shape our lives (p. 3).

Design thinking then leads to other design approaches. User-centered design focuses on how people, usually customers, will use the objects and documents we design. Participatory design likewise makes community members actual participants in the design research and development, and human-centered design considers issues of ***accessibility*** and general human welfare. Such is design *thinking*, in a nutshell: a process of finding communicative approaches to design problems, based on the people who are going to be using that document.

But how does this all fit with technical communication? To answer that, we must look at the way that technical communication developed as a field and enterprise. Technical communication began as a ***profession*** between World War I and II, somewhere around 1920. Every military product had both a part number and an instruction manual tied to a system of ***documentation***. By the beginning of the Cold War, practitioners could craft documentation quickly for the use of hundreds of thousands of soldiers, bureaucrats and service-people. Sadly, this proliferation of documentation led to abuse. In 1963, Malden Grange Bishop in *Billions for Confusion* claimed that, in the boom years of the Cold War, con men made fraudulent fortunes by cutting pages out of old military equipment manuals and pasting them into new manuals. This kind of rough surgery succeeded because neither the contracting officers nor the other writers knew enough about typography to tell the difference between the pasted-in pages. Clearly, the government and manufacturers spent time and money creating documentation nobody read. Communication from this period tended toward what might better be called *design thoughtlessness*.

Prior to the invention of the graphic user interfaces, including the use of color, image, and animation on screens of various resolutions and sizes, a technical writer dealt with words almost exclusively, or at most with their arrangement into sections or subsections in what became known as an "information architecture." The *strategic* aspect of documentation was left to management or engineers, including the visual and tactile aspects of a document, such as the size and weight of the paper, the binding, and data tables; a graphic designer managed the scientific illustrations, charts, and graphs. If the production values were low enough or if the institution's standard style sheet was specific enough, a document might skip the art department and go directly to a printer or publisher. The technical writer had no need for skills or technologies beyond those for basic writing and an ability to adhere to institutional style sheets.

With the advent of computers, technical communicators had role of designer thrust upon them: Within a decade, technical writers had to transition from typewriters to websites. Today's technical *communication* includes not just words and ***editing***, but other modes of communication variably called *design*, *information design*, *communication design*, *interface design*, and, most broadly, *user experience design*—titles that cover the various levels of design described above. This makes technical communication, by whatever title, a field that requires constant retraining.

A variety of design theorists and historians of design have charted the ***history*** of design as it relates to production and consumption of texts. The most significant design scholars tend to be polymaths, as interested in the liberal arts (such as history) as they are in the social sciences. They are as familiar with good design culture as they are with design practices and research methodologies. For example, Saul Carliner (2003) proposed that designers must consider the affective, the physical, and the cognitive aspects of information design; John Gage (1999), an art historian, offered an exhaustive discussion of the social and cultural value of color; Karen Schriver (1997), one of the best-known document designers, conducted robust original research in establishing those features most valued by practitioners; and Elizabeth Tebeaux, a historian of technical communication, has pointed out that even early technical documents had features that we associate with technical documents today, including the use of white space, lists, tables, and graphic illustrations.

While this is only a handful of scholars on design, the subject continues to grow in interest and impact. We would do well to build our design skills and see design as an integral part of technical communication.

References

Bishop, M. G. (1963). *Billions for confusion*. McNally and Laughlin.

Buchanan, R. (1992). Wicked problems in design thinking. *Design Issues*, *8*(2), 5–21.

Carliner, S. (2003). Physical, cognitive, and affective: A three-part framework for information design. In M. J. Albert & B. Mazur (Eds.), *Content and complexity: Information design in technical communication* (pp. 39–58). Lawrence Erlbaum.

Forty, A. (1986). *Objects of desire: Design and society since 1750*. Thames and Hudson.

Gage, J. (1999). *Color and meaning: Art, science, and symbolism*. University of California Press.

Goldman, S. & Kabayadondo, Z. (2017). Introduction. In S. Goldman & Z. Kabayadondo (Eds.), *Taking design thinking to school: How the technology of design can transform teachers, learners, and classrooms* (pp. 3–19). Routledge. https://doi.org/10.4324/9781317327585.

Kimball, M. & Hawkins, A. (2008). *Document design: A guide for technical communicators*. Bedford-St. Martin's.

Kostelnick, C. (1994). From pen to print: The new visual landscape of professional communication. *Journal of Business and Technical Communication*, *8*(1), 91–117. https://doi.org/10.1177/1050651994008001004.

Kostelnick, C. & Roberts, D. (1998). *Designing visual language*. Allyn & Bacon.

Langlands, A. (2018). *Cræft: An inquiry into the origins and true meaning of traditional crafts*. W. W. Norton.

Lupton, E. (2004). *Thinking with type: A critical guide for designers, writers, editors, and students*. Princeton Architectural Press.

Norman, D. (2013). *The design of everyday things*. Basic Books.

Norman, D. (2018). *Design thinking: A useful myth*. https://jnd.org/design_thinking_a_useful_myth/.

Oxford University Press. (n.d.). Design. In *Oxford English Dictionary*. Retrieved August 9, 2021, from www.oed.com.

Salen, K. (2000). Surrogate multiplicities: In search of the visual voice-over. In G. Swanson (Ed.), *Graphic design and reading* (pp. 77–90). Allworth.

Schriver, K. (1997). *Dynamics in document design: Creating texts for readers*. John Wiley.

Tebeaux, E. (2008). Technical writing in English Renaissance shipwrightery: Breaching the shoals of orality. *Journal of Technical Writing and Communication*, *38*(1), 3–25. https://doi.org/10.2190/TW.38.1.b.

Warde, B. (1999). The crystal goblet, or printing should be invisible. In S. Heller & R. Poyner (Eds.), *Looking closer 3: Classic writings on graphic design* (pp. 56–59). Allworth.

Wilson, S. & Zamberlan, L. (2015, Spring). Design for an unknown future: Amplified roles for collaboration, new design knowledge, and creativity. *Design Issues*, *31*(2), 3–15. https://doi.org/10.1162/DESI_a_00318.

10. Documentation

David Farkas
University of Washington

Within the field of technical communication, the term *documentation* generally refers to writing that describes the features and functions of complex systems. Since the 1960s, it has been most widely used in the computer industry or, more broadly, the industries called "information and computer technology," or ICT. Within ICT, the term *documentation* and the role of technical communication have in large measure evolved together. This evolution can be examined through influential books such as those by Thomas F. Walton (1968), Edmund H. Weiss (1985), and William K. Horton (1990), and, more systematically, through the serial publications of the Society for Technical Communication (1953–present). However, for certain historical projects, such as this chapter, this evolution can be advantageously traced through the direct report by someone whose adult life has been co-extensive with the transformation of our culture by ICT and who was closely tied to the field of technical communication.

In 1964, the new high school that opened for my senior year in Clifton, New Jersey, included an administrative computing room with multiple refrigerator-sized units sporting big tape drives. We were surprised to receive report cards with grades that were printed out by the computer, not hand written in ink by our teachers. The computer also "powered" our language lab. About a decade later, as a graduate student, I learned that some folks were keyboarding their dissertations, not with a typewriter, but using WYLBUR, a mainframe-based text editor. Then, during my academic career in technical communication, I became closely tied to the world of computer documentation (aka "software user assistance") and interacted in various ways with high-tech companies, small and large, including IBM and Microsoft and, later, Amazon and Facebook (now Meta). Because this historical review of the term *documentation* is primarily a personal ***history***, I have cross-checked my recollections with ICT veterans Saul Carliner, Lori Fisher, Jo Ann Hackos, and Joe Welinske.

Outside of technical communication, documentation has generally referred to a text (or set of texts) that "furnishes evidence and ***information***" (Oxford University Press, n.d.), often within a legal or regulatory context. If you wish to export a work of art from a nation, their customs officials will require appropriate documentation. To meet ISO 9001 standards for quality control in manufacturing, you must be able to provide documentation that verifies how your manufacturing processes are carried out. Documentation is now a contested concept due to resistance from transgender people and undocumented immigrants to dominant cultural narratives and governmental policies (Caminero-Santangelo, 2016; National Center for Transgender Equality, 2023).

DOI: https://doi.org/10.37514/TPC-B.2023.1923.2.10

Etymologically, the Latin root of *documentation* is *docere*, meaning to teach. However, documentation as a synonym for teaching is obsolete (Oxford University Press, n.d.). So, for example, no one now says, "She documented her students in mathematics." But the idea that documentation provides instruction is inherent to the modern use of the term.

In ICT, the term *documentation* is deeply intertwined with the term *manual*. *Manual*, much like its sibling term *handbook*, originally connoted a relatively small, easily handled volume that directly supports some kind of specific use. Another close relative is *guide*, when used in the context of such publications as *The Complete Guide to a Successful and Secure Retirement* (Swedroe & Grogan, 2021).

In past centuries, manuals were often devotional. They guided you through the process of prayer. One of the most popular works of the renowned Renaissance Humanist Desiderius Erasmus is *Enchiridion Militis Christiani*. This guide to moral improvement was known to English speakers as *The Manual of a Christian Knight* (Tyndale's 1533 translation). Later, the term *manual* became more centered on physical tasks—for example, Hans Busk's (1858) *Rifleman's Manual: Or Rifles and How to Use Them*. While a true manual in the modern sense, Busk's procedural (how to) content is phrased with indicative paragraphs rather than the numbered steps and imperative sentences that are used now.

With growing industrialization, the general ***public*** became familiar with various kinds of manuals, including automobile maintenance manuals, which evolved greatly from their early origins (Ford Motor Company, 1919, as cited in Crabbe, 2012). The instructions that come with consumer products of all kinds are, except for their narrower scope and brevity, very similar to manuals (Crabbe, 2012; Leitz, 1937, as cited in Crabbe, 2012). World War II saw the production of a great many military manuals and a dramatic growth in the number of individuals engaged in technical communication (O'Hara, 2001). The rapid expansion of consumer culture after World War II resulted in a great proliferation of instruction booklets for the assembly and use of all kinds of products, technical and non-technical. It also led to the development of technical manuals for the professional servicing of such consumer products as television sets (Early Television Museum, 2023; Sams, 1958).

The core form of the manual, in ICT and elsewhere, is a hierarchy of task-oriented procedures sequenced in some approximation of likely use. So, for example, the manual's introduction is the top node of the hierarchy, and chapters such as those explaining how to create, format, save, and print documents are at the second level. These chapters have sections, and perhaps subsections, that contain the individual procedures. Such manuals are very often termed "user's guides."

Computer documentation in the early days of the computer industry existed primarily in print and was a broad umbrella term for myriad texts related to the ***design***, installation, operation, and repair of a (mainframe) computer system. In the 1960s and 1970s, computer use expanded from ***research*** facilities to the corporate world and, finally, with the advent of the microcomputer, to hobbyists, small business owners, and everyone else. However, the computer documentation of this era, which was often prepared by developers rather than technical communicators, paid only inconsistent attention to the needs of computer users (meaning end users—those who simply wanted to get work done). Often, users were carelessly assumed to have sufficient expertise to make good use of a reference manual that did no more than define commands. They might or might not get a true user's guide and/or a tutorial (a set of lessons that impart basic conceptual ***knowledge***, provide instructions for fundamental tasks, and promote retention). At times, user manuals were merely modified design documents consisting in large part of nearly useless product specifications (Waite, 1984).

Gradually, however, the computer manual and manual set—for both hardware and software— evolved to meet the needs of a broad range of users. Furthermore, documentation increasingly became the purview of technical communication professionals. The contrasting terms *task oriented* and *systems oriented* (or *product oriented*) would appear to usefully distinguish between documentation written to support the work that users needed to do vs. documentation that focused on the system itself. Robert G. Waite (1984), an IBM information developer, identified the introduction of the System 36 midrange computer, in 1983, as the beginning of IBM's commitment to task-oriented documentation.

Apple Writer II was the most important software application of the hugely successful Apple II Plus microcomputer (sold 1979–1982). The Apple Writer II "operating manual" (Apple Computer, 1981), despite its rudimentary formatting, is an instance of competent, user-friendly software documentation. Most of the manual is a hybrid of tutorial documentation and a user's guide (with procedures formatted as numbered steps). The manual also includes a reference section.

Some of the best early end-user documentation took the form of article-length tutorials written by technical journalists in computer magazines published for the owners of particular microcomputers. A magazine for owners of the Tandy TRS 80 might feature an article such as "Chaining Short Texts with Scripsit," because the standard-issue documentation was especially inadequate in explaining this feature.

Largely under the influence of Microsoft, the standard print documentation set for a wide range of software products came into being in the 1980s. The primary components were three thick volumes: an extensive tutorial (sometimes with a supplementary floppy disk); a fully comprehensive, task-oriented user's guide; and the (less often used) command reference. There were also many third-party computer books that followed the tutorial or user's guide model.

Even when documentation existed primarily in print, online documentation had a significant role. Online documentation goes back as far as the "man page" command reference that was part of the early releases of the ubiquitous UNIX operating system in the early 1970s ("Man page," 2023; McIlroy, 1987). Man pages were essentially pages of a large manual with simple formatting. They could therefore be displayed on a character-based computer terminal (monitor).

Online documentation has inherent advantages over print. Users do not need to reach for a physical manual, and there are sophisticated means of integrating the online content into the software's user interface. Even when online documentation offered nothing more than the digital equivalents of the print table of contents and index, access to content was usually faster (Kearsley, 1988).

Online documentation evolved throughout the 1980s in a diverse manner. Pop-up annotations, often referred to as "field-level help," displayed brief explanations and instructions on specific elements of the user interface, especially where users needed to input text to complete online forms. Many software companies developed high-quality online tutorials that guided users through the core features of the product or in some cases were comprehensive.

Microsoft's Windows operating system, replacing DOS, was very widely adopted starting in 1990 with the release of Windows 3.0. With this release came the WinHelp (.hlp) help development platform (1990–1995). WinHelp offered technical communicators many implementation options, including a multi-level hyperlinked table of contents, an online index, search, a Back button to return to previously visited topics, a "browse sequence" enabling users to follow a pathway of help topics chosen by the help author, and hyperlinking from one help topic to another. Most important, WinHelp enforced uniformity across the help systems of the many software products that ran in Windows. Now, Windows users who opened a help system already knew how it worked because they had seen something similar in other products. WinHelp, however, was not well suited to tutorial documentation and had little impact on tutorials.

The central component of most WinHelp help systems was a hierarchy of task-oriented procedures with "overview topics"—not so different from the print user's guide where paragraphs of overview information would typically precede and introduce a cluster of procedures. There was also a command reference which, if context sensitivity was implemented, allowed users to display relevant topics directly from the user interface (Boggan et al., 1996).

Little by little, online help eclipsed the print user's guide in large parts of the computer world. Software companies began to favor thin "Getting Started" manuals that explained only core features of the product. Users were expected to transition to help as they made greater use of the product.

In addition, online tutorials eclipsed print tutorials during the 1980s. A problem that plagued print tutorials was that one mistake could throw the user out of synch with the next step in the tutorial, often making it impossible for the user to advance further. Online tutorials largely prevented this problem. If a user made an error, the tutorial would block the error, emit a "ding" sound, and in many instances indicate the correct action. Not only was online documentation, both help and tutorial, more functional than the various forms of print documentation, but adding one or two more floppy disks was vastly cheaper than printing and shipping thick books. Even so, the very considerable development effort required for online tutorials led to their gradual decline. Today's YouTube tutorials are relatively easy to develop but, like print tutorials, are not integrated with the software.

Newer forms of user assistance emerged that looked and behaved still less like print than did WinHelp (Boggan et al., 1996; Knabe, 1995). Wizards sidestepped the product's primary user interface and instead walked the user panel by panel through a limited number of tasks. Apple Guide and Microsoft Coaches/Cue Cards dramatically superimposed a series of prompts on the regular user interface as the user worked through a task. Certain Microsoft help topics actually operated the user interface—as some do today. These newer forms of user assistance were collectively termed "performance support," because reading and acting were closely tied and because the focus was on enabling users to get work done, with less concern for promoting full understanding and retention (Farkas, 1998).

While essentially just a new generation of the old "field-level" pop-up annotations, Apple's Balloon Help and Microsoft's Tool Tips—both embraced by computer users—were more capable and looked and felt like something new, something we would not so readily term documentation.

Increasingly, the term *documentation*, with its roots in print, seemed to apply less well to user-facing online content. While the phrase "online documentation" remained—and still remains—in wide use, it is an older and even tired-sounding term. Technical communicators in the computer industry are likely to describe their work function as "user assistance" and "user support." One will likely hear "I write online help and other kinds of user assistance" (or "other kinds of user content") rather than "I write online help and other kinds of documentation." Responding to this change in the industry, I changed the name of my course "Computer Documentation" to "Software User Assistance" in 2005.

A further shift in the history of documentation/user assistance has been the rapid growth of community support in ICT. Increasingly, software vendors leave it to users to answer each other's questions, although forum moderators and community managers often write FAQ and other documentation for the forum and point forum visitors to the relevant content that exists outside of the forum (Frith, 2014).

One area within ICT where the term *documentation* remains strong is software development. To pick just one example, Amazon Web Services, which primarily produces content for developers, identifies this content as documentation (https://docs.aws.amazon.com). The Agile software ***project management*** process calls for less extensive documentation than traditional processes (e.g., waterfall), but it is documentation just the same (Nispel, 2018; Rüping, 2003). On the other hand, in 2016, the Committee for Software and System Engineering of ISO, the International Organization for Standards, replaced the term *documentation* with *information for users* (ISO, 2023). So the full story of the term *documentation* in the world of ICT has yet to be written.

References

Apple Computer. (1981). *Apple Writer II operating manual.*

Boggan, S., Farkas, D. K. & Welinske, J. (1996). *Developing online help for Windows 95.* International Thompson Computer Press.

Busk, H. (1858). *The rifleman's manual: Or rifles and how to use them.* (2nd ed.). Charles Noble.

Caminero-Santangelo, M. (2016). *Documenting the undocumented: Latino/a narratives and social justice in the era of Operation Gatekeeper.* University Press of Florida. https://doi.org/10.5744/florida/9780813062594.001.0001.

Crabbe, S. (2012). Constructing a contextual history of English language technical writing. *trans-kom, 5*(1), 40–59. http://www.trans-kom.eu/bd05nr01/trans-kom_05_01_03_Crabbe_Technical_Writing.20120614.pdf.

Early Television Museum. (2023). *Postwar American television: Television technical data—postwar black and white.* https://www.earlytelevision.org/postwar_tv_schematic_diagrams.html.

Erasmus, D. (1533). *Enchiridion militis Christiani and in English The manual of a Christian knight, replenished with the most wholesome precepts . . .* (1905 reprint). Methuen & Co.

Farkas, D. K. (1998). Layering as a "safety net" for minimalist documentation. In J. M. Carroll (Ed.), *Minimalism beyond the Nurnberg Funnel* (pp. 247–274). MIT Press.

Ford Motor Company. (1919). *Ford manual for owners and operators of Ford cars and trucks.*

Frith, J. (2014). Forum moderation as technical communication: The social web and employment opportunities for technical communicators. *Technical Communication, 61*(3), 173–184.

Horton, W. K. (1990). *Designing and writing online documentation: Help files to hypertext.* Wiley.

ISO International Standards Organization. (2023). *Technical Committees ISO/IEC JTC 1, Information technology*. https://www.iso.org/committee/45020.html.

Kearsley, G. (1988). *Online help systems: Design and implementation*. Ablex.

Knabe, K. (1995, May 5–11). Apple Guide: A case study in user-aided design of online help. In *Proceedings of ACM CHI'95 Conference on Human Factors in Computing Systems* (Vol.2, pp. 286–287). https://doi.org/10.1145/223355.223677.

Leitz, E. (1937). *Leitz directions Leica Camera*. Ernst Leitz GmbH.

Man page. (2023). In *Wikipedia*. https://en.wikipedia.org/w/index.php?title=Man_page&oldid=969638945.

McIlroy, M. D. (1987). *A research Unix reader: Annotated excerpts from the programmer's manual, 1971–1986.* (Computer science technical report 139). Bell Laboratories. https://www.cs.dartmouth.edu/~doug/reader.pdf.

National Center for Transgender Equality. (2023). https://transequality.org.

Nispel, M. (2018). *Creating documentation in an Agile scrum environment.* Nispel; Amazon.

O'Hara, F. M. (2001). A brief history of technical communication. In *Proceedings of 48th Annual Conference of the Society for Technical Communication* (pp. 500–504).

Oxford University Press. (n.d.). Documentation. In *Oxford English Dictionary*. Retrieved April 8, 2021, from www.oed.com.

Rüping, A. (2003). *Agile documentation: A pattern guide to producing lightweight documents for software projects.* Wiley.

Sams. (1958). *Photofact folder 4 (schematic repair manual) for the Zenith A2221Y.* https://www.earlytelevision.org/pdf/zenith_a2221_sams_393-4.pdf.

Swedroe, L. E. & Grogan, K. (2021). (2nd ed.). *The complete guide to a successful and secure retirement*. Harriman House.

Waite, R. G. (1984). Organizing computer manuals on the basis of user tasks. In *Proceedings of the 31st International Technical Communication Conference* (pp. WE-38–WE-40).

Walton, T. F. (1968). *Technical manual writing and administration.* McGraw-Hill.

Weiss, E. H. (1985). *How to write a usable user manual.* ISI Press.

11. Editing

Angela Eaton
Angela Eaton & Associates

Within the field of technical communication, *editing* is quite a common term. Almost 20 years ago, three quarters of responding members from the Society for Technical Communication indicated that editing others was an important job function (Dayton, 2004, pp. 86–87). Today, most undergraduate and graduate programs in technical communication have a technical editing course. Editing is represented in our professional societies, such as the International Society of Managing and Technical Editors; the Society for Technical Communication has the Technical Editing Special Interest Group.

The *Oxford English Dictionary* states that *edit*'s etymology is partially a back-formation from *editor* and partially from the Latin *ēditus,* "to bring forth, to produce, to utter, to tell, relate, to declare, to publish (writings), to display, show" (Oxford University Press, n.d.). The *Oxford English Dictionary* defines *edit*: "To prepare an edition of written work by (an author) for publication, by selecting and arranging the contents, adding commentary, etc" (Oxford University Press, n.d., Definition 1.a.). The first example of this definition in use was in 1699.

Edit quickly evolved to mean more generally "to prepare (a piece of writing, copy for a newspaper or magazine, etc.) for publication or use by correcting, condensing, or otherwise modifying it" (Oxford University Press, n.d., Definition 1.b.). The first example of this definition in use was in 1867.

Editing, however, became even more complicated over the last 100 years. Sub-definitions were added to show what new discoveries could be edited: a television or radio program (1913), computer code (1958), a digital image (1971), and genes (1969). This extension of the term *editing* to multiple fields and careers is its first problem: It causes problems with search terms, complicating searches for jobs, educational programs, and ***research*** literature.

To distinguish technical communication's editing from all the other types, we typically use the term *technical editing*. Technical editing "does not have a well-established definition" (Flanagan, 2019, p. 15); definitions have been grouped into ***technology***-based, ***rhetoric***-based, ***actor***- or ***activity***-based, discipline-based, and levels-based definitions (Flanagan, 2019). One of the best definitions of technical editing is "the planning, analysis, restructuring, and language changes made to other people's technological or scientific documents in order to make them more useful and accurate for their intended audiences" (Murphy, 2010, p. 1).

As there are so many types of editors, technical editing job searches typically have a few inappropriate ads mixed in with relevant ones. An indeed.com search for "technical editor" in Dallas, Texas on February 26, 2021 provided mixed

DOI: https://doi.org/10.37514/TPC-B.2023.1923.2.11

opportunities. Of the first 15 ads, less than half were technical writing and/or editing of written documents, five were editing other media (such as film), and three were not even close—two 9-1-1 operators and a "labor editor" whose job appeared to be keeping timecards. For a Monster.com search for "editor" in Dallas, TX, three of the first five ads were for technical editors, along with one video editor and, somehow, a principal engineer with no ***documentation*** responsibilities. While these two searches on the two largest job-finding websites are hardly exhaustive, they indicate that any search for a technical editing position involves time eliminating extraneous job listings. This isn't news—in their 2011 technical editing textbook, Nicole Amare et al. note that every workplace has its own nomenclature for job titles.

The widely applied term also makes it difficult to find educational programs and publications on technical editing. The University of Chicago has an Editing Certificate, but it has to do with the publishing industry, preparing students for positions as acquisitions editors and managing editors.

Searching the ***research*** databases using the term *editing* presents similar problems to the job search. A Google Scholar search of the term on March 23, 2021 returned 5.74 million results. Of these, the top four articles pertained to film editing, image editing, editing software, and surface editing. When *technical editing* was specified, the number of results dropped to 24,700.

Is this issue with the term *edit* being applied to multiple fields likely to change? Frankly, no. It's simply too entrenched in our culture. We will just have to wade through extraneous job listings, educational searches, and research literature databases.

The second issue with the term *edit* comes from within technical editing. The field agrees fairly well on what the editing process accomplishes. Editing the text means "making it complete, accurate, correct, comprehensible, usable, and appropriate for the readers" (Rude & Eaton, 2010, p. 8).

However, where the field really disagrees is with how we envision the editing process. We have dozens of models that have been created over the last 45 years, models created from the authors' professional experiences and workplace.

Robert Van Buren and Mary Fran Buehler's (1980) *The Levels of Edit* is often cited as the first modern editing process. At the Jet Propulsion Lab, they created the levels to better describe what edits were available, along with time to complete and cost, so that their program managers could better plan. They first grouped all of the editorial tasks they could think of into separate categories, ending up with nine, such as editing for format, mechanical ***style***, policy (checks whether the new document contradicts any existing policies), and integrity (making sure the document is consistent). The levels then indicate how many of those listed edits will be conducted. The lightest level is Level 5, with only two of the nine types of editing performed. The most intense editing happens in a Level 1 edit, which contains all nine categories. In addition to providing a better understanding of the services available, showing clearly to editors and authors what edits to expect, and serving as a tool for budgeting, the levels of edit also serve as scaffolding for new editors and assist professional editors with planning schedules (Tarutz, 1992, p. 162).

There are, however, different ways to categorize levels of edits, and not even the textbooks agree on one process. The textbook *Technical Editing* defines three levels of edit: proofreading, copyediting, and comprehensive editing (Rude & Eaton, 2010). Proofreading is simply checking for errors introduced when a document moves from manuscript or draft form to printed form; it looks for the mistakes introduced by the graphic designer when laying the document out in ***design*** software. Proofreading was a more important stage when old-fashioned printing presses were used, which used humans to lay out the actual letters in a frame. As a human-based process, printing made a lot of opportunity for introduction of errors. Now that most graphic designers are taking computer text and cutting and pasting it into a different software program, there is less of an opportunity for introduced errors, but proofreading is still a necessary editing step. At the next level, copyediting "check[s] for correct spelling, punctuation, and grammar; for consistency in mechanics, such as capitalization, from one part of the document to the next, and for document accuracy and completeness" (Rude & Eaton, 2010, p. 9). Last, comprehensive editing "evaluates how well the content, organization, visual design, and style of the document support comprehension" (Rude & Eaton, 2010, p. 203).

The Amare et al. textbook (2011) also uses three levels of edit: editing for correctness, ***visual*** readability, and effectiveness (p. 12). Those levels, however, don't correspond to the levels of proofreading, copyediting, and comprehensive editing used in the *Technical Editing* textbook (Rude & Eaton, 2010). The correctness edit involves fixing grammatical and mechanical errors (similar to proofreading and copyediting), while the effectiveness edit deals with all rhetorical issues, and is defined as "substantive editing for content issues such as organization, sentence, structure, style, logic, and meaning" (similar to comprehensive editing; Amare et al., 2011, p. 12). The middle level of edit, the edit for visual readability, however, is completely different, entirely about formatting and page design, including color issues, white space, bulleting, and all graphics.

Other texts categorize for both types of edits and levels of edits. For example, Judith A. Tarutz's textbook (1992) describes four major types of edits—developmental, preliminary, copy and literary, and production. She adds a chapter on the levels of edit, providing her own levels: what is found by turning pages, skimming, skimming and comparing, reading, analyzing, and testing and using (p. 165). Similarly, Donald H. Cunningham and colleagues (2020) use both approaches, types and levels, to classify editing practices, but they also introduce a third, scope. For them, the types are substantive editing and copyediting. Substantive editing covers editing for organization, navigation, completeness, accuracy, and style as well as effective visuals and page design. Copyediting usually focuses on correcting errors in grammar, mechanics, typography, alignment, and punctuation; correcting formatting inconsistencies in headings, tables of contents, etc.; and ensuring adherence to style sheets and style manuals. Proofreading, which they fold into copyediting, is a late-stage check for errors—especially those introduced during the editing process. Cunningham et al.'s levels of editing reflect the amount of time, attention, and effort entailed during substantive editing (minimal, moderate, extensive) or copyediting (light, standard, heavy). Finally, the scope of the editing can be global (throughout the document) or local (in one part of it).

Even a study which surveyed authors who had been edited by professional editors still turned up baffling definitions and levels of editing (Eaton et al., 2008). Of the more than 400 respondents, only 26 percent defined editing in terms of all three types of editing (proofreading, copyediting, and comprehensive). Only 50 percent of respondents' definitions included comprehensive editing at all. In other words, not even those who have been edited define the process as an editor would.

What are the negative outcomes of not having consistent terms to describe editing? For potential clients, not knowing about the editorial process, particularly that comprehensive editing exists, really limits their ability to envision how an editor might help them. It limits the editor's ability to sell their services.

For practitioners, this means having to explain to every new client what model of editing they are following. Skipping the explanation can result in mismatched expectations and conflict. Practitioners will also need to learn the editorial process at each workplace. For teachers, having so many models, we have to use what mirrors our experience the best, what we find most helpful. For researchers, these different models negatively affect planning studies: we don't have large groups of students who have been trained using the same techniques.

Are these problems with the term *editing* very serious? The use of "editing" to describe multiple professional activities is inconvenient for people who must take more time to find job opportunities or relevant articles, but ultimately not serious. But editing having multiple processes is the larger issue. I predict that no matter how well the field defines its editing process, we will always have to explain the editorial process every time we work with a new client.

References

Amare, N., Nowlin, B. & J. H. Weber. (2011). *Technical editing in the 21st century*. Prentice Hall.

Cunningham, D. H, Malone, E. A. & Rothschild, J. M. (2020). *Technical editing: An introduction to editing in the workplace*. Oxford University Press.

Dayton, D. (2004). Electronic editing in technical communication: The compelling logics of local contexts. *Technical Communication*, *51*, 86–101.

Eaton, A., Brewer, P. E., Portewig, T. C. & Davidson, C. R. (2008). Examining editing in the workplace from the author's point of view: Results of an online survey. *Technical Communication*, *55*(2), 111–139.

Flanagan, S. (2019). The current state of technical editing research and the open questions. In S. Flanagan & M. J. Albers (Eds.), *Editing in the modern classroom* (pp. 15–46). Routledge. https://doi.org/10.4324/9781351132756-2.

Murphy, A. J. (Ed.). (2010). *New perspectives in technical editing*. Baywood.
Oxford University Press (n.d.). Edit. In *Oxford English Dictionary*. Retrieved February 27, 2021, from https://www.oed.com/.
Rude, C. D. & Eaton, A. (2010). *Technical editing* (5th ed.). Pearson.
Tarutz, J. A. (1992). *Technical editing: The practical guide for editors and writers*. Hewlett Packard.
University of Chicago. (n.d.). *Non-credit certificate program in editing*. https://graham-school.uchicago.edu/academic-programs/professional-development/editing.
Van Buren, R. & Buehler, M. F. (1980). *The levels of edit* (2nd ed.). Jet Propulsion Laboratory California Institutes of Technology.

12. Ethics

Steven B. Katz
Clemson University

To newcomers in the field of technical communication, the term *ethics*, and the phrase "ethics in technical communication," may seem superfluous if not oxymoronic. The phrase may seem superfluous because technical communication is by definition *technical*, and many people believe that ***technology*** does not have ethics (think of how many times people have argued that "guns don't kill people, people do."). Therefore, technical communication has nothing to do but simply communicate technical "facts," "truth." And *if* technology has ethics and values, they're those of the manufacturer or company or culture. The phrase "ethics in technical communication" may seem an oxymoron because the idea of allowing space in technical communication for considerations of human morals may appear both contradictory and a waste of time. In all cases, ethics themselves usually remain unarticulated.

In fact, ethical questions in ***rhetoric*** are as old as Plato and Aristotle, and as young as the field of technical communication (begun as a field of study in 1953 [Whitburn, 2009]). In technical communication, ethics entails different sets of moral concepts and values and associated practices. In its short ***history***, ethics in technical communication continue to evolve, with important keywords and concepts determining the direction of the field—in theory if not always practice. Whether acknowledged, these different concepts of ethics, like technical communication itself, are deeply rooted in epistemology, the study of ***knowledge***. One thing that these keywords and concepts have in common is that they ultimately devolve to one question: *What is the relationship between language and reality?* For example, is language a transparent window onto some objective reality? Or do authors to varying degrees use language to construct reality, co-construct it with readers?

The relationship between language and "reality" in a given context can have implications for the kind of ethical roles played by technical communicators. If authors are viewed as shaping reality to some extent through technical communication, their ethics become increasingly important. But if language does not matter in the perception and communication of what are regarded as "facts," then writers have little or no ethical responsibility for what they say (Katz & Linvill, 2017).

Reductively speaking, this latter view was held by Plato (1956), who believed that "Truth" existed not only outside language but outside the material world, in a transcendental realm of Ideal Forms. Plato's pupil Aristotle, who differed from his teacher in believing in observable empirical facts located in the physical world, was a little more forgiving. But Aristotle (1984) wished that language—in particular, ***style***—was unnecessary, "owing to a defect in our hearers" (emotions); he wished that facts could be communicated without style.

DOI: https://doi.org/10.37514/TPC-B.2023.1923.2.12

This position also was held by the inventors of modern ***science***, and by extension technical communication (Longo, 2000). Francis Bacon (1902, LI-LXII; 2000, XVIII), often called the father of modern scientific method, mistrusted the human senses and thus called for repeatable experiments and the verifiable replication of results, which rhetoric could be used to report "systematically." Thomas Sprat (1667) vehemently opposed the "flourish" and "digression" of rhetoric in science and urged the Royal Society to develop and practice a "plain style" of writing that would lead to a "faithful *Records*, of all the Works of *Nature*" (p. 61). Underlying this idea of the "plain" communication of facts, articulated a little later by John Locke (1975), was the notion of language as a "pipeline." In this view, the morality of the author is not as important as scientific method and facts plainly reported via "a conduit."

In this view, the author is invisible, and thus "ethically" the objectivity and accuracy of transmission are all that count. Although perhaps an ideal rarely achieved in science and technology given the multiple meanings of words, and even mathematics when considered as arguments, this ideal is the standard, default *ethical* position in traditional scientific and technical writing (Slack et al., 2006). In this standard view of technical communication, any consideration of author morality is minimized: Language and authors are just passive receivers and transmitters of ***information***—the so-called "information model of communication" (Katz & Miller, 1996; Waddell, 1996).

This view of language as a transmission line, a conduit for information gleaned objectively, placed on naïve senses, and printed directly upon the mind, reappears in several contemporary schools of ethics in technical communication, perhaps most notably "instrumentalism," which holds that technical communication is not rhetorical (Moore, 1996). The purpose of technical communication is not to persuade, but rather to simply convey information that serves corporations and society. One might be tempted to say that instrumentalism has no ethics at all, but this would be wrong on two accounts: 1) Any statement or position— any human endeavor (including this one)—uses language to persuade; 2) Instrumentalism itself, as its proponents argue, *is* ethical in its ideological commitment to capitalism (Moore, 2005). In this utilitarian philosophy of ethics in technical communication, the moral role of the author is present, but diminished. Perhaps one manifestation of this philosophy in technical communication is what Bradley Dilger (2006) calls "extreme usability," which "reduces user engagement, forbids considering the wider scope of culture, and limits the ends of usability to achievement of expediency" (p. 47).

Contrary to these conventional scientific or instrumental philosophies of language focused on communicating facts objectively for economic ends, there are several schools of contemporary ethics of technical communication that are *rhetorically* based. In these schools of technical writing, ethics, and thus authors, figure more prominently. The study of rhetorical ethics in technical communication can be said to have begun with Carolyn Miller's (1979) foundational work "A Humanistic Rationale for Technical Writing." In this essay, the question of the relation of *praxis*, or practice, to *phronesis*, wisdom or prudence, is *the* primary consideration. That is, the basis of ethical reasoning is not only the morality of the means (*praxis*) but also reasoning about ends (*phronesis*). Miller's essay rooted technical communication in the ancient and reviving discipline of classical rhetoric, finding there its humanistic as opposed to simply technical rationale. Miller's essay spawned many essays central to understanding ethics in technical communication, including the dangers of what Katz (1992) labels "the ethic of expediency," in which technological means becomes its own moral end.

Katz, in both the 1992 essay which explores one translation of a technical memo (Ward, 2014) about improving gassing vans prior to the Final Solution of death camps in WWII, and a follow-up essay on Hitler's *Mein Kampf* (Katz, 1993), discovered *phronesis* itself operating on an ideology of utility *in extremis*. This ideology is not limited to genocidal atrocities, and Katz points to a number of technical decisions in the 20th century that share not the political ideology of Nazism but the technological ideology of expediency. Paul Dombrowski (2000) applies Katz's concept of the ethic of expediency to a number of classic examples in technical communication, including the Three Mile Island communication disaster and the Challenger shuttle explosion. Later, Sam Dragga and Dan Voss (2001) employed the ethic of expediency, among other considerations, to question the "humanity" of the newly burgeoning study of graphics in technical communication.

Perhaps it is in the relation of *praxis* and *phronesis* that we find moral space for the introduction of other ethical concerns in technical communication. For example, the Society for Technical Communication (STC), the largest technical communication practitioner organization in the US, broadened the scope of its Code of Ethics to include professional principles beyond "objectivity," "accuracy," and "clarity." They include legality, honesty, confidentiality, fairness, professionalism, creativity, obligations to clients and employees, proper attribution, and use of employer time and equipment (STC, 1998).

Growing out of feminist critiques of gender bias in scientific and technical communication, "the ethics of care" rejects "ethics based on impersonal, abstract principles" (Dombrowski, 2000, p. 63). The ethics of care acknowledges and implements "women's ways of thinking" and emphasizes empathy and compassion in technical writing for the welfare of the people, which already was shifting theory, practice, and teaching away from being exclusively male-dominated "technological reasoning" (Brasseur, 1993; Lay, 1991; Sauer, 1993). Ecological ethics too, with their focus on environmental issues in the Anthropocene (Zylinska, 2014), also are a central focus in technical communication as rhetorical (Pilsch, 2017; Propen, 2018). In a discussion of ethics and expertise that would include all of these, Ashley Rose Mehlenbacher (2022) critiques Aristotle's concept of *phronesis* itself (pp. 7–19).

Echoing Rebecca Walton et al. (2019), in the "***social justice*** turn" and beyond, technical communication itself is seen as an important form of advocacy, addressing structural oppression, making ethics and social change the primary concern of technical communication (Colton et al., 2017; Colton & Holmes, 2018a). Ethics in technical communication pay new attention to equality for people "otherized" on the basis of race and ethnicity (Williams & Pimentel, 2012), queer and transgender identity (Edenfield et al., 2019; Fancher, 2018; Ramler, 2020), and incarceration (Stephens, 2018).

If readers were expecting this brief survey of *ethics* in technical communication *not* to return to concepts and practices like *truth*, *accuracy*, and *objectivity*, they will be disappointed. For there is a new school of ethics in technical communication, as in society at large, that is powerful because it is both pervasive and invisible. In it, accuracy, objectivity, and truth have been reborn in another keyword that has become what Kenneth Burke (1969) calls a "god-term"—one that organizes and dominates a way of seeing and thinking and behaving. That word is *transparency*. Not only in technical communication, but globally, *transparency* "is a buzzword . . . applied freely by government agencies, scientists, the media, and the public"; it mythically "assumes an ideal, objective unvarnished coding and decoding of information," constitutes "a metaphor for access and 'clarity' of communication," and "conceals the operations of rhetoric" (Hartzog & Katz, 2014). Transparency is a "happy vision" of communication and society (Han 2015).

In ***visual*** communication, Jay David Bolter and Diane Gromala (2005) demonstrate that transparency is "the myth of the windowpane." That myth is built on the metaphor of *perception* "as a clear glass." The myth and metaphor of transparency is found not only in graphic design but technical communication as a whole. One easy example is the computer screen. The screen seems transparent, a window that creates the illusion that the writer has direct and unfettered access to and control of the data, words, and meaning. But "phantom" hardware/software intervene: Not only do they necessarily underlie and co-construct meaning, but also, in emails for instance, they encode social status (.edu, .net, .com, etc.) and other data that belie the ostensible freedom (including privacy) that users believe; other values such as speed, productivity, and efficiency are ideologically embedded in the technology itself (Moses & Katz, 2006). Jared Colton and Steve Holmes (2018b) examine the assumed morality of "networked collaboration" in the face of proprietary rights, cookies, privacy, etc., and argue for rhetorical "virtue ethics" (equality, care, generosity, patience) in designing and programming new forms of digital communication.

The content of transparency *in language* is also created by and hidden in writing style; the best way of making transparency visible is to render it "opaque" through style analysis (Lanham, 2003). For example, in biotechnology communication with the ***public***, where transparency is hailed as a panacea, style analysis reveals contradictory motives in the language, including an unintended and unfortunate metaphor after the Titanic of biotechnology as "the tip of the iceberg"! (Katz, 2001). Style is like a "black box" where the "real content" of language might be revealed (Latour, 2007; Simon, 1999). For instance, a style analysis of the diction from the guidance document of the National Society of Genetic Counselors exposes a deep rift in that field concerning empathetic vs. objective communication with patients (Mebust & Katz, 2007)—a conflict partially resolved by rhetorical flexibility (Flach, 2019).

As a metaphor of a clear windowpane, transparency seems to reflect democratic values, and thus grounds for good governance. Transparency presents itself as a neutral medium or tool for communication. But there is no deliberation, no consideration of *praxis* and *phronesis* in transparency, only the myth of direct and open access, shiny diaphanous surfaces. Transparency is a contemporary word for "truth." This is the case in two technical reports prepared by the Canadian Biotechnology Advisory Committee that, *based on the information model of communication*, "argue . . . for transparency" in their discussion with the public about labeling GMO foods. But at the level of style, these reports are studded with two contradictory sets of words in the same description: "objective" visual and spatial imagery vs. "affective" appeals to social beliefs and subjective emotions (Katz, 2009).

Transparency also may cloak the profit motive, as seems to be evident in a debate between the British biotechnology firm Oxitec and scientists at the Max Planck Institute (MPI) in Plön, Germany, concerning the release of genetically modified mosquitoes on unwitting populations. Guy Reeves (2012) of MPI argued for transparency "not for its own sake" but as part of an "engagement approach" that "seeks to involve the public, stakeholders and local inhabitants of release areas . . . by making all scientific content available"; Camilla Beech (2012) of Oxitec, on the other hand, argued that transparency is letting the public "see" only the "relevant" (and nonproprietary) "information"—ironically what Molly Hartzog and Steven Katz (2014) call "selective transparency." Thus, transparency can *conceal* data in support of any other economic, political, scientific, or technical end, "frame" discussion (Heidegger, 1977; Katz & Rhodes, 2010), and so become what Kenneth Burke (1966) calls "terministic screens" that not only "reflect" but also "select," and thus "deflect" as much as reveal (p.45). Like conspiracy theories, claims of transparency can obviate the need for more, good evidence (Rice, 2020); transparency can be weaponized against opponents (see Ridolfo & Hart-Davidson, 2019). And like "the ethic of expediency" (Katz, 1992), transparency can become an ethical end in itself.

Technical communication began (at least for some) as an instrumental discipline. Turns to rhetoric, ***feminism***, care, social justice, and racial and ethnic equality have reframed the discussion of ethics in technical communication. Yet in the wider sphere in which technical communication operates, the old values of objectivity, accuracy, and open access have been reinstantiated in transparency as *the* communication ethic. As such, "the ethic of transparency" (re)presents 1) the same epistemological problems of Truth, and validity of empirical ***knowledge***, found in Platonic philosophy and traditional science; 2) rhetorical ambiguity regarding *phronesis* and the moral contribution of practicing technical writers; and 3) an ongoing ethical challenge to the field of technical communication, and society as a whole.

References

Aristotle. (1984). Rhetoric. In J. Barnes (Ed.), *The complete works of Aristotle* (Vol. 2). (pp. 2152–2269). (W. R. Roberts, Trans.). Princeton University Press. https://doi.org/10.1515/9781400835850-017.

Bacon, F. (1902). *The new organon: Or true directions concerning the interpretation of nature* (J. Devey, Ed.). https://oll.libertyfund.org/titles/bacon-novum-organum (Original work publication 1620).

Bacon, F. (2000). *The advancement of learning*. Paul Dry. (Original work publication 1605)

Beech, C. (2012, March 22). Editorial response: GM mosquitoes: A perspective from Oxitec. Malaria World. https://malariaworld.org/blogs/gm-mosquitoes-perspective-oxitec.

Bolter, J. D. & Gromala, D. (2005). *Windows and mirrors: Interaction design, digital art, and the myth of transparency*. MIT Press.

Brasseur, L. (1993). Contesting the objectivist paradigm: Gender issues in the technical and professional communication classroom. *IEEE Transactions on Professional Communication, 36*(3), 114–123. https://doi.org/10.1109/47.238051.

Burke, K. (1966). *Language as symbolic action: Essays on life, literature, and method*. University of California Press.

Burke, K. (1969). *A grammar of motives*. University of California Press. https://doi.org/10.1525/9780520341715.

Colton, J. S. & Holmes, S. (2018a). A social justice theory of active equality for technical communication. *Journal of Technical Writing and Communication, 48*(1), 4–30. https://doi.org/10.1177/0047281616647803.

Colton, J. S. & Holmes, S. (2018b). *Rhetoric, technology, and the virtues*. Utah State University Press.

Colton, J. S., Holmes, S. & Walwema, J. (2017). From NoobGuides to #OpKKK: Ethics of anonymous' tactical technical communication. *Technical Communication Quarterly, 26*(1), 59–75.

Dilger, B. (2006). Extreme usability and technical communication. In J. B. Scott, B. Longo & K. V. Willis (Eds.), *Critical power tools: Technical communication and cultural studies* (pp. 47–69). SUNY Press.

Dombrowski, P. (2000). *Ethics in technical communication*. Allyn and Bacon.

Dragga, S. & Voss, D. (2001, July). Cruel pies: The inhumanity of technical illustrations. *Technical Communication, 48*(3), 265–274.

Edenfield, A. C., Colton, J. S. & Holmes, S. (2019). Always already geopolitical: Trans healthcare and global tactical technical communication. *Journal of Technical Writing and Communication, 49*(4), 433–445. https://doi.org/10.1177/0047281619871211.

Fancher, P. (2018). Embodying Turing's Machine: Queer, embodied rhetorics in the history of digital computation. *Rhetoric Review, 37*(1), 90–104. https://doi.org/10.1080/07350198.2018.1395268.

Flach, M. (2019). *Exploring patient perceptions and misconceptions: Beliefs regarding hereditary cancer*. (Unpublished master thesis). University of South Carolina.

Han, B. C. (2015). *The transparency society* (Erik Butler, Trans.). Stanford University Press.

Hartzog, M. & Katz, S. B. (2014, May 28). *The appeal to transparency in the regulation of genetically modified insects* [Presentation]. Second Annual Governance of Emerging Technologies: Law, Policy, Ethics, Sandra Day O'Connor College of Law, Scottsdale, AZ, United States.

Heidegger, M. (1977). *The question concerning technology and other essays* (W. Lovitt, Trans.). Harper & Row.

Katz, S. B. (1992). The ethic of expediency: Classical rhetoric, technology, and the Holocaust. *College English, 54*(3), 255–275. https://doi.org/10.2307/378062.

Katz, S. B. (1993). Aristotle's rhetoric, Hitler's program, and the ideological problem of praxis, power, and professional discourse as a social construction of knowledge.

Journal of Business and Technical Communication, *7*(1), 37–62. https://doi.org/10.1177/10 50651993007001003.

Katz, S. B. (2001). Language and persuasion in biotechnology communication with the public: How to not say what you're not going to not say and not say it. *AgBioForum*, *4*(2). https://agbioforum.org/language-and-persuasion-in-biotechnology-communication-with-the-public-how-to-not-say-what-youre-not-going-to-not-say-and-not-say-it/.

Katz, S. B. (2009). Biotechnology and global miscommunication with the public: Rhetorical assumptions, stylistic acts, ethical implications. In H. Grady & G. Hayhoe (Eds.), *Connecting people with technology: Issues in professional communication* (pp. 167–175). Baywood Publications. https://doi.org/10.4324/9781315224909-19.

Katz, S. B. & Linvill, C. (2017). Lines and fields of ethical force in scientific authorship: The legitimacy and power of the Office of Research Integrity. In H. Yu & K. M. Northcut (Eds.), *Scientific communication: Practices, theories, and pedagogies* (pp. 39–63). Routledge Press. https://doi.org/10.4324/9781315160191-3.

Katz, S. B. & Miller, C. R. (1996). The low-level radioactive waste siting controversy in North Carolina: Toward a rhetorical model of risk communication. In C. G. Herndl & S. C. Brown (Eds.), *Green culture: Environmental rhetoric in contemporary America* (pp. 111–140). University of Wisconsin Press.

Katz, S. B. & Rhodes, V. W. (2010). Beyond ethical frames of technical relations: Digital being in the workplace world. In R. Spilka (Ed.), *Digital literacy for technical communication: 21st century theory and practice* (pp. 230–256). Routledge Press.

Lanham, R. A. (2003). *Analyzing prose* (2nd ed.). Continuum.

Lanham, R. A. (2006). *The economics of attention: Style and substance in the age of information*. University of Chicago Press.

Latour, B. (2007). *Reassembling the social: An introduction to Actor-Network-Theory*. Oxford University Press.

Lay, M. M. (1991). Feministic theory and the redefinition of technical communication. *Journal of Business and Technical Communication*, *5*, 384–370. https://doi.org/10.1177/1050651991005004002.

Locke, J. (1975). *An essay concerning human understanding* (P. H. Nidditch, Ed.). Oxford/ Clarendon. (Original work publication 1689)

Longo, B. (2000). *Spurious coin: A history of science, management, and technical writing*. SUNY Press.

Mebust, M. R. & Katz, S. B. (2007). Rhetorical assumptions, rhetorical risks: Communication models in genetic counseling In B. Heifferon & S. Brown (Eds.), *Rhetoric of healthcare: Essays toward a new disciplinary inquiry* (pp. 91–114). Hampton Press.

Mehlenbacher, A. R. (2022). *On expertise: Cultivating character, goodwill, and practical wisdom*. Pennsylvania State University Press. https://doi.org/10.1515/9780271093130.

Miller, C. R. (1979). A humanistic rationale for technical writing. *College English*, *40*(6), 610–617. https://doi.org/10.2307/375964.

Moore, P. (1996). Instrumental discourse is as humanistic as rhetoric. *Journal of Business and Technical Communication*, *10*(1), 100–118. https://doi.org/10.1177/10506519960100010 05.

Moore, P. (2005). From cultural capitalism to entrepreneurial humanism: Understanding and re-evaluating critical theory. In L. Carter (Ed.), *Market matters: Applied rhetoric studies and free market competition* (pp. 53–76). Hampton Press.

Moses, M. G & Katz, S. B. (2006). The phantom machine: The invisible ideology of email (A cultural critique). In J. B. Scott, B. Longo & K. V. Willis (Eds.), Critical power tools: Technical communication and cultural Studies (pp. 71–105). SUNY Press.

Pilsch, A. (2017). Invoking darkness: Skotison, scalar derangement, and inhuman rhetoric. *Philosophy and Rhetoric, 50*(3), 336–355. https://doi.org/10.5325/philrhet.50.3.0336.

Plato. (1956). Phaedrus. (W. C. Helmbold & W. G. Rabinowitz, Trans.). Bobbs-Merrill.

Propen, A. D. (2018). *Visualizing posthuman conservation in the age of the Anthropocene*. Ohio State University Press. https://doi.org/10.26818/9780814213773.

Ramler, M. (2020). Queer usability. *Technical Communication Quarterly, 30*(4), 345–358. https://doi.org/10.1080/10572252.2020.1831614.

Reeves, R. G. (2012, February 28). Guest editorial: Scientific standards and the release of genetically modified insects for vector control. *Malaria World*. https://malariaworld.org/blogs/guest-editorial-scientific-standards-and-release-genetically-modified-insects-vector-control.

Rice, J. (2020). *Awful evidence: Conspiracy theory, rhetoric, and acts of evidence*. The Ohio State University Press. https://doi.org/10.26818/9780814214350.

Ridolfo, J. & Hart-Davidson, W. (Eds.). (2019). *Rhet ops: Rhetoric and information warfare*. University of Pittsburgh Press. https://doi.org/10.2307/j.ctvqc6hmj.

Sauer, B. (1993). Sense and sensibility in technical documentation: How feminist interpretation strategies can save lives in the nation's mines. *Journal of Business and Technical Communication, 9*, 63–83. https://doi.org/10.1177/1050651993007001004.

Simon, H. A. (1999). *The sciences of the artificial* (3rd ed.). MIT Press.

Slack, J. D., Miller, D. J. & Doak, J. (2006). The technical communicator as author: Meaning, power, authority. In J. B. Scott, B. Longo & K. V. Wills (Eds.), *Critical power tools: Technical communication and cultural studies* (pp. 25–46). SUNY Press.

Society for Technical Communication (STC). (1998). *Ethical principles*. https://www.stc.org/about-stc/ethical-principles/.

Sprat, T. (1667). *The history of the Royal-Society of London, for the improving of natural knowledge.* Royal Society of London.

Stephens, E. J. (2018). *Prisons, genres, and Big Data: Understanding the language of corrections in America's prisons* (Unpublished doctoral dissertation). Clemson University.

Waddell, C. (1996). Saving the Great Lakes: Public participation in environmental policy. In C. G. Herndl & S. C. Brown (Eds.), *Green culture: Environmental rhetoric in contemporary America* (pp. 141–165). University of Wisconsin Press.

Walton, R., Moore, K. R. & Jones, N. N. (2019). *Technical communication after the social justice turn: Building coalitions for action*. Routledge. https://doi.org/10.4324/9780429198748

Ward, M. (2014). *Deadly documents: Technical communication, organizational discourse, and the Holocaust—Lessons from the rhetorical work of everyday texts*. Baywood Publishing. https://doi.org/10.2190/DOC.

Whitburn, M. D. (2009). The first weeklong Technical Writers' Institute and its impact. *Journal of Business and Technical Communication, 23*(4), 428–447. https://doi.org/10.1177/1050651909338801.

Williams, M. F. & Pimentel, O. (2012). Editorial introduction: Race, ethnicity, and technical communication. *Journal of Business and Technical Communication*, *26*(3), 271–276. https://doi.org/10.1177/1050651912439535.

Zylinska, J. (2014). *Minimal ethics for the Anthropocene (Critical climate change)*. (T. Cohen & C. Colebrook, Eds.). Open Humanities Press. https://doi.org/10.3998/ohp.12917741.0001.001.

13. Feminisms

Erin Clark Frost
East Carolina University

Feminist approaches to technical and professional communication (TPC) can lead to more ethical engagements with users, communities, and other stakeholders—engagements that disrupt traditional understandings of gender, power, and discourse to the benefit of all involved. To appreciate the possibilities of feminist approaches, one must first understand the ***history*** of feminism (really feminisms) in the field of TPC.

Mary Lay's (1989) "Interpersonal Conflict in Collaborative Writing: What We Can Learn from Gender Studies" is widely regarded as the first explicit engagement of technical communication with gender studies. In this piece, Lay transfers gender studies ***knowledge*** of the ways gender perceptions affect relationships to the domain of technical writing and offers strategies for helping technical communication students to see the limitations of gender roles and better collaborate. However, her work was not initially taken up, as the field was still grappling with terminology and entry points for the sorts of critical studies that include feminisms.

A bit later, feminisms gained a foothold in technical communication through special issues, including a *Journal of Business and Technical Communication* (*JBTC*) special issue (5.4) in 1991, an *IEEE Transactions on Professional Communication* (*IEEE TPC*) special issue (35.4) in 1992, and special issues of *Technical Communication Quarterly* (*TCQ*) in 1994 (3.3) and 1997 (6.3). These special issues were critical to the advancement of feminist technical communication. As Isabelle Thompson (1999) noted in her qualitative content analysis spanning 1989 to 1997, "most journal articles about women and feminism in technical communication appeared in special issues devoted to those topics" (p. 155). Further, these special issues did not appear out of nowhere; Thompson argues that "The journals publishing the most articles about women and feminism are currently edited by women" (p. 163), and she shows that *JBTC* and *TCQ* outpaced the other journals in her corpus (*IEEE TPC*, *Journal of Technical Writing and Communication*, and *Technical Communication*) in terms of percentage of articles published about women and feminism. For more information about how Lay's 1989 article came about prior to the publication of these special issues, its author offers a history that also includes related information about women in the field of technical communication (Schuster, 2015).

Special issues devoted to feminisms and related topics have mostly disappeared since 1997, and by some measures, "interest in feminism and women's issues has declined over the past 15 years" (Thompson & Smith, 2006). However, feminist technical communicators now persist in doing feminist work in the

DOI: https://doi.org/10.37514/TPC-B.2023.1923.2.13

absence of discipline-sponsored forums, through individual articles and chapters (e.g., Hallenbeck, 2012; Jones, 2016; Koerber, 2002; Ledbetter, 2018; Mallette, 2017; Malone, 2010; Petersen, 2014, 2019; Raign, 2019; Rauch, 2012; Rohrer-Vanzo et al., 2016; Sullivan & Moore, 2013). The past five to seven years have also seen some book projects that engage feminisms, sex, or gender and technical communication as significant themes (e.g., Agboka & Matveeva, 2018; Koerber, 2013, 2018; Owens, 2015). All of the above and more contribute to some common themes in feminist approaches to technical communication, including 1) feminist historiographical work, 2) interventions into misogynist practices, and 3) attention to plurality, intersectionality, and interdisciplinarity. This last theme points toward the fact that increasingly intersectional approaches mean that feminist work is happening in a variety of contexts and may not always be apparent in keyword searches of titles and abstracts; it also represents perhaps the most important trajectory for advancement of feminist (including womanist, Black feminist, and queer feminist) work in the field.

Feminist historiographical work is paradoxically connected with ***professionalization***, which can serve as a code word for unmarked maleness, and it is a common topic among technical writers (Coppola, 2012; Davis, 2001; Faber & Johnson-Eilola, 2002; Kynell-Hunt & Savage, 2003; Savage, 1996, 1999, 2004, 2010). Some would say "technical writing finally became a genuine ***profession*** as wartime technologies were translated into peacetime uses" and "the demand for [technical writing] courses rose dramatically as the colleges were deluged with returning veterans after 1945" (Connors, 1982, p. 12). If this history is to be believed, then technical communication was growing up as a field just before the time when mainstream second-wave feminism was gaining power. The second wave, often said to have begun with the 1963 publication of Betty Frieden's *The Feminine Mystique* and certainly associated with the Civil Rights movement, shifted attention from suffrage to identity and gender roles. Many women (particularly but not only white women) began to question the notion that being a wife and mother was their only path to success. The second wave gave rise to various kinds of feminisms that were sometimes in conflict with one another; for example, cultural feminists' belief in valuing traditionally female roles could sometimes clash with liberal feminists' injunctions to respond to stereotyping with resistance. And it is at what is typically considered the end of the second wave that explicitly feminist interventions into formal technical communication literature began.

The reflective bent of the second wave shows up in technical communication through field historiographies. The 1992 special issue of *IEEE TPC* investigated the effects of gendered assumptions on understandings of rationality. In this issue, Elizabeth Tebeaux and Lay (1992) engage in a historiographical recovery of English Renaissance-era technical writing for women; Kathryn A. Neeley explicates a history of women mediators in the 18th and 19th centuries. Later, in the 1997 *TCQ* special issue, authors worked to recover histories of women technical communicators and question the absence of such histories. Indeed, Katherine T.

Durack (1997) begins by suggesting that women's work in technical communication has been overlooked because the field has been seen as the domain of men and because historians have tended to internalize this belief. Elizabeth Flynn (1997) and John F. Flynn (1997), among others, begin to remedy this situation by paying attention to the mapping of feminisms in technical communication and by engaging in the recovery of domestic ***sciences*** and technologies—like grocery shopping, cooking, and bread-making—as technical communication practices. More recently, Marie E. Moeller and I (2016) additionally point to uncritical recoveries as potential feminist problems in our analysis of liberation vis-a-vis cookbook ***rhetorics*** and connected critique of field narratives. That is, we suggest that feminist approaches to technical communication artifacts should be attentive to context and should avoid hailing entire ***genres***—particularly domestic genres—as necessarily liberatory.

The 1992 special issue of *IEEE TPC* has perhaps offered the largest trove of scholarship that directly addresses feminist interventions to misogynist practices. In that issue, Beverly Sauer (1992) argues that gendered assumptions about male ways of thinking have affected mine safety management. L. J. Rifkind and L. F. Harper (1992) assert a paradox between sexual harassment policies and the necessity of interpersonal relationships in the workplace, and S. Dell (1992) draws on communication theory in a rhetorical analysis of the "glass ceiling." Stephen A. Bernhardt (1992) and Deborah S. Bosley (1992) separately engage issues of gender in ***visual design***. Beyond this issue, M. Z. Corbett (1990), Dell (1990), and Jeanette Vaughn (1989) all provide examples of ways to address sexist language in technical ***documentation***. Others interrogate the intersections of gender and technologies (Aschauer, 1999; Brasseur, 1993; Lay,1993). Notably, Angela M. Haas et al. (2002) complicate constructions of women's and girls' relationships with technology and technical communication, arguing that it is dangerous to "presume that 'going online' somehow alleviates gender inequity and power imbalance" (p. 247).

Defining intervention work as rhetorical means that almost any feminist technical communication work could be thought of as an intervention. An important entry into this body of work, then, is scholarship that addresses the language of the field. It is no accident that some of this intervention work looks inward, as does Sauer (1992) when she uses literature published by the Mine Safety and Health Administration to demonstrate the importance of training technical writers to understand how gendered assumptions about male rationality can influence the epistemological underpinnings of technical documentation. Likewise, the 1994 issue of *TCQ* showcased work—especially the articles by Jo Allen (1994), Bosley (1994), and Susan Mallon Ross (1994)—that continues a conversation about the unmarked maleness of the field. Allen and Bosley point to ways of challenging and making apparent otherwise implicit misogyny. I (2016) recommend apparency as a specific approach to intervening in technical rhetorics (including those within the field) that privilege unmarked maleness through efficiency rhetorics; apparent feminism advocates putting a face on feminisms, hailing non-feminist allies, and doing the rhetorical work to show how efficiency (and other terms like it) are often used to quell diverse approaches and perspectives.

Interest in paying attention to a greater plurality of feminisms, and especially to addressing issues of intersectionality, has become increasingly important to feminist technical communicators and apparent in their work. In particular, these concerns have overlapped with ***social justice*** movements. Much of this work has been made possible by Haas' (2012) argument for intersectional approaches to race, rhetoric, and technology. Since then, a number of works have been published that engage with feminisms and gender studies approaches to technical communication as part of a larger decolonial agenda to incorporate cultural studies and social justice into the field (De Hertogh, 2018; Jones et al., 2016; Moeller & Frost, 2016; Novotny & Hutchinson, 2019; Petersen & Moeller, 2016; Petersen & Walton, 2018; Smith, 2014). Notably, feminist technical communication scholars who embrace plurality and intersectionality increasingly combat the isolation and potential of myopic viewpoints of individual scholarship by co-authoring and engaging in other forms of scholarly ***collaboration***—often without institutional support for such endeavors.

While recent work has been able to explicitly name intersectional feminisms as both goals and approaches, a number of scholars laid the groundwork for this with important ***research*** on the subjectivities of technical communication and the importance of feminist methods (Coletta, 1992; Dragga, 1993; Sauer, 1993; Tebeaux, 1993). As just some examples, Gail Lippincott (2003) examines Ellen Swallow Richards' rhetorical development of ethos, Lee Brasseur (2005) shines a light on Florence Nightingale's persuasive use of rose diagrams, and Jeffrey T. Grabill (2007) focuses on the penetration of information technologies into everyday lives as he encourages emancipatory action. E. P. Boyer and T. G. Webb (1992) and M. de Armas Ladd and M. Tangum (1992) look to diversity and difference as guiding principles in feminist thought in technical communication.

The special issues described above were especially important in laying the groundwork for plurality, intersectionality, and interdisciplinarity. *TCQ* issue 3.3 expanded upon feminist approaches to technical communication with an issue that "explores gender as a social force that shapes and is shaped by professional communication practices and readerships" (LaDuc & Goldrick-Jones, 1994, p. 246). In this special issue, Linda LaDuc and Amanda Goldrick Jones (1994) invoke the power of feminism's ability to take on multiple theoretical and political positions, "forsaking the comfort of even a single feminist method or 'truth stance'" (p. 249). Laura J. Gurak and Nancy L. Bayer (1994) and Sauer (1994) describe a variety of feminist methodological approaches (and resulting implications) to their subjects rather than limiting their investigations to a single methodological approach. This variety of methodological approaches opens the door to rich interdisciplinarity in feminisms' contributions to technical communication.

The 1991 special issue of *JBTC* promotes a cultural turn in technical communication, providing foundations for work in feminisms and cultural studies and addressing the relationship between these two. (This cultural turn was not initially taken up, as suggested by the later return to the idea by J. Blake Scott et al. [2006].) In the 1991 *JBTC* special issue, Lay (1991) suggests a redefinition of technical communication that considers cultural issues, most notably issues of gender. Lay relies on technical communicators' understandings of social constructionism to combat and make visible scientific positivism in technical communication artifacts. Diane D. Brunner (1991) encourages recognizing that "we and our students operate within a culture in which domination/subordination is produced and reproduced" (p. 409) and that, embodied as we are, this creates ideologies in which some people are affirmed and others are cast out. Others in the issue advocate revision to static conceptions of female cultures and resistance to auto-colonization (Carrell, 1991; Flynn et al., 1991) and explicitly advocate for interdisciplinary work to support feminisms (Flynn et al., 1991).

Finally, in the 1994 special issue of *TCQ*, Ross looks to sources outside the discipline for insight, pushing for ***intercultural*** studies such as her own on the interactions between a Mohawk community and the Environmental Protection Agency. She provides an example of how feminist concern with other injustices—namely, racism and environmental oppressions—can inform broader understandings of the applicability of feminisms to a field like technical communication.

Feminisms and social justice agendas, in other words, are symbiotic—and they allow for the inclusion of queer, race-based, and (dis)ability studies approaches to technical communication. Through plural, intersectional, and interdisciplinary lenses, feminisms address structural oppressions—and more—that exist in technical communication scholarship and practice. For example, Cecilia D. Shelton (2019) emphasizes the confluence of Black feminisms and social justice work. Her dissertation offers a Techné of Marginality that emphasizes the value of Black subjectivities and experiences and employs digital activism as a medium to help technical communicators to "recognize the ways in which Black communities, and particularly Black women, have always, already done the unpaid labor that builds the communication infrastructures for equity, inclusion, and freedom" (p. 1). Temptaous T. Mckoy (2019) offers amplification rhetorics as a theoretical framework describing Black discursive and communicative practices that technical communicators can model their work on in order to center the lived experiences and epistemologies of Black people and other historically marginalized groups. Indeed, you can see feminist collaborative work that decenters positions of power in action by reading the Afterword of this keyword collection, in which Kristen R. Moore, Lauren E. Cagle and Nicole Lowman describe the process of a citation check intended to help the collection be as inclusive, accessible, and intersectional as possible.

The future of feminisms in technical communication is both plural and clear: Feminist technical communicators are devoted to decentering traditional centers of power in favor of radical, inclusive, and diverse feminist praxes.

References

Agboka, G. & Matveeva, N. (Eds.). (2018). *Citizenship and advocacy in technical communication*. Routledge. https://doi.org/10.4324/9780203711422.

Allen, J. (1994). Women and authority in business/technical communication: An analysis of writing features, methods, and strategies. *Technical Communication Quarterly*, *3*, 271–292. https://doi.org/10.1080/10572259409364572.

Bernhardt, S. A. (1992). The design of sexism: The case of an army maintenance manual. *IEEE Transactions on Professional Communication*, *35*(4), 217–221. https://doi.org/10.1109/47.180282.

Bosley, D. S. (1992). Gender and visual communication: Toward a feminist theory of design. *IEEE Transactions on Professional Communication*, *35*(4), 222–229. https://doi.org/10.1109/47.180283.

Bosley, D. S. (1994). Feminist theory, audience analysis, and verbal and visual representation in a technical communication writing task. *Technical Communication Quarterly*, *3*, 293–307. https://doi.org/10.1080/10572259409364573.

Boyer, E. P. & Webb, T. G. (1992). Ethics and diversity: A correlation enhanced through corporate communication. *IEEE Transactions on Professional Communication*, *35*(1), 38–43. https://doi.org/10.1109/47.126938.

Brady Aschauer, A. (1999). Tinkering with technological skill: An examination of the gendered uses of technologies. *Computers and Composition*, *16*(1), 7–23. https://doi.org/10.1016/S8755-4615(99)80003-6Brasseur, L. (1993). Contesting the objectivist paradigm: Gender issues in the technical and professional communication curriculum. *IEEE Transactions on Professional Communication*, *36*(3), 114–123. https://doi.org/10.1109/47.238051.

Brasseur, L. (2005). Florence Nightingale's visual rhetoric in the rose diagrams. *Technical Communication Quarterly*, *14*, 161–182. https://doi.org/10.1207/s15427625tcq1402_3.

Brunner, D. D. (1991). Who owns this work? The question of authorship in professional/academic writing. *Journal of Business and Technical Communication*, *5*, 393–411. https://doi.org/10.1177/1050651991005004004.

Carrell, D. (1991). Gender scripts in professional writing textbooks. *Journal of Business and Technical Communication*, *5*, 463–468. https://doi.org/10.1177/1050651991005004007.

Coletta, W. J. (1992). The ideologically biased use of language in scientific and technical writing. *Technical Communication Quarterly*, *1*(1), 59–70. https://doi.org/10.1080/10572259209359491.

Connors, R. (1982). The rise of technical writing instruction in America. *Journal of Technical Writing and Communication*, *12*(4), 329–352. https://doi.org/10.1177/004728168201200406.

Coppola, N. (2012). Professionalization of technical communication: Zeitgeist for our age. *Technical Communication*, *59*(1), 1–7.

Corbett, M. Z. (1990). Clearing the air: Some thoughts on gender-neutral writing. *IEEE Transactions on Professional Communication*, *33*(1), 197–200. https://doi.org/10.1109/47.49063.

Davis, M. T. (2001). Shaping the future of our profession. *Technical Communication*, *48*(2), 139–144.

de Armas Ladd, M. & Tangum, M. (1992). What difference does inherited difference make? Exploring culture and gender in scientific and technical professions. *IEEE*

Transactions on Professional Communication, 35(4), 183–188. https://doi.org/10.1109/47.158986.

De Hertogh, L. B. (2018). Feminist digital research methodology for rhetoricians of health and medicine. *Journal of Business and Technical Communication, 32*(4), 480–503. https://doi.org/10.1177/1050651918780188.

Dell, S. A. (1990). Promoting equality of the sexes through technical writing. *Technical Communication, 37*, 248–251.

Dell, S. A. (1992). A communication-based theory of the glass ceiling: Rhetorical sensitivity and upward mobility within the technical organization. *IEEE Transactions on Professional Communication, 35*(4), 230–235. https://doi.org/10.1109/47.180284.

Dragga, S. (1993). Women and the profession of technical writing: Social and economic influences and implications. *Journal of Business and Technical Communication, 7*, 312–321. https://doi.org/10.1177/1050651993007003002.

Durack, K. T. (1997). Gender, technology, and the history of technical communication. *Technical Communication Quarterly, 6*, 249–260. https://doi.org/10.1207/s15427625tcq0603_2.

Faber, B. & Johnson-Eilola, J. (2002). Migrations: Strategic thinking about the future(s) of technical communication. In B. Mirel & R. Spilka (Eds.), *Reshaping technical communication: New directions and challenges for the 21st century* (pp. 135–148). Lawrence Erlbaum.

Flynn, E. A. (1997). Emergent feminist technical communication. *Technical Communication Quarterly, 6*, 313–320. https://doi.org/10.1207/s15427625tcq0603_6.

Flynn, E. A., Savage, G., Penti, M., Brown, C. & Watke, S. (1991). Gender and modes of collaboration in a chemical engineering design course. *Journal of Business and Technical Communication, 5*, 444–462. https://doi.org/10.1177/1050651991005004006.

Flynn, J. (1997). Toward a feminist historiography of technical communication. *Technical Communication Quarterly, 6*, 321–329. https://doi.org/10.1207/s15427625tcq0603_7.

Frieden, B. (1963). *The feminine mystique*. Norton.

Frost, E. A. (2016). Apparent feminism as a methodology for technical communication and rhetoric. *Journal of Business and Technical Communication, 30*(1), 3–28. https://doi.org/10.1177/1050651915602295.

Grabill, J. T. (2007). *Writing community change: Designing technologies for citizen action*. Hampton.

Gurak, L. J. & Bayer, N. L. (1994). Making gender visible: Extending feminist critiques of technology to technical communication. *Technical Communication Quarterly, 3*, 257–270. https://doi.org/10.1080/10572259409364571.

Haas, A. M. (2012). Race, rhetoric, and technology: A case study of decolonial technical communication theory, methodology, and pedagogy. *Journal of Business and Technical Communication, 26*(3), 277–310. https://doi.org/10.1177/1050651912439539.

Haas, A., Tulley, C. & Blair, K. (2002). Mentors versus masters: Women's and girls' narratives of (re)negotiation in Web-based writing spaces. *Computers and Composition, 19*, 231–249. https://doi.org/10.1016/S8755-4615(02)00128-7.

Hallenbeck, S. (2012). User agency, technical communication, and the 19th-century woman bicyclist. *Technical Communication Quarterly, 21*(4), 290–306. https://doi.org/10.1080/10572252.2012.686846.

Jones, N. N. (2016). The technical communicator as advocate: Integrating a social justice approach in technical communication. *Journal of Technical Writing and Communication, 46*(3), 342–361. https://doi.org/10.1177/0047281616639472.

Jones, N. N., Moore, K. R. & Walton, R. (2016). Disrupting the past to disrupt the future: An antenarrative of technical communication. *Technical Communication Quarterly, 25*(4), 211–229. https://doi.org/10.1080/10572252.2016.1224655.
Koerber, A. (2002). Postmodernism, resistance, and cyberspace: Making rhetorical spaces for feminist mothers on the web. *Women's Studies in Communication, 24*, 218–240. https://doi.org/10.1080/07491409.2001.10162435.
Koerber, A. (2013). *Breast or bottle? Contemporary controversies in infant-feeding policy and practice*. University of South Carolina Press. https://doi.org/10.2307/j.ctv6wgk8f.
Koerber, A. (2018). *From hysteria to hormones: A rhetorical history*. The Pennsylvania State University Press. https://doi.org/10.1515/9780271081571.
Kynell-Hunt, T. & Savage, G. (2003). *Power and legitimacy in technical communication: The historical and contemporary struggle for professional status*. Routledge.
LaDuc, L. & Goldrick-Jones, A. (1994). The critical eye, the gendered lens, and "situated" insights—Feminist contributions to professional communication. *Technical Communication Quarterly, 3*(3), 245–256. https://doi.org/10.1080/10572259409364570.
Lay, M. (1989). Interpersonal conflict in collaborative writing: What we can learn from gender studies. *Journal of Business and Technical Communication, 3*(2), 5–28. https://doi.org/10.1177/105065198900300202.
Lay, M. M. (1991). Feminist theory and the redefinition of technical communication. *Journal of Business and Technical Communication, 5*, 348–370. https://doi.org/10.1177/1050651991005004002.
Lay, M. M. (1993). Gender studies: Implications for the professional communication classroom. In N. R. Blyler & C. Thralls (Eds.), *Professional communication: The social perspective* (pp. 215–229). Sage.
Ledbetter, L. (2018) The rhetorical work of YouTube's beauty community: Relationship- and identity-building in user-created procedural discourse. *Technical Communication Quarterly, 27*(4), 287–299. https://doi.org/10.1080/10572252.2018.1518950.
Lippincott, G. (2003). Rhetorical chemistry: Negotiating gendered audiences in nineteenth-century nutrition studies. *Journal of Business and Technical Communication, 17*(1), 10–49. https://doi.org/10.1177/1050651902238544.
Mallette, J. C. (2017). Writing and women's retention in engineering. *Journal of Business and Technical Communication, 31*(4), 417–442. https://doi.org/10.1177/1050651917713253.
Malone, E. A. (2010). Chrysler's "Most beautiful engineer": Lucille J. Pieti in the pillory of fame. *Technical Communication Quarterly, 19*(2), 144–183. https://doi.org/10.1080/10572250903559258.
Mckoy, T. T. (2019). *Y'all call it technical and professional communication, we call it #fortheculture: The use of amplification rhetorics in Black communities and their implications for technical and professional communication studies* [Unpublished doctoral dissertation]. East Carolina University.
Moeller, M. E. & Frost, E. A. (2016). Food fights: Cookbook rhetorics, monolithic constructions of womanhood, and field narratives in technical communication. *Technical Communication Quarterly, 25*(1), 1–11. https://doi.org/10.1080/10572252.2016.1113025.
Neeley, K. A. (1992). Woman as mediatrix: Women as writers on science and technology in the eighteenth and nineteenth centuries. *IEEE Transactions on Professional Communication, 35*(4), 208–216. https://doi.org/10.1109/47.180281.
Novotny, M. & Hutchinson, L. (2019). Data our bodies tell: Towards critical feminist action in fertility and period tracking applications. *Technical Communication Quarterly, 28*(4), 332–360. https://doi.org/10.1080/10572252.2019.1607907.

Owens, K. H. (2015). *Writing childbirth: Women's Rhetorical agency in labor and online*. Southern Illinois University Press.

Petersen, E. J. (2014). Redefining the workplace: The professionalization of motherhood through blogging. *Journal of Technical Writing and Communication, 44*(3), 277–296. https://doi.org/10.2190/TW.44.3.d.

Petersen, E. J. (2019). The "reasonably bright girls": Accessing agency in the technical communication workplace through interactional power. *Technical Communication Quarterly, 28*(1), 21–38. https://doi.org/10.1080/10572252.2018.1540724.

Petersen, E. J. & Moeller, R. M. (2016). Using antenarrative to uncover systems of power in mid-20th century policies on marriage and maternity at IBM. *Journal of Technical Writing and Communication, 46*(3), 362–386. https://doi.org/10.1177/0047281616639473.

Petersen, E. J. & Walton, R. (2018). Bridging analysis and action: How feminist scholarship can inform the social justice turn. *Journal of Business and Technical Communication, 32*(4), 416–446. https://doi.org/10.1177/1050651918780192.

Raign, K. R. (2019). Finding our missing pieces—Women technical writers in ancient Mesopotamia. *Journal of Technical Writing and Communication, 49*(3), 338–364. https://doi.org/10.1177/0047281618793406.

Rauch, S. (2012). The accreditation of Hildegard von Bingen as medieval female technical writer. *Journal of Technical Writing and Communication, 42*(4), 393–411. https://doi.org/10.2190/TW.42.4.d.

Rifkind, L. J. & Harper, L. F. (1992). Cross-gender immediacy behaviors and sexual harassment in the workplace: A communication paradox. *IEEE Transactions on Professional Communication, 35*(4), 236–241. https://doi.org/10.1109/47.180285.

Rohrer-Vanzo, V., Stern, T., Ponocny-Seliger, E. & Schwarzbauer, P. (2016). Technical communication in assembly instructions: An empirical study to bridge the gap between theoretical gender differences and their practical influence. *Journal of Business and Technical Communication, 30*(1), 29–58. https://doi.org/10.1177/1050651915602292.

Ross, S. M. (1994). A feminist perspective on technical communicative action: Exploring how alternative worldviews affect environmental remediation efforts. *Technical Communication Quarterly, 3*(3), 325–342. https://doi.org/10.1080/10572259409364575.

Sauer, B. A. (1992). The engineer as rational man: The problem of imminent danger in a non-rational environment. *IEEE Transactions on Professional Communication, 35*(4), 242–249. https://doi.org/10.1109/47.180286.

Sauer, B. A. (1993). Sense and sensibility in technical documentation: How feminist interpretation strategies can save lives in the nation's mines. *Journal of Business and Technical Communication, 7*(1), 63–83. https://doi.org/10.1177/1050651993007001004.

Sauer, B. A. (1994). Sexual dynamics of the profession: Articulating the ecriture masculine of science and technology. *Technical Communication Quarterly, 3*(3), 309–323. https://doi.org/10.1080/10572259409364574.

Savage, G. J. (1996). Redefining the responsibilities of teachers and the social position of the technical communicators. *Technical Communication Quarterly, 5*(3), 309–327. https://doi.org/10.1207/s15427625tcq0503_4.

Savage, G. J. (1999). The process and prospects for professionalizing technical communication. *Journal of Technical Writing and Communication, 29*(4), 355–381. https://doi.org/10.2190/7GFX-A5PC-5P7R-9LHX.

Savage, G. J. (2004). Tricksters, fools, and sophists: Technical communication as postmodern rhetoric. In T. Kynell-Hunt & G. J. Savage (Eds.), *Power and legitimacy in technical communication: Strategies for professional status* (Vol. Two, pp. 167–193). Baywood.

Savage, G. J. (2010). Program assessment, strategic modernism, and professionalization politics: Complicating Coppola and Elliot's "relational model." In M. N. Hundleby & J. Allen (Eds.), *Assessment in technical and professional communication* (pp. 161–168). Baywood. https://doi.org/10.2190/AITC10.

Schuster, M. L. (2015). My career and the "rhetoric of" technical writing and communication. *Journal of Technical Writing and Communication, 45*(4), 381–391. https://doi.org/10.1177/0047281615585754.

Scott, J. B., Longo, B. & Wills, K. V. (2006). *Critical power tools: Technical communication and cultural studies.* State University of New York Press.

Shelton, C. D. (2019). *On edge: A techné of marginality* [Unpublished doctoral dissertation]. East Carolina University.

Smith, A. W. (2014). Porn architecture: User tagging and filtering in two online porn communities. *Communication Design Quarterly, 3*(1), 17–22. https://doi.org/10.1145/2721882.2721885.

Sullivan, P. & Moore, K. (2013). Time talk: On small changes that enact infrastructural mentoring for undergraduate women in technical fields. *Journal of Technical Writing and Communication, 43*(3), 333–354. https://doi.org/10.2190/TW.43.3.f.

Tebeaux, E. (1993). Technical writing for women of the English Renaissance. *Written Communication, 10*(2), 164–199. https://doi.org/10.1177/0741088393010002002.

Tebeaux, E. & Lay, M. (1992) Images of women in technical books from the English Renaissance. *IEEE Transactions on Professional Communication, 35*(4), 196–207. https://doi.org/10.1109/47.180280.

Thompson, I. (1999). Women and feminism in technical communication: A qualitative content analysis of journal articles published in 1989 through 1997. *Journal of Business and Technical Communication, 13*(2), 154–178. https://doi.org/10.1177/1050651999013002002.

Thompson, I. & Smith, E. O. (2006). Women and feminism in technical communication—An update. *Journal of Technical Writing and Communication, 36*(2), 183–199. https://doi.org/10.2190/4JUC-8RAC-73H6-N57U.

Vaughn, J. (1989). Sexist language—Still flourishing. *The Technical Writing Teacher, 16*(1), 33–40.

Walton, R., Moore, K. & Jones, N. (2019). *Technical communication after the social justice turn: Building coalitions for action.* Routledge. https://doi.org/10.4324/9780429198748.

14. Genre

Brent Henze
East Carolina University

The word *genre* comes from the French, meaning "kind." *Genre* was used as early as 1770 to name "a particular style or category of works of art; esp. a type of literary work characterized by a particular form, style, or purpose" (Oxford University Press, n.d.).

Just as the related term *genus* names a broad category or "kind" into which more *specific* members can be grouped (for example, horses and zebras are two *species* in the equine *genus*), a genre is a categorization: Diverse specimens sharing some quality are part of a genre defined by that quality.

You may be familiar with *genre* as a term that describes recognizable, repeated forms of literary expression (e.g., *sonnets*, or *Elizabethan sonnets*; *mystery novels*, or *young adult detective serials*). Technical communicators and educators often use *genre* similarly to identify common types of technical writing, such as ***proposals***, *instruction manuals*, and *sales letters*. This familiar usage helps us name and group individual texts, and conversely, it signals characteristics that ***audiences*** expect to find in a text. But as helpful as it is for classifying regularities of already-written texts, this usage is less helpful for guiding or explaining the composition of new messages.

To better tackle these matters, technical communication turns to 20th century ***rhetorical*** theory. Building on earlier work that related genres to types of rhetorical situation (Bitzer, 1968), Carolyn Miller (1984) famously described genres as "typified rhetorical actions based in recurrent situations" (p. 159). This insight gave rise to the rhetorical genre studies (RGS) model that dominates genre scholarship in technical communication today. At root, genres are particular *kinds* of communication, expressed in recurring *contexts*, used to accomplish particular *purposes* shared by writers and their *audiences*.

RGS scholarship has undergone numerous shifts since the 1970s, and several good summaries are available (e.g., Artemeva, 2006; Bawarshi & Reiff, 2010; Henze with Miller & Carradini, 2016; Kain, 2005). RGS helps us to understand what's happening when writers and readers communicate through the mediation of various kinds of text. In the RGS model, a genre is a way of understanding characteristic activities that happen in a particular context. Although a genre may in fact have a characteristic form or ***style***, these emerge as a result of "genred" ***activity***—the repeated responses of actual writers in routine or repeating practical contexts. What's important is the activity, not the form.

Genres may be regular, recognizable, authoritative, and even apparently stable, but they are also generative, creative, mutable, open-ended, dynamic, and

DOI: https://doi.org/10.37514/TPC-B.2023.1923.2.14

efficient. Catherine Schryer (2000) defines genres as "constellations of regulated, improvisational strategies" (p. 450): They're "regulated" because it's not just the author but the relevant social context that determines whether a communication will be legible, yet "improvisational" since context gives authors an indefinite range of choices. Writers learn to work within a set of genres that community members have validated through repeated use. These genres not only help the writer to identify situationally appropriate types of rhetorical response, but also create a rhetorical space for invention.

This notion of genres in dynamic tension is important because it helps to explain why even the most apparently stable genres still change over time and permit variations. Experienced writers, after all, don't simply follow templates; they respond to *exigencies* (circumstances that provoke an action), they account for *context* (the variables of circumstance, timing, and relationship that surround a communication), and they create *content* that has meaning in particular cultures (including institutional and professional cultures).

Technical communication often occurs in complex institutional settings, and in technical contexts, social dynamics include the many ways institutions act as agents in discourse. For example, technical communicators often do not "author" their own texts: Instead, they're parts of a larger system of content generation, repurposing, ***editing***, production, and distribution. In this system, the individual writer might be little aware of the ultimate rhetorical purposes of a text they create. The locus of rhetorical activity is just as likely to be an institution, a user-responsive system (e.g., context-sensitive help), or some other ***actor***.

Just as the complexity of rhetorical contexts has altered the priorities of genre work in technical communication, so too does genre look different in the heavily mediated contexts of technical communication. After all, even an individually authored text is the product of editors, publishers, and other intermediaries, not just its "author." But in many technical communication contexts, the extent of this mediation is even more profound. For example, the technical writers who create a context-sensitive help system for a computer program may compose discrete chunks of text that appear on users' screens. But the appearance, order, and timing of those texts are governed by user behavior (such as clicking a "help" button or entering an erroneous command). The text is also mediated by programming that neither writer nor user created. The "document" is not a fixed product; it's an emergent experience produced in response to user input, using content prepared by a technical writer, and mediated by programming.

Since the recognizable conventions of genres result from accumulated rhetorical performances, genres can evolve over time and vary across contexts. Genres might seem "stabilized-for-now," as Schryer (1993, p. 204) puts it, but over time they adopt some of the variations introduced by writers responding to their exigencies. For example, Charles Bazerman (1988) describes the evolution of ***scientific research*** articles over centuries in response to the changing social dynamics and rhetorical contexts of experimental science.

Genres can also hybridize as writers combine strategies from multiple genres to tackle new problems. Carolyn Rude (1995), for example, showed how the decision-making report genre adapted strategies taken from proposals, experimental reports, and persuasive essays. Far from being mere *constraints* or *rules* to be followed, genres are more like a toolbox of handy strategies that can be applied to conventional tasks, but also remixed, repurposed, and modified in response to novel rhetorical challenges.

Change happens very quickly in many technical contexts. In these fluid contexts, some genres might change so rapidly that formal and stylistic conventions between the "generations" of a genre are negligible. Simply examining two examples of the same genre—say, two weather forecasts, or two error reports, separated by a few years and a few iterations of media—might yield few obvious similarities. The equivalency of these genre performances resides in their communicative context, the "social action" that the texts engage in, despite the many differences in *how* the texts do what they do.

Because technical communication situations are often distributed and complex, the individual text is often less salient than groups of interacting texts: for example, the sequence of CFP, inquiry letter, grant application, budget, impact report, and other genres associated with grant seeking. Technical communication ***research*** has studied how genres relate to one another in sets (Devitt, 1991, 2004), systems (Bazerman, 1994; Russell, 1997; Yates & Orlikowski, 2002), repertoires (Orlikowski & Yates, 1994), ecologies (Spinuzzi & Zachry, 2000), and other assemblages.

Rather than operating independently, genres often function together in sequences of recognized discourse "moves." To understand a genre is to appreciate its rhetorical ecosystem, including other genres and the various actors and relationships surrounding it. Foundational scholarship drawing upon ***activity theory*** and actor-networks, especially that of David Russell (1997), Clay Spinuzzi and Mark Zachry (2000), and Spinuzzi (2003, 2008), has examined how complex and distributed communities and networks get things done by sharing resources, including genres. Natasha Jones (2016), for example, shows how members of the Innocence Project Northwest adapted the communication genres circulating among Innocence Project chapters to accomplish local goals. The community's genres, including weekly team meetings, client-completed questionnaires, and Facebook posts, not only "help[ed] coordinate and promote collaboration," but also helped the community to "shape a cohesive identity and common goals" (Jones, 2016, p. 310).

Individually and in assemblages, genres can not only help actors to get work done, but they are also part of the joint processes of enculturation, disciplinary learning and reproduction, and sense-making that enable participants to coordinate activity. In a sense, genres function as vectors, carrying elements of a discourse along the various branches and turnings of a complex activity system or network.

Finally, technical communication scholars study how new practitioners are enculturated into their disciplines and professions, in part, by way of genres. Though technical genres are still routinely taught in introductory courses, scholarly opinion about the efficacy of teaching genres is mixed. Some scholars (e.g., Freedman, 1993; Freedman et al., 1994) doubt that genres can be explicitly taught in the classroom, since genre use is responsive to exigency and context, and classrooms are not authentic contexts for these genres. Others, including Amy Devitt, counter that the classroom can provide effective preparation for future technical genre use. Although the classroom doesn't offer exigencies identical to those in professional settings, Devitt (2009) argues that teachers can introduce genre principles that prepare students to improvise in response to the exigencies they encounter in later workplace contexts.

Teaching students *about genre* (rather than teaching particular genres) can help them become more versatile, savvy communicators and observers of their disciplines and workplaces, and thus better able to acquire disciplinary skills and awareness quickly once they're in the workplace. As Anis Bawarshi and Mary Jo Reiff (2010) describe it, genres function as "learning strategies or tools for accessing unfamiliar writing situations" (p. 191). Devitt (2004) proposes a ***pedagogy*** based upon "meta-awareness of genres, as learning strategies rather than static features" of text (p. 197).

As content production becomes increasingly divisible from distribution and consumption, technical communicators are less likely to "author" whole, stable units of end-user text. They may also find themselves becoming more involved in the components of ***documentation*** or ***information*** systems that are harder to recognize as writing or communication: components like interface ***design***, content reuse, ***translation***, and distribution.

The shift in technical communication scholarship toward studies of larger information systems, networks, and genre ecologies reflects the new realities of our field. Just as the characteristic genres continue to evolve, we can expect our genre theory to continue to expand and hybridize as researchers study and theorize contemporary genres and communication practices in complex networks, systems, and institutions.

References

Artemeva, N. (2005). A time to speak, a time to act: A rhetorical genre analysis of a novice engineer's calculated risk taking. *Journal of Business and Technical Communication, 19*, 389–421. https://doi.org/10.1177/1050651905278309.

Artemeva, N. (2006). Approaches to learning genres: A bibliographical essay. In N. Artemeva & A. Freedman (Eds.), *Rhetorical genre studies and beyond* (pp. 9–99). Inkshed Publications.

Artemeva, N., Logie, S. & St-Martin, J. (1999). From page to stage: How theories of genre and situated learning help introduce engineering students to discipline-specific communication. *Technical Communication Quarterly, 8*(3), 301–316. https://doi.org/10.1080/10572259909364670.

Bawarshi, A. S. & Reiff, M. (2010). *Genre: An introduction to history, theory, research, and pedagogy.* Parlor Press.

Bazerman, C. (1988). *Shaping written knowledge: The genre and activity of the experimental article in science.* University of Wisconsin Press.

Bazerman, C. (1994). Systems of genre and the enactment of social intentions. In A. Freedman & P. Medway (Eds.), *Genre and the new rhetoric* (pp. 79–99). Taylor & Francis.

Bitzer, L. F. (1968). The rhetorical situation. *Philosophy and Rhetoric, 1*, 1–14.

Devitt, A. J. (1991). Intertextuality in tax accounting: Generic, referential, and functional. In C. Bazerman & J. Paradis (Eds.), *Textual dynamics of the professions: Historical and contemporary studies of writing in professional communities* (pp. 336–355). University of Wisconsin Press.

Devitt, A. J. (2004). *Writing genres.* Southern Illinois University Press.

Devitt, A. J. (2009). Teaching critical genre awareness. In C. Bazerman, A. Bonini & D. Figueiredo (Eds.), *Genre in a changing world* (pp. 337–351). The WAC Clearinghouse; Parlor Press. https://doi.org/10.37514/PER-B.2009.2324.2.17.

Freedman, A. (1993). Show and tell? The role of explicit teaching in the learning of new genres. *Research in the Teaching of English, 27*(3), 222–251.

Freedman, A., Adam, C. & Smart, G. (1994). Wearing suits to class: Simulating genres and simulations as genre. *Written Communication, 11*(2), 193–226. https://doi.org/10.1177/0741088394011002002.

Henze, B., with Miller, C. & Carradini, S. (2016). *Technical communication.* Genre Across Borders. http://genreacrossborders.org/research/technical-communication.

Jones, N. N. (2016). Found things: Genre, narrative, and identification in a networked activist organization. *Technical Communication Quarterly, 25*(4), 298–318. https://doi.org/10.1080/10572252.2016.1228790.

Kain, D. J. (2005). Constructing genre: A threefold typology. *Technical Communication Quarterly, 14*(4), 375–409. https://doi.org/10.1207/s15427625tcq1404_2.

Miller, C. R. (1984). Genre as social action. *Quarterly Journal of Speech, 70*(2), 151–167. https://doi.org/10.1080/00335638409383686.

Orlikowski, W. J. & Yates, J. (1994). Genre repertoire: The structuring of communicative practices in organizations. *Administrative Science Quarterly, 39*(4), 541–574. https://doi.org/10.2307/2393771.

Oxford University Press (n.d.). Genre (1.b.), n. In *Oxford English Dictionary.* Retrieved May 2, 2023, from https://www.oed.com.

Rude, C. D. (1995). The report for decision-making: Genre and inquiry. *Journal of Business and Technical Communication, 9*(2), 170–205. https://doi.org/10.1177/1050651995009002002.

Russell, D. R. (1997). Rethinking genre in school and society: An activity theory analysis. *Written Communication, 14*(4), 504–554. https://doi.org/10.1177/0741088397014004004.

Schryer, C. F. (1993). Records as genre. *Written Communication, 10*(2), 200–234. https://doi.org/10.1177/0741088393010002003.

Schryer, C. F. (2000). Walking a fine line: Writing negative letters in an insurance company. *Journal of Business and Technical Communication, 14*(4), 445–497. https://doi.org/10.1177/105065190001400402.

Spinuzzi, C. (2003). *Tracing genres through organizations: A sociocultural approach to information design.* MIT Press. https://doi.org/10.7551/mitpress/6875.001.0001.

Spinuzzi, C. (2008). *Network: Theorizing knowledge work in telecommunications.* Cambridge University Press. https://doi.org/10.1017/CBO9780511509605.

Spinuzzi, C. & Zachry, M. (2000). Genre ecologies: An open system approach to understanding and constructing documentation. *Journal of Computer Documentation*, *24*(3), 169–181. https://doi.org/10.1145/344599.344646.

Yates, J. & Orlikowski, W. J. (2002). Genre systems: Structuring interaction through communicative norms. *Journal of Business Communication*, *39*, 13–35. https://doi.org/10.1177/002194360203900102.

15. History

Edward A. Malone
Missouri S&T

In technical communication, the term *history* may refer to a timestream of events or the past events themselves, as in the statement "technical communication cannot be understood as standing outside of history" (Longo & Fountain, 2013, p. 176). More often, though, the term refers to attempts to understand the past through such activities as ***researching***, interpreting, and narrating: for example, "if the field of technical communication is instrumental communication, communication that gets things accomplished, so must its history" (Brockmann, 1998, p. 386). The following questions drive many of the conversations that scholars have about the study of technical communication history: Why should we study history? What history should we study? How should we study history?

On the question of why we should study history, Gerald Savage (1999/2003) suggested that historical studies of technical communication help to legitimize the ***profession*** of technical communication by contributing to the development of collective historical consciousness. For those who view technical communication as a humanistic endeavor, the study of history is one way of humanizing the practice of technical communication (Rutter, 1991/2004). Many scholars have offered project-specific justifications for studying history. Edward A. Malone (2007) classified some of these justifications into four categories: invention, precedent, distance, and context. We may study the past to discover (invent) ideas and find inspiration; we may look for past analogues (precedents) that help us persuade others to make decisions or take action; we may use a historical perspective (distance) to help an ***audience*** view a situation with greater objectivity; or we may gain a better understanding of our work by investigating the past events (context) that gave rise to and continue to influence the work. These four categories are not exhaustive, but they describe some of the major uses of history in our discipline. (For additional uses of history, see Brockmann, 1998, pp. 385–395; Connor, 1991; and Malone & Wright, 2012.)

Studying history can also improve our production and consumption of scholarship. All topics in technical communication have a history, and sometimes that history extends back several decades in technical communication journals, yet too many new articles in our field have literature reviews that cover only post-2000 works or (conversely) a few dated works from the 1990s. A literature review can be a form of historiography that interprets the evolution of scholarly interest. The history of scholarship on a topic may suggest novel avenues of research even as it undercuts claims of novelty. When we consume scholarship, a well-developed historical consciousness can help us evaluate cited sources critically, readily noticing when older sources are being used inappropriately.

DOI: https://doi.org/10.37514/TPC-B.2023.1923.2.15

The same kind of ***knowledge*** can help us evaluate claims about the discipline and decide whether to repeat those claims. For example, technical communication is often claimed to be a relatively new field, discipline, or profession. As Yvonne Cleary (2016) wrote, "technical communication is a new occupational field, relative to more traditional occupations such as medicine and law" (p. 126). Such statements stretch the meaning of "new" and are potentially misleading to students and others. Technical communication is, of course, an ancient practice. Technical documents have been produced throughout history around the world (e.g., Ding, 2020; Raign, 2019, 2022). And while technical communication may be "a new concept in China" (Yu, 2011, p. 72), "a new occupational field in Ireland" (Cleary, 2016, p. 127), and "a new profession" in Finland (Suojanen, 2010, p. 54), it is a well-established academic discipline and profession in the United States. The first university courses in technical writing were created at the beginning of the 20th century; the first academic degree programs in technical writing and ***editing*** were created in the 1950s (Connors, 1982/2004; Kynell, 2000). Full-time technical writers and editors in the modern sense had existed before World War II; these occupations grew quickly during and after the war, and since 1943 the job title "technical writer" has been included in the U.S. Department of Labor's *Dictionary of Occupational Titles* (Malone, 2011). A profession of technical communication began to emerge in the 1950s when technical writers, editors, illustrators, managers, and librarians formed professional associations, created codes of ***ethics***, published journals, and held conferences (Malone, 2011). Thus, we cannot say that technical communication is a "new" field, discipline, or profession in the United States.

Similarly, some claims should not be repeated without heavy qualification: for example, that "the history of technical writing still has not been written" (Moran & Tebeaux, 2012, p. 58) or that "we have no history to show our sustained existence in the world—just a collage of articles and a few monographs" (Tebeaux, 2014, p. 253). The situation is not as dire as these statements suggest. Scholars have been researching and writing about the history of technical communication since at least the 1950s. They have contributed many historical studies to the profession's body of knowledge, as documented in bibliographic essays by R. John Brockmann (1983), William Rivers (1994/1999), Edward A. Malone (2007), and Michael Moran and Elizabeth Tebeaux (2011, 2012). Their output has included more than a few book-length studies, such as the monographs by Brockmann (1998, 2002, 2004), Bernadette Longo (2000), Mark Ward (2014), Dirk Remley (2014), and Carol Siri Johnson (2016) and the edited collections by Teresa Kynell and Michael Moran (1999) and Miles Kimball and Charles Kostelnick (2017). Other disciplines, too, have shown an interest in technical writing history (e.g., Formisano & Van Der Eijk, 2009). Our discipline does not have a textbook or reference work that provides an overview of technical communication from ancient times to the present, but however useful such a work might be, it would still be just another thread in a tapestry of diverse perspectives on our history. (On historiography as tapestry weaving, see Brockmann, 1998, p. 3.)

What history should we study? Some scholars believe that the proper subject of historical study in our discipline is communication, usually in the form of documents. In their own research, they analyze historical texts and ***visuals***. They use the phrase "history of technical communication" to mean mainly writing in the past and seldom stray far from the artifacts they are studying. (For an example of this focus, see Tebeaux, 2014, pp. 253–258.) Other scholars in the field explore a broader range of history-related topics, such as the lives and careers of technical communicators (e.g., Hayhoe, 2017); technologies related to writing, ***designing***, and publishing (e.g., Durack, 2003a); the teaching of technical communication (e.g., Sullivan, 2012); the project of professionalization (e.g., Hallier & Malone, 2012); oral technical communication (e.g., Brockmann, 1998, pp. 99–116; Pochatko, 2017); communicative rituals in mathematics (Fiss, 2020); transmedia storytelling (Malone, 2019); and the subfield of technical editing (e.g., Cunningham et al., 2019, pp. 1–19; Malone, 2006; Warren, 2010). They may analyze technical documents as well, but they do not limit their focus to these artifacts.

A number of scholars have attempted to classify historical studies by historical period or theme (Kynell & Moran, 1999; Malone, 2007; Rivers, 1994/1999), but such classification systems inevitably break down because many historical studies cover material from more than one century or country, focus on more than one theme or topic, or include history as part of a larger discussion of a topic (e.g., Brasseur, 2003).

Over the decades, technical communication scholars have advocated for greater inclusiveness in historical research. Noting that historical studies before 1983 usually focused on "celebrated authors and scientists" as technical writers, Brockmann (1983) called for more studies of the "common man" as a technical writer (pp. 155–156). To investigate the work of uncelebrated and often anonymous technical communicators, a researcher must inspect unpublished (and often handwritten) documents, such as letters and memoranda; drafts of ***proposals***, reports, and drawings; job descriptions and personnel files; and other records in corporate archives and libraries' special collections. About 15 years after Brockmann's important contribution, Katherine Durack (1997) called for more historical studies of female technical communicators and their work, a project that required a reconsideration of what counts as technical communication. Thanks in part to her efforts, documents such as cookbooks, sewing patterns, and childcare manuals are more likely now than in the past to be recognized and appreciated as technical communication. Since 1997, there has been a steady stream of historical studies about female technical communicators and their work (e.g., Durack, 1998, 2003a, 2003b, 2003c; Hallenbeck, 2012; Lippincott, 2003a, 2003b, 2003c; Malone, 2010, 2013, 2015a, 2015b; Petersen, 2016; Raign, 2019; Rauch, 2012).

Because most of these studies are about American or British subjects, however, Emily Petersen (2017) has challenged historians of technical communication to heed ***international/intercultural communication***, giving special attention to "women of color and women of the Global South" (pp. 1, 25). India is one promising site for this research agenda, and oral history interviews may be the best research method (Petersen, 2017, p. 17). (For examples of oral history interviews with women in technical communication, see Lewenstein, 1987; Malone, 2014; Swent, 1989.) During her own interviews with female technical writers in India, Petersen (2017) gleaned information about the men and women who founded that country's decades-old technical writing industry. The first account of India's technical writing industry was published in a technical communication journal nearly 60 years ago (Sampath & Murthy, 1966).

How should we study history? Researchers studying the history of technical communication must use primary sources, such as the accident report of a historic train wreck, a map of the train's route and a timetable of its stops, interviews with passengers and bystanders, and even the train itself, but researchers must also conduct a thorough literature review and use relevant secondary sources, such as a documentary film or journal articles about the historic train accident or studies of other train wrecks. Beyond these basic working principles, several technical communication scholars have proposed multistep approaches to conducting historical research for either academic publication (Battalio, 2002; Connor, 1993; Kynell & Seely, 2002; Tebeaux & Killingsworth, 1992) or immediate workplace application (Longo & Fountain, 2013; Shirk, 2000/2004). These approaches emphasize the importance of understanding context, such as relevant details about the time period in which a document was created, the organization that created it, and its intended audience.

Sometimes, a researcher in technical communication may borrow a historiographic approach from another discipline, such as textual studies or literary studies. For example, W. Tracy Dillon (1997) explained how the methods of new historicism—a form of criticism once popular in literary studies—might be used to study historical technical documents. This approach is political and cultural as well as self-reflective. If the nature of historicity is such that every historical study is infused with the subjective ideologies of those who produced it (Jones & Walton, 2018, p. 253), then acknowledging our own ideologies as historians may be more honest and helpful than claiming—or giving the impression of—too much objectivity.

Another promising approach to historiography is the use of antenarratives. The history that has already been written is a history that privileges some people and activities over others, often unfairly. A dominant narrative in this history tends to drown out other narratives as it creates and maintains its own homogeneity. One way to rescue the nondominant (usually unnoticed or forgotten) stories in our history is by telling "a disruptive 'before' story that seeks to destabilize and unravel aspects of the tightly woven dominant narrative about who we are as a field, what we do, where our work occurs, and what we value" (Jones et al., 2016, p. 212). By interrogating previous historical studies, and (re)examining historical evidence, researchers can sometimes lend new, stronger voices to nondominant stories. This approach is ultimately future oriented: "Antenarratives open up a space that invites reinterpretation of the past so as to suggest—and enable—different possibilities for the future" (Jones et al., 2016, p. 212).

For example, Miriam Williams (2010) and others have started important conversations about issues of race and ethnicity in historical technical communication, but thus far no one has written about (or even mentioned) the role of African Americans in the profession-building activities of technical communicators in the 1950s and 1960s, yet there is evidence of significant contributions and presence—for example, the safety posters by technical illustrator John H. Terrell in the 1950s ("His Cartoons," 1956); a 1956 article by Herbert Augustus in *Technical Writing Review*; an *Ebony* magazine cover story about La Bonnie Bianchi, the first African American woman to graduate with a master's degree in technical writing in 1960 ("Woman Engineer," 1961); three technical writing textbooks by radio engineer Rufus P. Turner in the mid-1960s; and the accomplishments of David J. Chesnut (see Figure 15.1), the first African American fellow of the Society for Technical Communication. Investigating and recovering this part of our history might help to change perceptions about the field.

Although there is already a large body of literature about technical communication history, researchers still have plenty of work to do because writing the history of technical communication is an ongoing project that will never be finished and will always need reinterpretation and revision.

Figure 15.1. David J. Chesnut, first African American fellow of the Society for Technical Communication (STC). Photograph from the archives at STC headquarters in Fairfax, VA. Reprinted with permission.

References

Augustus, H. (1956). Simplicity in technical writing. *Technical Writing Review, 3*(4), 59–60.

Battalio, J. T. (2002). A methodology for streamlining historical research: The analysis of technical and scientific publications. *IEEE Transactions on Professional Communication, 45*(1), 21–39. https://doi.org/10.1109/47.988360.

Brasseur, L. E. (2003). *Visualizing technical communication: A cultural critique*. Routledge. https://doi.org/10.2190/VTI.

Brockmann, R. J. (1983). Bibliography of articles on the history of technical writing. *Journal of Technical Writing and Communication, 13*(2), 155–165. https://doi.org/10.2190/1LAR-4PLU-EKQ3-302W.

Brockmann, R. J. (1998). *From millwrights to shipwrights to the twenty-first century: Explorations in a history of technical communication in the United States*. Hampton Press.

Brockmann, R. J. (2002). *Exploding steamboats, senate debates, and technical reports: The convergence of technology, politics, and rhetoric in the Steamboat Bill of 1838*. Baywood. https://doi.org/10.2190/ESS.

Brockmann, R. J. (2004). *Twisted rails, sunken ships: The rhetoric of nineteenth century steamboat and railroad accident investigation reports, 1833–1879*. Baywood. https://doi.org/10.2190/TRS.

Cleary, Y. (2016). Community of practice and professionalization perspectives on technical communication in Ireland. *IEEE Transactions on Professional Communication, 59*(2), 126–139. https://doi.org/10.1109/TPC.2016.2561138.

Connor, J. J. (1991). History and the study of technical communication in Canada and the United States. *IEEE Transactions on Professional Communication, 34*(1), 3–6. https://doi.org/10.1109/47.68420.

Connor, J. J. (1993). Medical text and historical context: Research issues and methods in history and technical communication. *Journal of Technical Writing and Communication, 23*(3), 211–232. https://doi.org/10.2190/0P4Q-07X0-R2EV-WRD2.

Connors, R. J. (2004). The rise of technical writing instruction in America. In J. Johnson-Eilola & S. Selber (Eds.), *Central works in technical communication* (pp. 3–19). Oxford University Press. (Original work published 1982)

Cunningham, D. H., Malone, E. A. & Rothschild, J. M. (2020). *Technical editing: An introduction to editing in the workplace*. Oxford University Press.

Dillon, W. T. (1997). The new historicism and studies in the history of business and technical writing. *Journal of Business and Technical Communication, 11*(1), 60–73. https://doi.org/10.1177/1050651997011001004.

Ding, D. (2020). *The historical roots of technical communication in the Chinese tradition*. Cambridge Scholars Publishing.

Durack, K. T. (1997). Gender, technology, and the history of technical communication. *Technical Communication Quarterly, 6*(3), 249–260. https://doi.org/10.1207/s15427625tcq0603_2.

Durack, K. T. (1998). Authority and audience-centered writing strategies: Sexism in 19th century sewing machine manuals. *Technical Communication, 45*(2), 180–196.

Durack, K. T. (2003a). From the moon to the microchip: Fifty years of *Technical Communication*. *Technical Communication, 50*(4), 571–584.

Durack, K. T. (2003b). Instructions as "inventions": When the patent meets the prose. In T. C. Kynell-Hunt & G. J. Savage (Eds.), *Power and legitimacy in technical commu-*

nication. Volume I: The historical and contemporary struggle for professional status (pp. 15–37). Baywood. https://doi.org/10.2190/PL1C1.

Durack, K. T. (2003c). Observations on entrepreneurship, instructional texts, and personal interaction. *Journal of Technical Writing and Communication, 33*(2), 87–108. https://doi.org/10.2190/Y5VH-HAD2-PYT1-TR1N.

Fiss, A. (2020). *Performing math: A history of communication and anxiety in the American mathematics classroom*. Rutgers University Press. https://doi.org/10.36019/9781978820241.

Formisano, M. & Van Der Eijk, P. (Eds.). (2009). *Knowledge, text and practice in ancient technical writing*. Cambridge University Press.

Hallenbeck, S. (2012). User agency, technical communication, and the 19th-century woman bicyclist. *Technical Communication Quarterly, 21*(4), 290–306. https://doi.org/10.1080/10572252.2012.686846.

Hallier, P. A. & Malone, E. A. (2012). Light's "technical writing and professional status": Fifty years later. *Technical Communication, 59*(1), 29–31 and E13–E21.

Hayhoe, G. (Ed.). (2017, June). Legends of technical communication [Special issue]. *Intercom, 64*(6), 5–29.

His cartoons crusade for safety. (1956, March 4). *Philadelphia Inquirer Magazine*, p. 8

Johnson, C. S. (2016). *The language of work: Technical communication at Lukens Steel, 1810 to 1925*. Routledge.

Jones, N. N., Moore, K. R. & Walton, R. (2016). Disrupting the past to disrupt the future: An antenarrative of technical communication. *Technical Communication Quarterly, 24*(4), 211–229. https://doi.org/10.1080/10572252.2016.1224655.

Jones, N. N. & Walton, R. (2018). Using narratives to foster critical thinking about diversity and social justice. In A. M. Haas & M. F. Eble (Eds.), *Key theoretical frameworks: Teaching technical communication in the twenty-first century* (pp. 241–267). Utah State University Press. https://doi.org/10.7330/9781607327585.c010.

Kimball, M. A. & Kostelnick, C. (Eds.). (2017). *Visible numbers: Essays on the history of statistical graphics*. Routledge.

Kynell, T. C. (2000). *Writing in a milieu of utility: The move to technical communication in American engineering programs, 1850–1950* (2nd ed.). Ablex.

Kynell, T. C. & Moran, M. G. (Eds.). (1999). *Three keys to the past: The history of technical communication*. Ablex.

Kynell, T. C. & Seely, B. (2002). Historical methods for technical communication. In L. J. Gurak & M. M. Lay (Eds.), *Research in technical communication* (pp 67–92). Praeger.

Lewenstein, B. V. (1987, February 6). *Jane Stafford: Transcript of an interview*. National Association of Science Writers and Council for the Advancement of Science Writing Records, 1934–2014 (Collection 4448, Box 33). Cornell University, Carl A. Kroch Library, Ithaca, NY.

Lippincott, G. (2003a). Moving technical communication into the post-industrial age: Advice from 1910. *Technical Communication Quarterly, 12*(3), 321–342. https://doi.org/10.1207/s15427625tcq1203_6.

Lippincott, G. (2003b). Rhetorical chemistry: Negotiating gendered audiences in nineteenth-century nutrition studies. *Journal of Business and Technical Communication, 17*(1), 10–49. https://doi.org/10.1177/1050651902238544.

Lippincott, G. (2003c). "Something in motion and something to eat attract the crowd": Cooking with science at the 1893 World's Fair. *Journal of Technical Writing and Communication, 33*(2), 141–164. https://doi.org/10.2190/QXUU-WBAF-EWCX-VFMD.

Longo, B. (2000). *Spurious coin: A History of science, management, and technical Writing*. SUNY Press.

Longo, B. & Fountain, T. K. (2013). What can history teach us about technical communication? In J. Johnson-Eilola & S. A. Selber (Eds.), *Solving problems in technical communication* (pp. 165–186). The University of Chicago Press.

Malone, E. A. (2006). Learned correctors as technical editors: Specialization and collaboration in early modern European printing houses. *Journal of Business and Technical Communication, 20*(4), 389–424. https://doi.org/10.1177/1050651906290232.

Malone, E. A. (2007). Historical studies of technical communication in the United States and England: A fifteen-year retrospection and guide to resources. *IEEE Transactions on Professional Communication, 50*(4), 333–351. https://doi.org/10.1109/TPC.2007.908732.

Malone, E. A. (2010). "Chrysler's most beautiful engineer": Lucille J. Pieti in the pillory of fame. *Technical Communication Quarterly, 19*(2), 144–183. https://doi.org/10.1080/10572250903559258.

Malone, E. A. (2011). The first wave (1953–1961) of the professionalization movement in technical communication. *Technical Communication, 58*(4), 301–322.

Malone, E. A. (2013). Elsie Ray and the founding of STC. *Journal of Technical Writing and Communication, 43*(2), 121–143. https://doi.org/10.2190/TW.43.2.b.

Malone, E. A. (2014). *Teachers and practitioners of technical communication, 1940–1975 (oral history project)*. Missouri University of Science and Technology. http://web.mst.edu/~malonee/list.html.

Malone, E. A. (2015a). Eleanor McElwee and the Formation of IEEE PCS. *Journal of Technical Writing and Communication, 45*(2), 104–133. https://doi.org/10.1177/0047281615569480.

Malone, E. A. (2015b). Women organizers of the first professional associations in technical communication. *Technical Communication Quarterly, 24*(2), 121–146. https://doi.org/10.1080/10572252.2015.1001291.

Malone, E. A. (2019). "Don't be a Dilbert": Transmedia storytelling as technical communication during and after World War II. *Technical Communication, 66*(3), 209–229.

Malone, E. A. & Wright, D. (2012). The role of historical study in technical communication. *Programmatic Perspectives, 4*(1), 44–87.

Moran, M. G. & Tebeaux, E. (2011). A bibliography of works published in the history of professional communication from 1994–2009: Part 1. *Journal of Technical Writing and Communication, 41*(2), 193–214. https://doi.org/10.2190/TW.42.1.e.

Moran, M. G. & Tebeaux, E. (2012). A bibliography of works published in the history of professional communication from 1994–2009: Part 2. *Journal of Technical Writing and Communication, 42*(1), 57–86. https://doi.org/10.2190/TW.42.1.e.

Petersen, E. J. (2016). Beyond biography: Using technical and professional documentation to historically contextualize women's agency. *Rhetoric, Professional Communication, and Globalization, 9*(1), 55–77.

Petersen, E. J. (2017). Feminist historiography as methodology: The absence of international perspectives. *Connexions, 5*(2), 1–38.

Pochatko, A. M. (2017). The singer of technology: The oral-based origins of technical communication in the ancient world. *Journal of Technical Writing and Communication, 47*(4), 464–477. https://doi.org/10.1177/0047281616646751.

Raign, K. R. (2019). Finding our missing pieces—Women technical writers in ancient Mesopotamia. *Journal of Technical Writing and Communication, 49*(3), 338–364. https://doi.org/10.1177/0047281618793406.

Raign, K. R. (2022). The art of ancient Mesopotamian technical manuals and letters: The origin of instructional writing. *Technical Communication Quarterly, 31*(1), 44–61. https://doi.org/10.1080/10572252.2021.1915386.

Rauch, M. (2012). The accreditation of Hildegard von Bingen as Medieval female technical writer. *Journal of Technical Writing and Communication, 42*(4), 393–411. https://doi.org/10.2190/TW.42.4.d.

Remley, D. (2014). *Exploding technical communication: Workplace literacy hierarchies and their implications for literacy sponsorship*. Baywood. https://doi.org/10.2190/ETC.

Rivers, W. E. (1999). Studies in the history of business and technical writing: A bibliographical essay. In T. C. Kynell & M. G. Moran (Eds.), *Three keys to the past: The history of technical communication* (pp. 249–307). Ablex. (Original work published 1994)

Rutter, R. (2004). History, rhetoric, and humanism. In J. Johnson-Eilola & S. Selber (Eds.), *Central works in technical communication* (pp. 20–34). Oxford University Press. (Original work published 1991)

Sampath, P. & Murthy, V. S. (1966). Technical writing in India. *IEEE Transactions on Engineering Writing and Speech, 9*(2), 30–32. https://doi.org/10.1109/TEWS.1966.4322622.

Savage, G. (2003). The process and prospects for professionalizing technical communication. In T. Kynell-Hunt & G. J. Savage (Eds.), *Power and legitimacy in technical communication: Volume I: The historical and contemporary struggle for professional status* (pp. 137–165). Baywood. (Original work published 1999) https://doi.org/10.2190/PL1C7.

Shirk, H. N. (2004). Researching the history of technical communication: Accessing and analyzing corporate archives. In J. M. Dubinsky (Ed.), *Teaching technical communication: Critical issues for the classroom* (pp. 128–138). Bedford/St. Martin's. (Original work published 2000)

Sullivan, P. (2012). After the great war: Utility, humanities, and tracings from a technical writing class in the 1920s. *Journal of Business and Technical Communication, 26*(2), 202–228. https://doi.org/10.1177/1050651911430626.

Suojanen, T. (2010). Comparing translation and technical communication: A holistic approach. In T. Kinnunen & K. Koskinen (Eds.), *Translators' Agency* (pp. 47–60). Tampere University Press.

Swent, E. (1989). *Ian and Catherine Campbell, geologists: Teaching, government service, editing*. University of California.

Tebeaux, E. (2014). *The flowering of a tradition: Technical writing in England, 1641–1700*. Routledge. https://doi.org/10.2190/TFO.

Tebeaux, E. & Killingsworth, M. J. (1992). Expanding and redirecting historical research in technical writing: In search of our past. *Technical Communication Quarterly, 1*(2), 5–32. https://doi.org/10.1080/10572259209359496.

Turner, R. (1964a). *Grammar review for technical writers*. Holt, Rinehart and Winston.

Turner, R. (1964b). *Technical writer's and editor's stylebook*. H. W. Sams.

Turner, R. (1965). *Technical report writing*. Holt, Rinehart and Winston.
Ward, M., Sr. (2014). *Deadly documents: Technical communication, organizational discourse, and the Holocaust—Lessons from the rhetorical work of everyday texts*. Baywood. https://doi.org/10.2190/DOC.
Warren, T. L. (2010). History and trends in technical editing. In A. Murphy (Ed.), *New perspectives on technical editing* (pp. 29–50). Baywood. https://doi.org/10.2190/NPOC3.
Williams, M. (2010). *From Black Codes to recodification: Removing the veil from regulatory writing*. Baywood. https://doi.org/10.2190/FBC.
Woman engineer. (December, 1961). *Ebony*, pp. 87–92.
Yu, H. (2011). Integrating technical communication into China's English major curriculum. *Journal of Business and Technical Communication*, *25*(1), 6–94. https://doi.org/10.1177/1050651910380376.

16. Information

William Hart-Davidson
Michigan State University

Information is one of those terms that is widely used in both academic and popular discourse in ways that do not always relate to a more precise, technical definition. It can be helpful, in fact, to consider the various words that *information* is often paired with as a modifier in order to know how best to make sense of it. There are four especially helpful pairings for technical communicators to know: *information theory*, *information technology*, *information design*, and *information architecture*. These four terms mark points on a timeline of information's evolution in meaning as well as conceptual shifts in the work of technical communicators as it relates to information. Interestingly, in none of these pairings is the word *information* the neutral signifier that it can sometimes seem in popular usage, as when people ask for "just information." Rather, in each of the four cases, the term marks a site of consequential contestation over the nature of technical communication and the role technical communicators play in the social settings where their work unfolds.

This entry tracks the shifts in thinking about technical communication across the four pairings in four historical moments: information theory and technical communication as transmission, information technology and technical communication as ***translation***, information design and technical communication as transformation, and information architecture and technical communication as trans-disciplinary ***knowledge*** making. In each section, *information* serves as a compass point for a trajectory of further inquiry that, necessarily, exceeds the scope of this short essay.

Information theory is a mathematical formulation credited to MIT and Bell Labs scientist Claude Shannon. Published as a two-part article titled "A Mathematical Theory of Communication," Shannon's (1948) work contributed two key ideas that are foundational to both computing and telecommunications. The first is a means to reliably quantify how many binary digits are required to encode some amount of data, such as a text or voice message. The second idea, which applies to transmission of messages through a channel, is the means to reliably calculate the signal to noise ratio for the channel and to understand how the ratio varies given the channel bandwidth. Shannon's formulations of information entropy—the way the quality of a signal degrades under certain conditions—are the basis for compression and error-checking routines widely used today that allow for fast, clear, global communication (Collins, 2002). But Shannon's ideas have had more than instrumental influence. They also arguably underlay our current economic and political orientations to the term *information*, wherein we take it to be common sense that information is the valuable part of a signal (and

DOI: https://doi.org/10.37514/TPC-B.2023.1923.2.16

all other stuff is "noise"), and where the consistent reliable flow of information is understood to be vital but "information overload" is also a known threat. So how did a highly technical mathematical theory hatched in a telecommunications laboratory gain broad cultural cache?

In 1949, a colleague of Shannon's at Bell Labs, Warren Weaver, collaborated with Shannon to publish a book-length version of the original article under a slightly modified title: *The Mathematical Theory of Communication.* The move from "A" to "The" in the title signified an implicit argument about the generalizability of the ideas in the book. A model was born that would be taken up in many ***research*** and industry areas and applied to business and social affairs. The Shannon-Weaver model of communication also had a significant impact on technical communication, though not an uncontroversial one. To see why, a look at the model (Figure 16.1) is helpful.

Where is the work of technical communication in the Shannon-Weaver model? What is implied about the nature of that work? If we take this model from its original technical context and apply it more broadly to systems populated by humans, the technical communicator is most plausibly a "transmitter," a functional role that does not contribute any information value to the signal apart from error correction and compression, always with the risk of introducing rather than reducing information entropy. Not surprisingly, technical communication as a field has resisted this reduction to the value added by technical communication and has produced robust critiques of this "transmission model" of communication as well as alternative formulations that turn, in part, on alternative conceptions of "information" (c.f. Miller, 1979; Slack et al., 1993). Perhaps the most popular of these alternative formulations is the technical communicator as translator or, as once metaphorically represented in a since-retired Society for Technical Communication logo, a bridge.

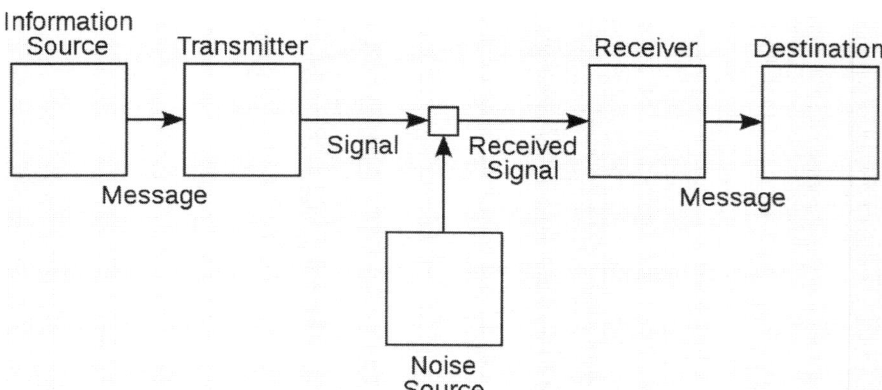

Figure 16.1. Shannon-Weaver Model of Communication.
Public domain image. Wikimedia Commons.

This metaphor is captured in the pairing *information technology*, a phrase whose meaning took shape as computer processors shrank in size and found their way into industrial and consumer products. Two areas of need arose that buoyed demand for technical communication: 1) experts in different knowledge domains such as health care and computing or agriculture and robotics needed to be able to understand one another, and 2) people who offer products and services needed to communicate technical information to a growing, global audience of consumers. Throughout the latter part of the 20th century, these two demands fueled the conception of information as a supporting product and, in some cases, as a companion service that had to be provided in order for increasingly technical products and services to be traded and used successfully.

Information in this model is not, by itself, inherently valuable. But without a manual, one might never learn to use a piece of expensive software. Or, without a documented programming interface, a software developer might not be able to connect one system with another. This view of information as knowledge to be translated gives rise to the role of technical communicator as a "bridge," either between disparate expert areas of knowledge or between an expert and a layperson. This model goes well with the way information is understood in the phrase *information technology*, wherein a technical device or object functions to do something useful without the user needing to "do the math" implied in information theory to derive the benefits. That is, the information in information technology—the representation of messages as quantities, calculations performed on those quantities, and the rapid communication of bits back and forth via microcircuitry—is "blackboxed" to the user. So your rice cooker that uses "fuzzy logic" may well use sophisticated computing algorithms, but as the cook you only need to add rice and water and push a button. You may also need a guide, and the manufacturer who developed the machine likely needed ***documentation*** from the company who manufactured the circuit board in the appliance.

The value of information products as ancillary also came into scrutiny by members of the field for the way it still positioned technical communicators not as creators of knowledge but as processors of it. This model left the hierarchies of expertise in place, even if it placed technical communicators in an important middle position between the originators of knowledge and those who needed to learn more. What changed, according to Johndan Johnson-Eilola (1996), was a shift accelerated by how we could interact on global networks such as the World Wide Web, a shift that predicated the re-ordering of the value of work across all of our categories of professional activity.

With the advent of the Web came a melding of what had previously been a quite clear split between the "product" and "information about the product"; with it, the value proposition that had supported the bridge model became far less clear (Hart-Davidson, 2001). Many of the most successful companies in the world began succeeding by selling information. And with Apple as perhaps the signature example, these companies would go on to develop service models that turned the old hierarchy upside down. Now, the device (such as the iPhone) was a gateway to a monthly service and a "content ecosystem" where the main commodities were information products. Today, the Society for Technical Communication's mission statement no longer refers to technical communicators as bridges or translators. Instead, it reads, "The Society for Technical Communication advances technical communication as the discipline of *transforming* [emphasis added] complex information into usable content for products, processes, and services."

Pairing "information" with the verb "design" offers one conceptual path to understanding technical communicators' work as transformative. As the kinds of products that technical communicators produced or helped to create—documents, websites, tutorials, infographics, videos, apps—evolved, so did a new understanding of information as raw and, potentially, re-usable material to create useful, usable products. The information of information design is malleable and valuable. It arrives from a variety of sources in a variety of formats and feeds any number of content streams where it might become part of a document, a tweet, an infographic, or a video. The value of the information can be measured in its potential, but is more often understood when an information asset is set in motion and users begin to engage with it. How much and what kinds of engagement an information asset accrues will determine how it might be repurposed and/or transformed further.

Karen Schriver's 1996 book *Dynamics in Document Design: Designing Texts for Readers* offered a thorough treatment of how technical communicators might realign their work such that it would be judged not by how documents looked but rather by what users of those documents did with them. While the focus of that book was on documents, the book is still in print today because it lays the groundwork for seeing the real value in information design not as visible in a product adhering to some technical standard or aesthetic benchmark. Rather, information design succeeds when the behavioral results of readers and users can be measured as outcomes.

The concept of information paired with design invites action from technical communicators across the full scope of the traditional ***rhetorical*** canons—invention, arrangement, style, memory, delivery. In this way, it differs dramatically from the transmission model, where technical communicators' only role was to smooth delivery largely using the tactics of ***plain language***. And, importantly, this work is never done, because there are always opportunities to make engagements richer, more satisfying, more effective, and, importantly, as Miriam Williams and Octavio Pimentel (2012) argue, more inclusive and inviting to other groups. And as Laura Gonzales (2018) has argued, this focus on transformation to facilitate inclusion also calls us to remediate our understanding of terms like *translation* that have been at the center of our work.

The work of technical communication today is often aligned with another professional area with "information" as a modifier: architecture. Information architecture names both a set of professional practices as well as an academic area of study with conferences and research journals and a professional identity with a well-defined career pathway. Rather than replace technical communication, information architecture, or IA, can be considered a complementary path to practicing technical communication skills and applying technical communication knowledge. In this pairing, information does not just exist *a priori*, nor do technical communicators or information architects wait for others to create it. Rather, information is seen as a potentiality to be maximized, realized, and capitalized.

Today, nearly everything we do—down to the most minute, involuntary gestures, such as eye-blinks or heartbeats—has the potential to become information stored in a system, fed to an algorithm, aggregated, analyzed, and visualized for our own or somebody else's use (Hart-Davidson & Grabill, 2012). That end-to-end conceptualization of an information lifecycle describes the scope of activity implied in the pairing of information and architecture. Technical communicators might realistically play a role in all of the phases where data becomes information and information becomes knowledge.

References

Collins, G. P. (2002). Claude E. Shannon: Founder of information theory. *Scientific American, 14*. https://www.scientificamerican.com/article/claude-e-shannon-founder/.

Gonzales, L. (2018). *Sites of translation: What multilinguals can teach us about digital writing and rhetoric*. University of Michigan Press. https://doi.org/10.2307/j.ctv65sx95.

Hart, H. & Conklin, J. (2006). Toward a meaningful model of technical communication. *Technical Communication, 53*(4), 395–415.

Hart-Davidson, W. (2001). On writing, technical communication, and information technology: The core competencies of technical communication. *Technical Communication, 48*(2), 145–155.

Hart-Davidson, W. & Grabill, J. (2012). The value of computing, ambient data, ubiquitous connectivity for changing the work of communication designers. *Communication Design Quarterly Review, 1*(1), 16–22. https://doi.org/10.1145/2448917.2448921.

Johnson-Eilola, J. (1996). Relocating the value of work: Technical communication in a post-industrial age. *Technical Communication Quarterly, 5*(3), 245–270. https://doi.org/10.1207/s15427625tcq0503_1.

Miller, C. R. (1979). A humanistic rationale for technical writing. *College English, 40*(6), 610–617. https://doi.org/10.2307/375964.

Schriver, K. A. (1996). *Dynamics in document design: Creating text for readers*. John Wiley & Sons, Inc.

Shannon, C. (1948, July and October). A mathematical theory of communication. *Bell System Technical Journal, 27*, 379–423, 623–656. https://doi.org/10.1002/j.1538-7305.1948.tb01338.x.

Shannon, C. E. & Weaver, W. (1949). *The mathematical theory of communication*. University of Illinois Press.

Slack, J. D., Miller, D. J. & Doak, J. (1993). The technical communicator as author: Meaning, power, authority. *Journal of Business and Technical Communication*, *7*(1), 12–36. https://doi.org/10.1177/1050651993007001002.

Society for Technical Communication. (n.d.). *Mission and Vision*. https://www.stc.org/about-stc/mission-a-vision/.

Williams, M. F. & Pimentel, O. (2012). Introduction: Race, ethnicity, and technical communication. *Journal of Business & Technical Communication 26*(3), 271–276. https://doi.org/10.1177/1050651912439535.

17. International/Intercultural Communication

Huiling Ding
North Carolina State University

Technical and professional communication (TPC) is produced in all languages and by people of all cultures, and TPC discourses are constantly moving across borders and between cultures. Thus, it is essential for all technical communicators to understand international and intercultural communication.

The term *international communication* was criticized for its nation-centric and Anglo-centric assumptions and its use of individual countries as the unit of analysis in studying global communication. Many factors have introduced both changes in and challenges to international communication, including but not limited to globalization, global trade, global cinema, global media, the rise of ***social media*** and the networked society, international education, transnational travel, contact zones, hybrid cultures, and the tendency to use the deficit model when examining communication and ***rhetorical*** practices in non-Western cultures (Castells, 1996; Mao, 2003; Singh & Doherty, 2004). To provide new nuanced analysis of communication across cultures, the term *intercultural communication* has become widely accepted today.

Early studies borrowed extensively from cultural heuristics and cultural dimension theories from intercultural communication and employed individual nation states as the unit of analysis (Marcus, 2005; Spyridakis & Fukuoka, 2002). Increasing attention has been shifting from sole dependence on, and oftentimes over-simplistic application of, cultural heuristics for individual nation states, which Ulrich Beck (2003) called "methodological nationalism," to alternative and non-nation-centric ways to conceive and analyze cultures at different levels (Ding, 2013; Hunsinger, 2006; Scott, 2006; Starke-Meyerring, 2005; Starke-Meyerring & Wilson, 2008; Sun, 2006, 2012; Thatcher, 2010).

The *Oxford English Dictionary* defines the term *intercultural* as "taking place between cultures, or derived from different cultures," with the prefix *inter-* meaning "between" and *cultural* meaning "of or relating to culture" (Oxford University Press, n.d.). Back in 1871, British anthropologist Sir Edward Burnett Tylor defined culture as a "complex whole which includes ***knowledge***, belief, art, morals, law, custom, and any other capabilities and habits acquired by man as a member of society" (p.1). For Clifford Geertz (1973), culture is "a historically transmitted pattern of meaning embodied in symbols" (p. 89). Geert Hofstede (1991) defined culture as "the collective programming of the mind that distinguishes the members of one group or category of people from others" (p. 5). Fred Dervin (2011) distinguished between "liquid" and "solid" interculturality by defining solid

DOI: https://doi.org/10.37514/TPC-B.2023.1923.2.17

interculturality as believing in "resolutely distinct human essences," which is featured by uncritical and systematic use of "the primordial and basic concept of culture" (pp. 38–39). Liquid interculturality, in contrast, emphasizes the "inter" as in interaction and interconnectedness rather than the "cultural".

Exploring "the cultural dimensions of globalization," Arjun Appadurai (1996) criticized the noun form of culture, which sees culture as a substance and uses nation states as the unit of analysis. He advocated shifting to the adjectival form of the word: *cultural*, which explores "the conscious mobilization of [situated] cultural differences in the service of a larger national or transnational politics" (p. 13).

Klaus B. Jensen (2011) defines three types of communication, namely, the embodied face-to-face communication, the technically reproduced mass communication, and networked communication enabled by digital technologies. Originating from intergroup communication, intercultural communication theories initially focused on embodied face-to-face communication before expanding their reach to networked communication (Chen, 2017). Four factors led to the development of the so-called global village and increasing intercultural communication: improvements in transportation ***technology*** and communication technologies, the economic globalization, and accelerated immigration (Samovar & Porter, 1997). Working together, these developments made possible technology-mediated intercultural communication, which increasingly takes place virtually among individuals.

To examine cultural variability in communication, different theories have been proposed to perform analysis at the societal level and at the individual level. Edward T. Hall (1976) proposed low-high context communication theory to examine direct and indirect communication practices. Hofstede (1980, 1991, 2001) identified six dimensions of cultural variability: individualism-collectivism, uncertainty avoidance, power distance, masculinity-femininity, long-term vs. short-term orientation, and indulgence vs. restraint. William B. Gudykunst and his co-authors (2005) categorized intercultural communication theories into five themes: effective outcome, accommodation and adaptation, identity negotiation, communication network, and acculturation and adjustment.

Various approaches have been taken to examine intercultural communication practices. Judith Martin and Thomas Nakayama (1997) identified three approaches to studying intercultural communication, namely, social science, interpretive, and critical approaches. The social science approach employs methods such as survey and observation to identify cultural variables and to describe and predict behavior. The interpretive approach, used mostly in sociolinguistics and anthropology, employs participant observation, field study, and ethnography to examine communication in different cultural contexts. The critical approach, in contrast, focuses on "macrocontext," namely, political and social structures, historical contexts, and power relations, in conducting textual analysis of cultural products (Martin & Nakayama, 1997, p. 35). Similarly, Gudykunst et al. (2005) emphasized the need for "indigenous theories developed by scholars outside the United States" and the inclusion of power in intercultural communication theories (p. 26).

Compared with the over-two-decade development of intercultural communication, the field of intercultural professional/technical communication began to develop only fairly recently because of the quick integration of the global economy and the globalization of the workplace. Many existing publications on intercultural technical communication still rely on intercultural communication theories such as cultural variables and face concepts in their analyses and focus on either interpersonal or organization communication processes (Constantinides et al., 2001; Gould, 2005; Marcus, 2005). This over-reliance on intercultural communication theories is particularly strong in pedagogical discussion of intercultural technical communication.

Early publications about pedagogical approaches took the ***information*** acquisition approach and emphasized the heuristic view of culture that sees culture as nation-centric traits (Andrews, 1996; Beamer, 1992; Chapel, 1997; DeVoss et al., 2002; Miles, 1997; Thrush, 1993; Tippens, 1993). In their analysis of professional and technical communication textbooks, both Libby Miles (1997) and Dànielle DeVoss et al. (2002) highlighted the limited, oversimplified, and problematic treatment of intercultural communication as problems to be overcome and the reliance on linear transmission models to teach such competencies. Another dominant theme in pedagogical experiments focuses on strategies to sensitize students to cultural differences. For instance, Emily Thrush (1993) calls for the teaching of cultural differences in communication strategies and an awareness of how such differences impact communication practices. Dora Tippens (1993) examines the problems of ethnocentrism, language barriers, and cultural differences in teaching intercultural communication and recommends strategies to modify existing assignments with intercultural elements. Han Yu (2011) explored the use of genre-based instruction to cultivate intercultural awareness and sensitivity in engineering students. To prepare students for intercultural technical communication tasks, Deborah Andrews (1996) suggested the integration of components such as contrastive rhetoric, ***translation***, internationalization, and localization, which attracted increasing scholarly attention with the rapid development of transnational corporations and multinational teams since the 1990s.

Globalization, localization, and translation are three important areas of focus for technical communication (Agboka, 2013; Aykin, 2005; Ding & Li, 2018; Gnecchi et al., 2011; Gonzales & Turner, 2017; Han et al., 2016; Maylath, 1997; Spyridakis et al., 1997; Yunker, 2003). Highlighting the complex and contested nature of the concept, Jan Scholte (2000) defined globalization as "a transformation of social geography marked by the growth of supraterritorial spaces" which "unfolded with unprecedented speeds and to unprecedented extents since the 1960s" (p. 8). Emphasizing the need to go beyond connections between nation-states, Doreen Starke-Meyerring (2005) defined globalization as "the increasing interdependence and integration of social, cultural, political, and economic processes across local, national, regional, and global levels" (p. 470). To help technical communication students develop global literacies, she called for the need to pay attention to digital networks, pluralized identities and blurred boundaries, interactions between diverse local and global discourses, and ideological contestation surrounding globalization.

Closely connected with the practice of intercultural technical communication, localization is defined as "the process of modifying products or services to accommodate differences in distinct markets" (Aykin, 2005, p. 5). Nuray Aykin's (2005) edited collection contains studies dealing with strategies for and issues of localization of various products such as documents, graphics, websites, and user interfaces (Aykin & Milewski, 2005; Horton, 2005; Marcus, 2005; Mayhew & Bias, 2005). Regarding graphics, Charles Kostelnick (1995) distinguished the global perspective from the culture-focused perspective. While the former tries to "invent an objective, universal language and to define such language through perceptual principles and empirical research," the latter asks ***designers*** to develop sensitivity to cultural contexts and beliefs to meet the needs of specific rhetorical situations (p. 184). In his popular book of localization for the software industry, Bert Esselink (2000) covered the issues of software engineering, software quality assurance, document translation, graphics localization, project evaluations, and ***project management***. Aykin (2005) and Esselink (2000) focused on business needs in localization and examined how producers in source cultures can use localization to better serve the needs of consumers in target cultures. For them, producers or service providers initiate and take charge of the localization processes, and markets in the target cultures receive and consume localized products. Starting in the early 1990s, scholars also worked with local scholars and programs to build localized courses and programs in Chinese universities (Barnum et al., 2001; Ding, 2019; Rainey et al., 2008).

In terms of translation, numerous scholars argue for the need to incorporate translation, including technical translation, into the technical communication curriculum (Ding & Li, 2018; Maylath,1997; Weiss, 1995). Timothy Weiss (1997), for instance, defined the role of professional communicators as that of "a translator who interprets contexts and formulates/reformulates communications" (p. 325). Brue Maylath and Emily Thrush (2000) identified several useful components related to translation, including cultural awareness, language awareness, and awareness of translation procedures. Multiple efforts have been made to give technical communication students opportunities to work with translation and localization students from European countries and to collaborate virtually with students from other cultures through bottom-up networked learning opportunities (Maylath, 1997; Starke-Meyerring & Wilson, 2008).

Beyond the three areas of globalization, localization, and translation, some efforts have been made to develop culturally appropriate empirical ***research*** methodologies in the study of intercultural technical communication. Barry Thatcher (2000) examined possible ways to balance differences with commonalities in designing more valid and ethical cross-cultural comparative studies. Advocating a methodology "situated within local cultures," Beth Kolko and Carolyn Wei (2003) explored possible ways to "incorporate an understanding of how culture, policy, and infrastructure affect patterns of Internet development" in the development of survey and interview tools in their study of information technology use patterns in technologically underdeveloped regions such as Uzbekistan (pp. 1–3). Huatong Sun (2012) proposed a sociocultural methodological framework of cultural usability to compare local uses of mobile messaging in the US and in China through questionnaire surveys, diary studies, qualitative interviews, and observations. Godwin Agboka (2013) explored the incorporation of ***social justice*** consideration and decolonial methodologies in studying cultural localization in disenfranchised cultural sites and discussed possible approaches in encouraging participatory localization. All these researchers stress the need to consider local cultural, political, and material contexts when designing empirical studies.

Scholars coming from non-western cultures have been examining intercultural technical communication practices from non-US-centric perspectives while introducing new insights about different source cultures (Fukuoka et al., 1998; Fukuoka & Spyridakis, 2000). Offering the Global South perspective, Sun (2012, 2020) explored the issues of culturally sensitive design of technologies and social media use across cultures, moving from designing usable and meaningful technology to designing usable, meaningful, and empowering social media technology. Huiling Ding (2013, 2014, 2020) investigated the transcultural ***risk communication*** about SARS and Zika by tracing both virtual and extra-institutional communication efforts made by experts, affected communities, and concerned citizens. The inclusion of intercultural studies focusing on cultures other than the US has added new perspectives and approaches to the field of intercultural technical communication.

While much progress has been made in the research on intercultural communication in the last few decades, we face new challenges today due to the rapid new developments in various areas, including artificial intelligence, data analytics, Industry 4.0, borderless digital labor platforms such as Amazon Mechanical Turk, increasing connectivity due to infrastructural improvement brought by 5G mobile technologies, the ongoing climate crisis, as well as the proliferation of chatbots, fake news, and social media tools. In addition, the continuous improvements in machine translation technologies make it easier for individuals to access and understand information written in other languages and to communicate with people speaking different languages.

Numerous contextual factors, including the ongoing pandemic of COVID-19, complicate the overarching picture of intercultural communication. These factors include the changing global geopolitical and technological landscapes, the shift from multilateralism and economic globalization to economic nationalism and protectionism (Frieden, 2019), the widening health and wealth disparity, and the ever-growing sociospatial inequities (Bhattacharya et al., 2017). How can technical communication scholars engage with these new technologies, developments, and challenges to shed light on possible approaches and strategies to improve intercultural communication efforts and to build new theories to guide such practices? What methodological and theoretical challenges will technical communication scholars encounter when engaging with these new practices? How can we revise and update our curriculum and pedagogical practices to help prepare students to become more effective intercultural communicators? As we move into a post-COVID world with accelerating automation and protectionism, technical communication scholars are in a unique position to engage with these new challenges and to explore possible entry points to help shape important conversations that will determine how the intercultural communities interact with one another in a world facing challenges on all fronts.

References

Agboka, G. Y. (2013). Participatory localization: A social justice approach to navigating unenfranchised/disenfranchised cultural sites. *Technical Communication Quarterly, 22*(1), 28–49. https://doi.org/10.1080/10572252.2013.730966.

Agboka, G. Y. (2014). Decolonial methodologies: Social justice perspectives in intercultural technical communication research. *Journal of Technical Writing and Communication, 44*(3), 297–327. https://doi.org/10.2190/TW.44.3.e.

Andrews, D. C. (1996). *International dimensions of technical communication*. Society for Technical Communication.

Appadurai, A. (1996). *Modernity at large: Cultural dimensions of globalization*. University of Minnesota Press.

Aykin, N. (2005). *Usability and internationalization of information technology*. Lawrence Erlbaum. https://doi.org/10.1201/b12471.

Aykin, N. & Milewski, A. E. (2005). Practical issues and guidelines for international information display. In N. Aykin (Ed.), *Usability and internationalization of information technology* (pp. 21–50). Lawrence Erlbaum. https://doi.org/10.1201/b12471.

Barnum, C. M., Philip, K., Reynolds, A., Shauf, M. S. & Mae, T. (2001). Globalizing technical communication: A field report from China. *Technical Communication Quarterly, 48*(4), 397–420.

Beamer, L. (1992). Learning intercultural communication competence. *Journal of Business Communication, 29*(3), 285–303. https://doi.org/10.1177/002194369202900306.

Beck, U. (2003). Toward a new critical theory with a cosmopolitan intent. *Constellations, 10*, 453–468. https://doi.org/10.1046/j.1351-0487.2003.00347.x.

Bhattacharya, A., Reeves, M., Lang, N. & Augustinraj, R. (2017). *New business models for a new global landscape*. https://www.bcg.com/publications/2017/globalization-new-business-models-global-landscape.

Castells, M. (1996). *The rise of the network society*. Blackwell Publishers.

Chapel, W. B. (1997). Developing international management communication competence. *Journal of Business and Technical Communication, 11*(3), 281–297. https://doi.org/10.1177/1050651997011003003.

Chen, L. (2017). Cultures, communication, and contexts of intercultural communication. In L. Chen (Ed.), *Intercultural Communication* (pp. 1–16). De Gruyter Mouton.

Constantinides, H., St.Amant, K. & Kampf, C. (2001). Organizational and intercultural communication: An annotated bibliography. *Technical Communication Quarterly, 10*(1), 31–59. https://doi.org/10.1207/s15427625tcq1001_2.

Dervin, F. (2011). A plea for change in research on intercultural discourses: A 'liquid' approach to the study of the acculturation of Chinese students. *Journal of multicultural discourses, 6*(1), 37–52.

DeVoss, D., Jasken, J. & Hayden, D. (2002). Teaching intracultural and intercultural communication: A critique and suggested method. *Journal of Business and Technical Communication, 16*(1), 69–95. https://doi.org/10.1177/1050651902016001003.

Ding, H. (2013). Transcultural risk communication and viral discourses: Grassroots movements to manage global risks of H1N1 flu pandemic. *Technical Communication Quarterly, 22*, 126–149. https://doi.org/10.1080/10572252.2013.746628.

Ding, H. (2014). *Rhetoric of global epidemic: Transcultural communication about SARS.* Southern Illinois University Press.

Ding, H. (2019). Development of technical communication in China: Program building and field convergence. *Technical Communication Quarterly, 28*, 223–237. https://doi.org/10.1080/10572252.2018.1551576.

Ding, H. (2020). Crowdsourcing, social media, and intercultural communication about Zika: Use contextualized research to bridge the digital divide in global health intervention. *Journal of Technical Writing and Communication, 50*, 141–166. https://doi.org/10.1177/0047281620906127.

Ding, H. & Li, X. (2018). Technical translation in China. In C. Shei & Z. Gao (Eds.), *Routledge handbook of Chinese translation* (pp. 537–550). Routledge. https://doi.org/10.4324/9781315675725-32.

Esselink, B. (2000). *A practical guide to localization.* John Benjamins Publishing Company. https://doi.org/10.1075/liwd.4.

Frieden, J. (2019). The backlash against globalization and the future of the international economic. In P. Diamond (Ed.), *The crisis of globalization: Democracy, capitalism and inequality in the twenty-first century* (pp. 43–52). I.B. Tauris. https://doi.org/10.5040/9781788316309.ch-002.

Fukuoka, W., Kojima, Y. & Spyridakis, J. (1998). Illustrations in user manuals: Preference and effectiveness with Japanese and American readers. *Technical Communication, 46*, 167–176.

Fukuoka, W. & Spyridakis, J. H. (2000). Japanese readers' comprehension of and preferences for inductively versus deductively organized text. *IEEE Transactions on Professional Communication, 43*, 355–367. https://doi.org/10.1109/47.888811.

Geertz, C. (1973). *The interpretation of culture: Selected essays.* Basic Books.

Gnecchi, M., Maylath, B., Mousten, B., Scarpa, F. & Vandepitte, S. (2011). Field convergence between technical writers and technical translators: Consequences for training institutions. *IEEE Transactions on Professional Communication, 54*(2), 168–184. https://doi.org/10.1109/TPC.2011.2121750.

Gonzales, L. & Turner, H. N. (2017). Converging fields, expanding outcomes: Technical communication, translation, and design at a non-profit organization. *Technical Communication, 64*(2), 126–140.

Gould, E. (2005). Synthesizing the literature on cultural values. In N. Aykin (Ed.), *Usability and internationalization of information technology* (pp. 79–121). Lawrence Erlbaum Associates.

Gudykunst, W. B., Carmen, M. L., Tsukasa, N. & Ogawa, N. (2005). Theorizing about intercultural communication: An introduction. In Gudykunst, W. B. (Ed.), *Theorizing About Intercultural Communication* (pp. 3–32). Sage Publications.

Hall, E. T. (1976). *Beyond culture*. Doubleday.

Han, T., Liu, L. & Gao, Z. (2016). Improve technical communication and localization to enhance the visibility of Chinese brands. *Science Reader*, *34*(17), 105–109.

Hofstede, G. (1980). *Culture's consequences*. Sage.

Hofstede, G. (1991). *Cultures and organizations: Software of the mind*. McGraw-Hill.

Hofstede, G. (2001). *Culture's consequences* (2nd ed.). Sage.

Horton, W. (2005). Graphics: The not quite universal language. In N. Aykin (Ed.), *Usability and internationalization of information technology* (pp. 157–188). Lawrence Erlbaum Associates.

Hunsinger, R. P. (2006). Culture and cultural identity in intercultural technical communication. *Technical Communication Quarterly*, *15*(1), 31–48. https://doi.org/10.1207/s15427625tcq1501_4.

Jensen, K. B. (2011). Introduction: The state of convergence in media and communication research. In K. B. Jensen (Ed.), *A handbook of media and communication research: Qualitative and quantitative methodologies* (pp. 1–14). Routledge.

Kolko, B. & Wei, C. Y. (2003). Internet use in Uzbekistan: Developing a methodology for tracking information technology implementation success. *Information Technologies and International Development*, *1*(2), 1–19. https://doi.org/10.1162/154475203322981932.

Kostelnick, C. (1995). Cultural adaptation and information design: Two contrasting views. *IEEE Transactions on Professional Communication*, *38*(4), 182–195. https://doi.org/10.1109/47.475590.

Mao, L. M. (2003). Reflective encounters: Illustrating comparative rhetoric. *Style*, *37*(4), 401–425.

Marcus, A. (2005). User interface design and culture. In N. Aykin (Ed.), *Usability and internationalization of information technology* (pp. 51–78). Lawrence Erlbaum.

Martin, J. & Nakayama, T. K. (1997). *Intercultural communication in contexts*. Mayfield Publishing Company.

Mayhew, D. & Bias, R. (2005). Cost-justifying usability engineering for cross-cultural user interface design. In N. Aykin (Ed.), *Usability and internationalization of information technology* (pp. 213–252). Lawrence Erlbaum.

Maylath, B. (1997). Writing globally: Teaching the technical writing student to prepare documents for translation. *Journal of Business and Technical Communication*, *11*(3), 339–353. https://doi.org/10.1177/1050651997011003006.

Maylath, B. & Thrush, E. (2000). Café, the ou lait? Teaching technical communicators to manage translation and localization. In P. J. Hager & H. J. Scheiber (Eds.), *Managing* Global Communication in Science and Technology (pp. 233–254). John Wiley & Son.

Miles, L. (1997). Globalizing professional writing curricula: Positioning students and re-positioning textbooks. *Technical Communication Quarterly*, *6*(2), 179–200. https://doi.org/10.1207/s15427625tcq0602_4.

Oxford University Press (n.d.). Intercultural. In *Oxford English Dictionary*. Retrieved May 15, 2023, from https://www.oed.com/

Rainey, K. T., Smith, H. J. & Barnum, C. M. (2008). Steps and missteps in negotiating a joint degree program with a Chinese university. In D. Starke-Meyerring & M. Wilson (Eds.), *Designing globally networked learning environments: Visionary partnerships, policies, and pedagogies* (pp. 67–86). Sense. https://doi.org/10.1163/9789087904753_006.

Samovar, L. A. & Porter, R. (1997). *Intercultural communication: A reader.* Wadsworth.

Scholte, J. A. (2000). *Globalization: A critical introduction.* St. Martin's.

Scott, J. B. (2006). Kairos as indeterminate risk management: The pharmaceutical industry's response to bioterrorism. *Quarterly Journal of Speech, 92,* 115–143. https://doi.org/10.1080/00335630600816938.

Singh, P. & Doherty, C. (2004). Global cultural flows and pedagogic dilemmas: Teaching in the global university contact zone. *TESOL Quarterly, 38*(1), 9–42. https://doi.org/10.2307/3588257.

Spyridakis, J. & Fukuoka, W. (2002). The effect of inductively versus deductively organized text on American and Japanese readers. *IEEE Transactions on Professional Communication,* 45(2), 99–114. https://10.1109/TPC.2002.1003692.

Spyridakis, J., Holmbeck, H. & Shubert, S. (1997). Measuring the translatability of simplified English in procedural documents. *IEEE Transactions on Professional Communication, 40,* 4–12. https://doi.org/10.1109/47.557512.

St.Amant, K. (2017). The cultural context of care in international communication design: A heuristic for addressing usability in international health and medical communication. *Communication Design Quarterly, 5*(2), 62–70. https://doi.org/10.1145/3131201.3131207.

Starke-Meyerring, D. (2005). Meeting the challenges of globalization: A framework for global literacies in professional communication programs. *Journal of Business and Technical Communication, 19*(4), 468–500. https://doi.org/10.1177/1050651905278033.

Starke-Meyerring, D. & Wilson, M. (2008). *Designing globally networked learning environments: Visionary partnerships, policies, and pedagogies.* Brill.

Sun, H. (2006). The triumph of users: Achieving cultural usability goals with user localization. *Technical Communication Quarterly, 15*(4), 457–481. https://doi.org/10.1207/s15427625tcq1504_3.

Sun, H. (2012). *Cross-cultural technology design: Creating culture-sensitive technology for local users.* Oxford University Press. https://doi.org/10.1093/acprof:oso/9780199744763.001.0001.

Sun, H. (2020). *Global social media design: Bridging differences across cultures.* Oxford University Press https://doi.org/10.1093/oso/9780190845582.001.0001.

Thatcher, B. L. (2000). Writing policies and procedures in a U.S./South American context. *Technical Communication Quarterly, 9*(4), 365–399. https://doi.org/10.1080/10572250009364706.

Thatcher, B. (2010). Editor's introduction to the first edition: Eight needed developments and eight critical contexts for global inquiry. *Journal of Rhetoric, Professional Communication, and Globalization, 1*(1). https://docs.lib.purdue.edu/rpcg/vol1/iss1/1.

Thrush, E. A. (1993). Bridging the gaps: Technical communication in an international and multicultural society. *Technical Communication Quarterly, 2*(3), 271—283. https://doi.org/10.1080/10572259309364541.

Tippens, D. (1993). Interculturalizing the technical communication course. *Journal of Technical Writing and Communication, 23*(4), 389–412. https://doi.org/10.2190/0L80-83WF-5MAA-TC8X.

Tylor, E. B. (1871). *Primitive culture: Researches into the development of mythology, philosophy, religion, language, art and custom.* J. Murray.

Weiss, T. (1995). Translation in a borderless world. *Technical Communication Quarterly, 4*(4), 407–423. https://doi.org/10.1080/10572259509364610.

Weiss, T. (1997). Reading culture: Professional communication as translation. *Journal of Business and Technical Communication, 11*(3), 321–339. https://doi.org/10.1177/1050651997011003005.

Yu, H. (2011). Integrating intercultural communication into an engineering communication service class. *IEEE Transactions on Professional Communication, 54*(1), 83–96. https://doi.org/10.1109/TPC.2010.2099830.

Yunker, J. (2003). *Beyond borders: Web globalization strategies*. New Riders Publishing.

18. Knowledge

Jason Swarts
North Carolina State University

As a concept, knowledge is central to technical communication. Technical communicators deliver knowledge (as in scientific and technical) in a form that readers can use. Technical communicators also produce knowledge, as insights about data, work processes, and ***user experiences***. This characterization of knowledge, as a thing that exists in the world, revealed through language, and as a thing created through the interaction of language with the world is central to understanding developments in the field of technical communication. The *Oxford English Dictionary* offers an accessible starting point. Two of its definitions for *knowledge* focus on how knowledge connects with technical communication.

First is *knowledge of*, or the act of knowing: "The apprehension of fact or truth with the mind; clear and certain perception of fact or truth; the state or condition of truth" (Oxford University Press, n.d.). Under this definition, knowledge is an act of ascertaining truth about the world with certainty and clarity. Technical communication has been portrayed as a way to do exactly this: reveal truth by allowing access to the world and what is truly there. The technical communicator does not get in the way of this transmission. This use of knowledge is positivistic in that it references a correct/formal process by which one acquires knowledge of the world. When used properly, language reveals the world without distortion.

A second definition of *knowledge* is more constructivist: "The fact or condition of having acquired a practical understanding or command of, or competence or skill in, a particular subject, language, etc., esp. through instruction, study, or practice" (Oxford University Press, n.d.). Here, knowledge is seen as something one acquires by engaging in actions that produce knowledge. Knowing involves intentionality, engagement, and situatedness. Language is the medium through which we express intentions and make sense of our engagements, making language essential to the creation of knowledge.

Technical communication has long grappled with these approaches to knowledge, as practitioners have sought to articulate their role in the process of knowledge creation. Some of the earliest forms of technical writing, technical descriptions from the late 15th century, on medicine and navigation, came about as ways to preserve knowledge that was experiential and detailed, knowledge that was difficult to transmit orally with any degree of comprehensiveness or reliability (Tebeaux, 1991, p. 61).

The need for transmittable knowledge grew alongside publication technologies that circulated content widely and helped professions enrich their knowledge base. These professionals required technical writing to capture developments

DOI: https://doi.org/10.37514/TPC-B.2023.1923.2.18

using specialized technical terms (Tebeaux, 1991, p. 106). The need served by technical writing in these contexts held constant throughout the development of technical writing as an area of instruction in the 19th century, where its purpose was to ensure clear transmission of specialized ***information*** among engineers (Connors, 1982). In these contexts, technical writing was "the skill of subduing language so that it most accurately and directly transmits reality" (Miller, 1979, p. 610), a relationship between technical writing and reality that Carolyn R. Miller (1979) called the "windowpane theory of language" (p. 611).

If technical writing is to be a windowpane on the world, then the writing itself must be highly formalized and words must be chosen carefully to be direct, to the point, and to mean one and only one thing (Britton, 1965, p. 114). This view on knowledge is prevalent today among practicing technical communicators who describe their work as "that of transferring information from those who have it (subject matter experts or SMEs) to those who need it . . . packaging that information to be more accessible and more readily understood by the user" (Hughes, 2002, p. 275). This position "implies that the source information 'exists' and someone 'has' that information" (Hughes, 2002, p. 275).

The function of technical communication to create knowledge by revealing truth is also captured in Jennifer Daryl Slack, David James Miller, and Jeffrey Doak's (1993) typology of technical communicator roles. Among the three roles, "transmitter" stands out as being linked most closely to a positivistic outlook on knowledge. A transmitter is one whose words frame knowledge in the world, reveal it, and move it from one place to another with little or no signal loss. In the second role, "translator," the technical communicator still encodes knowledge in a format that reflects the source, but they must now interact with receivers who actively decode that content. Meaning is negotiated (Slack et al., 1993, p. 20). The third role, "articulator," moves us closer to a constructivist concept of knowledge in technical communication, where more power is invested in the technical communicator and knowledge is recognized as something that is created through language and situated within a location and nexus of identities and positionalities. The articulator role becomes possible if we take the knowledge that technical communicators deal with to be socially constructed, rather than strictly revealed through objective and formal means.

This social, constructed view of knowledge parallels thinking in ***science*** and ***technology*** studies, such as David Bloor's (1976) work on the Strong Programme, which views social influence on scientific knowledge not just as the source of error but the source of success as well. Social conditions must inhere for any kind of knowledge to develop. A similar perspective is echoed in Ludwig Fleck's (1981) social explanations of scientific facts as well as, famously, Thomas Kuhn's (1996) discussions of "paradigms."

Knowledge construction is particularly evident where interpretations of the world intersect and disrupt what Richard Rorty (1979) describes as "normal discourse," or that use of language "which is conducted within an agreed-upon set of conventions about what counts as a relevant contribution" (p. 320). Normal discourse is kept in tension by the work of edification, the "project of finding new, better, more interesting, more fruitful ways of speaking" (Rorty, 1979, p. 360). Across these views of knowledge, language is understood to be constitutive of reality (Berger & Luckmann, 1967), of what we know and care to remember (Havelock, 1988). Technical writing in particular "becomes, rather than the revelation of absolute reality, a persuasive version of experience" (Miller, 1979, p. 616).

The swing toward constructivist notions of knowledge characterizes much of technical communication scholarship throughout the late 20th century. Marilyn Samuels (1985) describes this turn as one that characterizes technical communication as a creative enterprise, crafting "reality for special purposes" (p.11). The language of science is just an example. Other contexts, like the technological and political, can also reflect in technical communicators' choices of language. Those contexts and the languages associated with them reflect discursive norms within different domains of practice while also reinforcing norms of knowing and acting entailed by those discourses (Thralls & Blyler, 1993, pp. 254, 259). An example might be procedure writing, from a technological context, that positions users as those who must bend their expectations to fit a technology's ***design*** constraints (Norman, 2002).

Within this space opened up by a constructivist approach to technical communication, scholars saw ways to raise the profile of situated knowledges that accompanied ways of being in the world (e.g., Durack, 1997). Paul Dombrowski (1995) saw the move as a way of focusing on knowledge creation, especially forms of knowledge that have been "excluded, suppressed, and marginalized" (p. 265) as well as knowledge that has been misconstrued, ignored, or otherwise silenced (Jones, 2016). When knowledge is understood to be socially constructed, writers must give attention to forces of "knowledge legitimation (i.e., whose knowledge do we value, whose knowledge do we seek and solicit, and whose opinions do we include)" (Jones, 2016, p. 479). Mary Lay (1991) also saw value in resisting positivistic notions of knowledge to create room for feminist approaches that valued situated experience and collaborative, community-based ways of knowing, where knowledge is negotiated (p.356, 365), socially achieved (Winsor, 1990, p. 12), and strongly informed by lived experience (Jones, 2020).

A constructivist outlook on communication foregrounds the role of the receiver and acknowledges that knowledge is not passive (Winsor, 1990, p. 13). Instead, receivers actively interpret and create knowledge as they read (Redish, 1993). As a result, technical communicators increasingly think of themselves less exclusively as generators of knowledge and sometimes also as "information managers," who help bridge different "content spaces" (Regli, 1999, p. 32; see also Wilson & Herndl, 2007). A focus on the social as a source of knowledge production is also evident in the field's turn toward user involvement, as clients are deliberately integrated into the knowledge-creation process, whether through interviews, focus groups, usability testing, or other means (Johnson, 1997).

A constructivist approach to knowledge production is also foundational to technical communicators who position themselves as "knowledge workers," trading in the creation and circulation of knowledge within particular communities (Johnson-Eilola, 1996). More recently, scholars have looked at this knowledge work as supportive of users but also as supportive of knowledge communities within organizations (Hart-Davidson, 2013; see also Smart, 1999). Knowledge is what technical communicators facilitate, and they do so through their contact with different social actors that they help put into conversation (Read & Swarts, 2015). Knowledge is literally in and between the minds of the actors that we engage with in social settings and connect through language and text (Winsor, 2001).

This constructivist outlook on knowledge creation positions technical communicators as social agents of knowledge creation. Over time, the field has developed techniques and heuristics for generating this kind of social knowledge. Technical communication sees itself as a "problem-solving activity" (Johnson-Eilola & Selber, 2013, p. 3), and its practitioners solve problems by learning through the use of heuristics, which are "rough frameworks for approaching specific types of situations" (Johnson-Eilola & Selber, 2013, p. 4). There are heuristics for understanding audiences and users (Redish, 1993), usability (Mirel, 1998), ***project management*** (Dicks, 2003), content strategy (Halvorson & Rach, 2012), and information architecture (Rosenfeld & Morville, 2002), to name a few. But as Johndan Johnson-Eilola and Stuart A. Selber note, these heuristics must account for differences in the cultural, economic, and political contexts where they are applied.

Heuristics like audience and task analysis, user profiles, scenarios, content maps, and content plans are used to create knowledge, but in doing so, one must be aware of how those heuristics engage in a process of creating and recreating normal discourse that belongs to particular regimes of power (Thralls & Blyler, 1993, p. 254). Knowledge making through communication helps create a reality for those who use it—it is an ***ethical*** activity (Cooper, 2005, p. 37). The problem, as scholars in technical communication are coming to realize, is that while we respect the instrumental value and utility of standardized approaches to language use (see Moore, 1996), if we are not critical of our heuristics, they can overemphasize an ethos of efficiency and effectiveness, which flattens and simplifies readers and contexts of communication, at the expense of building local, situational knowledge that will be more complex and diverse than heuristics aimed at efficient data collection and processing will allow. The danger in the zealous pursuit of efficiency is precisely presented in Steven Katz's (1992) work on technical communication in Nazi Germany. And Natasha Jones and colleagues (2016) broadly characterize the issue this way:

> The official narrative of our field indicates that TPC is about practical problem solving: a pragmatic identity that values effectiveness. But this is not the whole story. The narrative should be reframed to

> make visible competing (i.e., a collection of nondominant) narratives about the work our field can and should do. (p. 212)

The values associated with effectiveness and efficiency are central to our pragmatic, disciplinary identity and are characteristically present in the heuristics that we use to create knowledge. Procedurally, we rely on our heuristics to create methodological distance from which we pretend to get a true view of the readers and contexts we are trying to reach. All the while, we may not realize how the heuristics are themselves constructions that reinforce ways of knowing and seeing from a particular vantage point. The danger is that if we do not acknowledge the partiality and positionalities from which we generate knowledge, we run a risk of essentialism by overlooking ways that culture is socially constructed and local (Agboka, 2012, p. 174). Heuristics and other tools, especially when deployed to understand other cultures, tend to treat culture as "a set of habits and traits that one can learn and regurgitate" (Agboka, 2012, p. 169). A better approach to knowledge creation is local and participatory (Agboka, 2013, p. 42; Longo, 2014, p. 24).

The meaning and pursuit of knowledge in technical communication continues to be a matter of importance for how we see ourselves and our work. New information and communication technologies, as well as new information environments, require technical communicators to face new demands for creating and sharing knowledge. Ongoing discussions about knowledge and knowledge creation will also help us become better at articulating our relationships to other fields and industries.

References

Agboka, G. (2012). Liberating intercultural technical communication from "large culture" ideologies: Constructing culture discursively. *Journal of Technical Writing and Communication, 42*(2), 159–181. https://doi.org/10.2190/TW.42.2.e.

Agboka, G. Y. (2013). Participatory localization: A social justice approach to navigating unenfranchised/disenfranchised cultural sites. *Technical Communication Quarterly, 22*(1), 28–49. https://doi.org/10.1080/10572252.2013.730966.

Berger, P. L. & Luckmann, T. (1967). *The social construction of reality: A treatise in the sociology of knowledge*. Anchor Books.

Bloor, D. (1976). *Knowledge and social imagery*. University Of Chicago Press.

Britton, W. E. (1965). What is technical writing? *College Composition and Communication, 16*(2), 113–116. https://doi.org/10.2307/354886.

Connors, R. J. (1982). The rise of writing instruction in America. *Journal of Technical Writing and Communication, 12*(4), 329–352. https://doi.org/10.1177/004728168201200406.

Cooper, M. M. (2005). Bringing forth worlds. *Computers and Composition, 22*(1), 31–38. https://doi.org/10.1016/j.compcom.2004.12.013.

Dicks, R. S. (2003). *Management principles and practices for technical communicators*. Longman.

Dombrowski, P. M. (1995). Post-modernism as the resurgence of humanism in technical communication studies. *Technical Communication Quarterly*, *4*(2), 165–185. https://doi.org/10.1080/10572259509364595.

Durack, K. T. (1997). Gender, technology, and the history of technical communication. *Technical Communication Quarterly*, *6*(3), 249–260. https://doi.org/10.1207/s15427625tcq0603_2.

Fleck, L. (1981). *Genesis and development of a scientific fact*. University of Chicago Press. https://doi.org/10.7208/chicago/9780226190341.001.0001.

Halvorson, K. & Rach, M. (2012). *Content strategy for the web*. New Riders.

Hart-Davidson, B. (2013). What are the work patterns of technical communication? In J. Johnson-Eilola & S.A. Selber (Eds.), *Solving problems in technical communication* (pp. 50–74). University of Chicago Press.

Havelock, E. (1988). *The muse learns to write: Reflections on orality and literacy from antiquity to the present*. Yale University Press.

Hughes, M. (2002). *Moving from information transfer to knowledge creation: A new value proposition for technical communicators*. Society for Technical Communication.

Johnson, R. R. (1997). Audience involved: Toward a participatory model of writing. *Computers and Composition*, *14*(3), 361–376. https://doi.org/10.1016/S8755-4615(97)90006-2.

Johnson-Eilola, J. (1996). Relocating the value of work: Technical communication in a post-industrial age. *Technical Communication Quarterly*, *5*(3), 245–270. https://doi.org/10.1207/s15427625tcq0503_1

Johnson-Eilola, J. & Selber, S. A. (Eds.). (2013). *Solving problems in technical communication*. University of Chicago Press.

Jones, N. N. (2016). Narrative inquiry in human-centered design: Examining silence and voice to promote social justice in design scenarios. *Journal of Technical Writing and Communication*, *46*(4), 471–492. https://doi.org/10.1177/0047281616653489.

Jones, N. N. (2020). Coalitional learning in the contact zones: Inclusion and narrative inquiry in technical communication and composition studies. *College English*, *82*(5), 515–526.

Jones, N. N., Moore, K. R. & Walton, R. (2016). Disrupting the past to disrupt the future: An antenarrative of technical communication. *Technical Communication Quarterly*, *25*(4), 211–229. https://doi.org/10.1080/10572252.2016.1224655.

Katz, S. B. (1992). The ethic of expediency: Classical rhetoric, technology, and the Holocaust. *College English*, *54*(3), 255–275. https://doi.org/10.2307/378062.

Kuhn, T. S. (1996). *The structure of scientific revolutions*. University of Chicago Press. https://doi.org/10.7208/chicago/9780226458106.001.0001.

Lay, M. (1991). Feminist theory and the redefinition of technical communication. *Journal of Business and Technical Communication*, *5*(4), 348–370. https://doi.org/10.1177/1050651991005004002.

Longo, B. (2014). Using social media for collective knowledge-making: Technical communication between the Global North and South. *Technical Communication Quarterly*, *23*(1), 22–34. https://doi.org/10.1080/10572252.2014.850846.

Miller, C. R. (1979). A humanistic rationale for technical writing. *College English*, *40*(6), 610–617. https://doi.org/10.2307/375964.

Mirel, B. (1998). "Applied constructivism" for user documentation alternatives to conventional task orientation. *Journal of Business and Technical Communication*, *12*(1), 7–49. https://doi.org/10.1177/1050651998012001002.

Moore, P. (1996). Instrumental discourse is as humanistic as rhetoric. *Journal of Business and Technical Communication*, *10*(1), 100–118. https://doi.org/10.1177/105065199601000 1005.

Norman, D. (2002). *The design of everyday things*. Basic Books.

Oxford University Press. (n.d.). Knowledge. In *Oxford English Dictionary*. Retrieved June 4, 2020, from https://www.oed.com.

Read, S. & Swarts, J. (2015). Visualizing and tracing: Research methodologies for the study of networked, sociotechnical activity, otherwise known as knowledge work. *Technical Communication Quarterly*, *24*(1), 14–44. https://doi.org/10.1080/10572252.201 5.975961.

Redish, J. C. (1993). Understanding readers. In C. M. Barnum & S. Carliner (Eds.), *Techniques for technical communicators* (pp. 15–41). Macmillian.

Regli, S. H. (1999). Whose ideas? The technical writer's expertise in inventio. *Journal of Technical Writing and Communication*, *29*(1), 31–40. https://doi.org/10.2190/73VW -YBUC-YHXW-WU0C.

Rorty, R. (1979). *Philosophy and the mirror of nature*. Princeton University Press.

Rosenfeld, L. & Morville, P. (2002). *Information architecture for the World Wide Web*. O'Reilly.

Samuels, M. S. (1985). Technical writing and the recreation of reality. *Journal of Technical Writing and Communication*, *15*(1), 3–13. https://doi.org/10.2190/V6M7-43G5-9PT7 -C5BH.

Slack, J. D., Miller, D. J. & Doak, J. (1993). The technical communicator as author: Meaning, power, authority. *Journal of Business and Technical Communication*, *7*(1), 12–36. https://doi.org/10.1177/1050651993007001002.

Smart, G. (1999). Storytelling in a central bank: The role of narrative in the creation and use of specialized economic knowledge. *Journal of Business and Technical Communication*, *13*(3), 249–273. https://doi.org/10.1177/105065199901300302.

Tebeaux, E. (1991). The evolution of technical description in Renaissance English technical writing, 1475–1640: From orality to textuality. *Issues in Writing*, *4*(1), 59–109.

Thralls, C. & Blyler, N. R. (1993). The social perspective and pedagogy in technical communication. *Technical Communication Quarterly*, *2*(3), 249–270. https://doi.org/10.1080 /10572259309364540.

Wilson, G. & Herndl, C. G. (2007). Boundary objects as rhetorical exigence: Knowledge mapping and interdisciplinary cooperation at the Los Alamos National Laboratory. *Journal of Business and Technical Communication*, *21*(2), 129–154. https://doi.org/10 .1177/1050651906297164.

Winsor, D. A. (1990). The construction of knowledge in organizations: Asking the right questions about the Challenger. *Journal of Business and Technical Communication*, *4*(2), 7–20. https://doi.org/10.1177/105065199000400201.

Winsor, D. A. (2001). Learning to do knowledge work in systems of distributed cognition. *Journal of Business and Technical Communication*, *15*(1), 5–28. https://doi.org/10 .1177/105065190101500101.

19. Literacy

Kelli Cargile Cook
Texas Tech University

The term *literacy* is so commonplace that few sources bother to define it. *Literacy*, in lay terms, means "the ability to read and write." The term *literacy*, according to David Barton (2007), did not appear in dictionaries until 1924; when it did, it was simply defined as "educated." Over time, the definition of literacy has evolved. The United Nations Educational, Scientific and Cultural Organization (UNESCO; 2005), an agency that has offered international literacy support for decades, offers this more complex definition:

> Literacy is the ability to identify, understand, interpret, create, communicate and compute, using printed and written materials associated with varying contexts. Literacy involves a continuum of learning in enabling individuals to achieve their goals, to develop their knowledge and potential, and to participate fully in their community and wider society. (p. 40)

For technical and professional communication—a discipline dedicated to goal-oriented, contextually relevant communication—literacy can serve as a powerful framework for understanding the practices of both technical communicators and their ***audiences***.

Literacy practices are embedded in social situations: "the meaning of literacy depends on the social institutions in which it is embedded . . . [and] . . . the particular practices of reading and writing that are taught in any context depend upon such aspects of social structure as stratification . . . and the role of educational institutions" (Street, 1984, p. 8). Similarly, Gerald J. Savage (2003) writes that "no set of institutional or social arrangements, no body of knowledge, values, or beliefs is an essence. All have ***histories*** and arise from historical exigencies" (p. 3). These statements are particularly true when discussing *literacy* as a keyword in technical and professional communication. This literacy story begins in a social setting: English departments, embedded in higher education, organizations themselves fraught with systemic imbalances.

Historical scholars suggest the origins of technical and professional communication ***pedagogy*** arose from engineering and agricultural students' need for better workplace writing and speaking skills (Connors, 2004; Kynell, 2000; Longo, 2000). Instruction was frequently outsourced to departments of English, where these students read and critiqued literature. This outsourcing came with its own problems. On the surface, these courses were designed to improve students' functional literacy—their abilities to read and write—but on English teachers' terms:

DOI: https://doi.org/10.37514/TPC-B.2023.1923.2.19

"If engineers wanted English instruction, they would have to accept literature along with writing, because the English graduate schools of the time were not producing anything but literary scholars—who wanted work" (Connors, 2004, p. 7). Robert J. Connors (2004) documents several problems that ultimately led to failure in these early 20th-century classrooms: English faculty tended to focus on composition and critique of literature as a means of improving students' functional literacy; inexperienced, junior faculty were most assigned to teach these courses; and cooperation between English and engineering faculty was minimal, at best (pp. 7–8). Complicating problems, "academic literary professionals felt alienated from 'real world' matters, and indeed cultivated that alienation as a virtue, setting themselves apart from business and industrial concerns and upholding values they took to be higher than those of what they viewed as philistine commercial interests" (Russell, 1993, p. 86). Describing technical writing instruction occurring at the end of the 20th century, Mary Sue Garay (1998) depicts this attitude among English faculty as the "filthy lucre bias" against physical labor and applied workplaces (p. 4).

The "filthy lucre bias" not only impacted how technical and professional communication programs evolved in English departments over time, but it also affected how scholars in the field approached pedagogy. To an English department audience unconvinced of the value of the technical writing course, "the common opinion [is] that the undergraduate technical writing course is a 'skills' course with little or no humanistic value" (Miller, 1979, p. 610). Carolyn Miller (1979) counters this opinion and argues for technical writing as an acceptable humanities offering. Her argument concludes with this recommendation for technical writing pedagogy: Rather than focusing on writing skill sets, it should focus on contextualizing skills within social settings and considering the ethical implications of technical writers within those settings (p. 617).

Miller's ***rhetoric*** shifted the focus away from workplace skills to a more palatable English department goal: a literate study grounded in humanism. Her turn from "skills" to "literacy" provided a more solid foundation on which to build and ***assess*** programs in technical, scientific, and professional communication in the late 1980s and 1990s (p. 617). It was in these programs that scholars in the late 20th century and early 21st century began to explore and open the boundaries of literacy in technical and professional communication pedagogy. Among the scholars who pushed these boundaries was Billie Wahlstrom (1997), whose essay revisits traditional definitions of literacy and explores how those definitions must be expanded to include new configurations of community and the agency students possess within those communities:

> Too often . . . technical communication educators have abdicated the larger obligation to help students become responsible citizens and ethical workers in favor of focusing on smaller topics such as teaching the skill sets our graduates need to get successful jobs. We have opted for functional literacy instead of designing true

> teaching and learning environments that enable students to build layered literacies. Functional literacy may help our students to get jobs, but in this era only a broader set of literacies will enable students to develop fully as competent communicators, ethical agents of change, and engaged citizens. (p. 130)

Wahlstrom's (1997) concept of layered literacies inspired me to consider how best to articulate the layered literacies technical and professional communication students needed (Cargile Cook, 2002). Reflecting on Wahlstrom's (1997) call for "literacies [that] are not isolated but integrated and situated through a complex of classroom goals and activities" (Cargile Cook, 2002, p. 6), I wrote,

> Two problems face technical communication instructors as they construct learning communities with integrated, situated, and multiple literacy-learning opportunities. The first is the lack of a concise identification of literacies that technical communicators should possess. This problem does not result from lack of literature on the literacies that technical communicators should acquire; rather it results from the breadth of that literature. The second problem is the lack of understanding about how these multiple literacies can be integrated, situated, or, as Wahlstrom advocates, layered into programs, courses, and specific course activities. (Cargile Cook, 2002, p. 6).

My response to these problems is to synthesize the breadth of the existing literature into six "literacies" that could be "layered" into multiple configurations within varied lessons, units, and courses in professional communication. I identify the following literacies: basic, rhetorical, social, technological, ethical, and critical. These literacies, I argue, are important because they provide students with more than functional literacy: "By focusing on these literacies rather than on specific workplace skills, technical communication instructors may better prepare students for many workplaces and prepare them for lifelong learning, not learning for a specific vocation" (Cargile Cook, 2002, p. 24).

As opposed to this broad approach to literacies, Stuart Selber (2004) delves more deeply into computer literacy, calling for students to gain the "multiliteracies," which he places in three categories: "functional, critical, and rhetorical" (p. 24):

> The functional layer implies access to—and control over—technologies that can support the educational goals of students, help them manage their computer-based activities, and help them resolve their technological impasses. The critical layer implies access to computer technologies for the purposes of critique, and not just one platform. . . . And the rhetorical layer implies access to robust computer environments that can support the technical side

> of interface design, which includes the collaborative production of rapid prototypes and visual images, not to mention actual interfaces that function. (Selber, 2004, p. 192)

Although he focuses primarily on computer literacies, Selber (2004) proposes an extensive framework for literacy programs, beginning with students' introduction to the functional uses of computer hardware and software and extending to broader systemic change within institutional settings. Whether literacy instruction is combined into a single course or divided within a curricular series, Selber (2004) argues that his framework provides "direction and structure for teachers of writing and communication who work in departments of English" (p. 29).

However, literacies, even when defined as "layered" or "multi-," do not take into account multiple, tacit knowledges that simply reading and writing cannot encompass, such as those gained through extended practice within specific cultural settings. As examples of these practices, consider the challenges of learning to play a musical instrument, to lay bricks, fold a parachute, or weave a cloth with only the guidance of the printed word. Shirley Brice Heath (1980) notes that even print media themselves have had a paradoxical effect on literacy: While it opened literate practices for many, it "also made possible new kinds of control over the people" (Heath, 1980, p. 124). Furthermore, scholars like Cynthia L. Selfe (1999) warned that "federally sponsored literacy programs . . . can actually contribute to the ongoing problems of racism, sexism, poverty, and illiteracy in the United States" (p. 12). In UNESCO's Expert Meeting on Literacy (2005), this problem was further elaborated:

> Literacy may be a means of domination, for example when it is taught to promote particular ideologies or where new readers are served a diet of propaganda. More subtly, literacy promotion often serves to socialise learners into the dominant social discourse, rather than opening up new opportunities of expression and creative diversity. (p. 15)

Concerns about the use of "literacy" standards and measures to create and maintain institutionalized biases appear in other disciplines too. Literacy historians, such as Carl F. Kaestle (1985), have examined how historical assumptions about literacy have resulted in cultural biases used to disempower marginalized groups. Such beliefs include assumptions that upper classes are more literate than lower, that white people are more literate than people of color, that Protestants are more literate than Catholics, and that Northerners are more literate than Southerners (Kaestle, 1985, p. 22). These cultural stereotypes, frequently unquestioned and unrecognized by those in power, have had devastating consequences when they are enacted in educational and legal decisions (see Cook-Gumperz's [2006] discussion of the "ideology of literacy" in education, Prendergast's [2002] analysis of the "economy of literacy" in Supreme Court rulings, and Jones & Williams' [2018] analysis of literacy tests as "technologies of disenfranchisement").

Such critiques of literacy and literacy standards are especially poignant since 2020. In the throes of a global pandemic, protesters lined American streets decrying invidious discrimination and police brutality. Black, indigenous, and other people of color have asked the privileged among us to witness, to listen, to read, and to take note of their lives. Is it not time, then, to question our use of certain keywords like *literacy*? Is it time to retire this term, adopted originally in our field to appease literature faculty but used systematically in many disciplines to establish and maintain cultural superiority? Are we ready, as a field, to reassess our pedagogies and our programs in this light? And, if so, what is the new keyword that should take its place? The answer is as complex as the questions. Terms like "skills," "competencies," and "standards" have been used as frequently as "literacies" in technical and professional scholarship (Carliner, 2001; Gillis, 2006; Hart-Davidson, 2001; Pringle & Williams, 2005; Rainey et al., 2005; Whiteside, 2003). These terms, more situated in practical workplaces, do not carry the negative cultural and historical connotations of "literacy," nor, unfortunately, do they carry the positive connotations of an engaged citizen advocating change. Perhaps, a better term for the pedagogical aims is simply "***knowledges***," a word that connotes all the capabilities we desire for our students: the know-hows, know-whens, and know-whys of technical and professional communication as well as the know-whats it takes to be an engaged citizen and good human in the world.

References

Barton, D. (2007). *Literacy: An introduction to the ecology of written language*. Blackwell Publishing. https://eprints.lancs.ac.uk/id/eprint/68942.

Cargile Cook, K. (2002). Layered literacies: A theoretical frame for technical communication pedagogy. *Technical Communication Quarterly*, *11*(1), 5–29. https://doi.org/10.1207/s15427625tcq1101_1.

Carliner, S. (2001). Emerging skills in technical communication: The information designer's place in a new career path for technical communicators. *Technical Communication*, *48*(2), 156–175.

Cook-Gumperz, J. (2006). *The social construction of literacy* (Vol. 25). Cambridge University Press. https://doi.org/10.1017/CBO9780511617454.

Connors, R. J. (2004). The rise of technical writing instruction in America. In J. Johnson-Eilola & S. A. Selber (Eds), *Central works in technical communication* (pp. 3–19). Oxford University Press. (Original work published 1982)

Garay, M. S. (1998). Of work and English. In M.S. Garay & S. A. Bernhardt (Eds.), *Expanding literacies: English teaching and the new workplace.* (pp. 3–20). SUNY Press.

Gillis, T. (Ed.). (2006). *The IABC handbook of organizational communication: A guide to internal communication, public relations, marketing and leadership* (Vol. 2). John Wiley & Sons.

Hart-Davidson, W. (2001). On writing, technical communication, and information technology: The core competencies of technical communication. *Technical Communication*, *48*(2), 145–155.

Heath, S. B. (1980). The functions and uses of literacy. *Journal of Communication*, *30*(1), 123–133. https://doi.org/10.1111/j.1460-2466.1980.tb01778.x.
Jones, N. N. & Williams, M. F. (2018). Technologies of disenfranchisement: Literacy tests and Black voters in the US from 1890 to 1965. *Technical Communication*, *65*(4), 371–386.
Kaestle, C. F. (1985). The history of literacy and the history of readers. In E. S. Edmund (Ed.), *Review of research in education* (pp. 11–53). https://doi.org/10.2307/1167145.
Kynell, T. C. (2000). *Writing in a milieu of utility: The move to technical communication in American engineering programs, 1850–1950* (No. 12). Greenwood Publishing Group.
Longo, B. (2000). *Spurious coin: A history of science, management, and technical writing*. SUNY Press.
Miller, C. R. (1979). A humanistic rationale for technical writing. *College English*, *40*(6), 610–617. https://doi.org/10.2307/375964.
Prendergast, C. (2002). The economy of literacy: How the Supreme Court stalled the civil rights movement. *Harvard Educational Review*, *72*(2), 206–230. https://doi.org/10.17763/haer.72.2.l8112t70x6klx6j0.
Pringle, K. & Williams, S. (2005). The future is the past: Has technical communication arrived as a profession? *Technical Communication*, *52*(3), 361–370.
Rainey, K. T., Turner, R. K. & Dayton, D. (2005). Do curricula correspond to managerial expectations? Core competencies for technical communicators. *Technical Communication*, *52*(3), 323–352.
Russell, D. R. (1993). The ethics of teaching ethics in professional communication: The case of engineering publicity at MIT in the 1920s. *Journal of Business and Technical Communication*, *7*(1), 84–111. https://doi.org/10.1177/1050651993007001005.
Savage, G. J. (2003). Toward professional status in technical communication. In T. Kynell-Hunt & G. J. Savage (Eds.), *Power and legitimacy in technical communication*, *1* (pp. 1–12). Baywood.
Selber, S. (2004). *Multiliteracies for a digital age*. SIU Press.
Selfe, C. L. (1999). *Technology and literacy in the 21st century: The importance of paying attention*. SIU Press.
Street, B. V. (1984). *Literacy in theory and practice* (Vol. 9). Cambridge University Press.
United Nations Educational, Scientific and Cultural Organization (UNESCO). (2005). *Aspects of literacy assessment: Topics and issues from the UNESCO expert meeting*. https://unesdoc.unesco.org/ark:/48223/pf0000140125.
Wahlstrom, B. (1997). Teaching and learning communities: Locating literacy, agency, and authority in a digital domain. In S.A. Selber (Ed.), *Computers and Technical Communication: Pedagogical and Programmatic Perspectives Series*, *3* (pp. 129–146). Greenwood Publishing Group.
Whiteside, A. L. (2003). The skills that technical communicators need: An investigation of technical communication graduates, managers, and curricula. *Journal of Technical Writing and Communication*, *33*(4), 303–318. https://doi.org/10.2190/3164-E4V0-BF7D-TDVA.

20. Medical/Health Communication

Christa Teston
Ohio State University

As a practice, medical/health communication (M/HC) existed long before the field of technical communication (TC). In fact, Barbara L. Harris (1991) identified Hippocrates' "Corpus Hippocraticum," a treatise that modeled how to describe patients' case histories concisely and precisely, as one of "Western Civilization's Earliest Technical Documents." Since then, and especially in recent years, M/HC has become a significant domain of TC, with ***information*** shared both between medical professionals and between doctors and their patients in a host of in-person, print, and digital ***genres***. Yet classical sources can guide how today's TC scholars approach M/HC; for example, the following tenets inspired by Aristotle's *Nicomachean Ethics*: (1) The art of medicine is a model for ethical communication, and (2) "Good health" was (for the Greeks) an indicator of a "good life." In other words, the corporeal conditions that mark someone as "healthy" (or sick) were used to make judgments that tend to confer extra-corporeal advantage. So long as "virtues of the body" are intimately tethered to "virtues of the soul" (Jaeger, 1957, p. 57), M/HC will remain an ethical and political enterprise that has enormous consequences for individuals and publics.

Contemporary M/HC reflects a cross-pollination of ideas between and among scholars in such fields as social studies of ***science***, science and ***technology*** studies, behavioral science, ***history*** of medicine, medical humanities, communication studies, and TC itself (to name but a few). Intellectual overlap among rhetoric of science, medical rhetoric, and the emergence of TC as a discipline constitutes the bedrock of contemporary M/HC scholarship in TC. It's important to note that this scholarship is distinct from other approaches to medical and/or health communication. The field of health communication, for example, is a rich, stand-alone area of study (typically housed within communication departments) that has its own, unique disciplinary ancestry (see Lynch & Zoller, 2015).

During the early 1990s, TC publications treated M/HC largely as textual phenomena that, when analyzed critically, could shed light on cultural practices, beliefs, and values (see Brasseur & Thompson, 1995; Connor, 1993; Harris, 1991). At around that same time, TC scholars interrogated scientific communication, which similarly involved analyses of textual artifacts, for what they might tell us about specific disciplinary practices and the ethical-sociopolitical construction of ***knowledge***, more generally (Bazerman, 1988; Condit, 1990; Paradis, 2019; Zappen, 1991). Analyses of scientific texts from a TC perspective yielded new constructs for unpacking how medical texts—as both practical and professional documents—perform important rhetorical work. In fact, Jessica M. Eberhard

DOI: https://doi.org/10.37514/TPC-B.2023.1923.2.20

(2012) has argued that TC's "history of collaboration with the applied sciences" and its "attention to workplace writing genres" resulted in the emergence of the rhetoric of medicine (p. 1). The iterative emergence of the rhetoric of medicine and TC's interest in M/HC is further evidenced by Barbara Heifferon and Stuart Brown's (2000) special issue on medical rhetoric in *Technical Communication Quarterly*, which was, according to Eberhard (2012) "the first ever collection of articles fathered [sic] under the name 'medical rhetoric'" (p. 14). Other prominent special issues include Ellen Barton's (2005) special issue on the discourse of medicine in *Journal of Business and Technical Communication*, Amy Koerber and Brian Still's (2008) special issue on online health communication in *Technical Communication Quarterly*, Christina Haas' (2009) special issue on writing and medicine in *Written Communication*, and Lisa Melonçon and Erin Frost's (2015) special issue on the rhetorics of health and medicine in *Communication and Design Quarterly*.

Today, disciplinary and analytic overlap between humanistic traditions that tend toward critique (e.g., rhetorical criticism, critical disability studies, critical race studies) and more socially scientific fields (e.g., sociology, anthropology, political science) continues. Beyond its inherent transdisciplinarity, determining the scope of M/HC is further complicated by that pesky slash between "medical" and "health." Generally speaking, *medical* communication could be characterized as communicative practices, processes, and products within the domain of medical science, while *health* communication includes a more expansive material-discursive corpus that, in tandem with sociocultural contexts, indexes what it means to be healthy (or not). But tensions between medicine and health have a long and sordid history. That tension is all the more amplified when we inquire about M/HC's goals. Are M/HC communicators working toward cure? Or care? Is the goal of M/HC to achieve some idealized standard of how *the* (not *a*) healthy human body ought to look and act?

Adjacent fields of study such as disability studies have asked similar ends/means questions that often result in critiques of M/HC for its unabashed pursuit of cure (often at the expense of care), which, according to such critiques, advances normative ideologies about human bodies. Building from such cure vs. care critiques, I'd argue that what animates the productive power of the slash between medicine and health, at least as it concerns TC, is amplified attention to how power operates—in all its (intersectional) forms.

Practicing medicine or performing health requires a constellation of suasive evidences, many of which are textual inscriptions. Historiographic or archival studies offer one means to uncover some of these evidences. For example, Carolyn Skinner (2012) studied "the incompatible rhetorical expectations for women and for physicians" in the 19th century (p. 307), Lee E. Brasseur and Torri L. Thompson (1995) critiqued the "gendered ideologies" in medical manuals used during the Renaissance, and Carol Berkenkotter and Cristina Hanganu-Bresch (2011) conducted archival ***research*** of admissions records for a 19th-century asylum. In addition, TC scholars have attempted to trace how power circulates by investigating exigent M/HC documents within both forensic and deliberative situations. These include Susan Popham's (2014) examination of juvenile mental health records, Mary Lay Schuster et al.'s (2013) analysis of court case documents regarding end-of-life decisions, and Carolyn Schryer et al.'s (2012) discourse analysis of dignity interviews. TC researchers in M/HC have also examined medical record-keeping (Popham & Graham, 2008; Scott, 2014; Varpio et al., 2007) and whether said records accurately reflect concerns and contributions from patients and their caretakers (Breuch et al., 2016). Other TC scholars have chosen to study M/HC's writing practices and processes (see Heifferon, 2005; Opel & Hart-Davidson, 2019; Willerton, 2008).

But it's not always evident from textual products, practices, and processes how economies, geographies, race, gender, sex, and politics (to name only a few) intersect and influence who or what counts as "healthy." Intersectional power differentials are often legitimized, if not enabled, by medicalized institutions and technologies in less visible ways (Moore et al., 2018; Teston, 2016). Consider, for example, the computational code that structures genetic tests' results (Condit, 2018; Kirkscey, 2019; Sidler & Jones, 2008; Teston, 2018), or medical professionals' implicit biases (Hernández & Dean, 2020; Liz, 2020; Segal, 2005). These less visible sites of rhetorical power, while difficult to isolate and analyze from a purely textual vantage point, have serious consequences on M/HC. One way TC researchers have sought to better understand how extra-textual medicalized "discourses and practices" (Lupton, 2002, p. 95) affect individuals is to wed patient-centered care with human-centered ***design*** (Bellwoar, 2012; Gouge, 2017; Melonçon, 2017)—especially as it concerns informed consent (Bivens, 2017; Kim et al., 2008).

Capturing how power circulates beyond the text has led TC scholars to consider a wider range of M/HC artifacts, perhaps best described as information ecologies—e.g., oral, gestural, textual, ***visual***, and/or statistical forms of communication, the boundaries of which often bleed into one another and therefore require multiple methodological approaches. Many scholars in TC have sought to unspool how power operates in M/HC's information ecologies through site-based ***research*** methods, as exemplified by Fountain's (2014) rich analyses of the anatomy laboratory, Debra Burleson's (2014) interviews with hospitalists, S. Scott Graham and Carl Herndl's (2013) observational study of a pain management team, Elizabeth L. Angeli's (2015) robust *in situ* analyses of emergency medical services professionals' reliance on memory in their workplace writing, and Ellen Barton and Susan Eggly's (2009) observations of how physicians pitch to cancer patients the opportunity to participate in a clinical trial.

Integral to each of these projects is the generalizable finding that medicalized power matrices are often occluded by bureaucratic regimes that prevent individuals from accessing the means by which they might not just survive but thrive (Barton et al., 2018; Lynch, 2009; Scott, 2002). That is, such M/HC projects uncover how the medical ***profession*** cultivates and maintains a sense of (hegemonic) expertise through what Colleen Derkatch (2016) might call "boundary work" (see also Stone, 1997). Medicine's ethos is frequently "distributed and mediated" (Sánchez, 2020) via symbolic representations such as figures, graphs, medical images, and other forms of visual evidence (Graham, 2009; Longo et al., 2007; Welhausen, 2015; Wise, 2018). But ethos is also negotiated, if not challenged, behind the scenes, as evidenced by (anti)vaccination controversies (Campeau, 2019; Lawrence, 2020; Scott, 2016), or "do-it-yourself" argumentation tactics employed by holistic health coaches (Gigante, 2018).

Fueled by the desire to design more democratic if not equitable medical or health spaces, some TC researchers have waded into digital or online communities where M/HC circulates—i.e., spaces where ethos and expertise are negotiated in real time, (presumably) beyond the constraints of medicalized bureaucracies (Ding, 2009; Freeman & Spyridakis, 2009; Moeller, 2015; Segal, 2009; Spoel, 2008). For example, Lori Beth De Hertogh (2018) pairs TC frameworks with a feminist digital research methodology in a five-year case study of an online childbirth community. Given users' vulnerability to health and medical misinformation in online spaces such as these, Rebecca K. Britt and Kristen Nicole Hatten (2016) propose an "e-health communication competence scale." Similarly, Abigail Bakke (2019) examines the risks of misinformation in a Parkinson's disease online community, and Amy Roundtree (2017) studies "health-related Facebook usage of people not designated as patients" (p. 300). As new communication technologies emerge, it's likely that more TC researchers will pursue projects related to telemedicine (continuing the work of Mirel et al., 2008) and how so-called "smart" devices are marketed as a way to improve care coordination and communication (see Alaiad & Zhou, 2017), especially in developing countries.

Transdisciplinary variety in M/HC scholars' theoretical frameworks and methodological approaches will undoubtedly continue in response to changing sociopolitical and economic conditions—including the effects of environmental degradation on human health, global pandemics, health consumerism, and how to treat "invisible injuries," like those sustained during pervasive military imperialism around the world (Lindsley, 2015). Such evolutions may further blur disciplinary territory between, say, M/HC and consumer science, disability studies, political science, economics, environmental studies, and interdisciplinary approaches to human vulnerability.

Looking toward the future, it is important to recognize transnational medical and health precarities, which have been enabled by the rise of power among the Global Right. Those who teach, research, and practice M/HC in the US might expand their investigative repertoire to account for "non-native-English speakers" (Koerber & Graham, 2017; see also Bloom-Pojar, 2018; Ding, 2009, 2020; Gonzales et al., 2018; Walton & DeRenzi, 2009), or the ways immigrants and asylum seekers, for example, are disproportionately affected by medicalized patienthoods (see Cedillo, 2020; Rose et al., 2017). A word of caution, though: These M/HC projects ought to be pursued in a way that is neither exploitative nor extractive. Intellectual bridges should be built between TC and Indigenous methodologists, for example, who are careful to critique the ways academic research—especially as it concerns medicine and health—has been used exploitatively to deny basic human rights via biocitizenship (Happe et al., 2018; see also TallBear, 2013; Washington, 2006).

Through these and other ongoing disciplinary evolutions, it's possible to imagine that the communicative hegemony associated with "medicine" and "health" might more forcefully be reckoned with. Toward that end, it is important that those who study M/HC's practices represent a wider range of diverse identities and desires, as embodied in the work of Avery Edenfield, who has published extensively on ***social justice***, power, and the need to queer tactical technical communication (Edenfield, 2019; Edenfield, Colton & Holmes, 2019; Edenfield, Holmes & Colton, 2019), and Modupe Yusuf (2022), a rising star in M/HC, whose dissertation examines the circulation of mobile health information among women and children in Nigerian communities. Ideally, the outcome of such diversification will make TC scholars who study M/HC an important resource for clinicians who serve diverse ***publics***. TC scholars who study and practice M/HC ought to continue to work toward catalyzing public policy such that it does more than reify Aristotelean (and neoliberal) assumptions about the relationship between good health and good living.

References

Alaiad, A. & Zhou, L. (2017). Patients' adoption of WSN-based smart home healthcare systems: An integrated model of facilitators and barriers. *IEEE Transactions on Professional Communication*, *60*(1), 4–23. https://doi.org/10.1109/TPC.2016.2632822.

Angeli, E. L. (2015). Three types of memory in emergency medical services communication. *Written Communication*, *32*(1), 3–38. https://doi.org/10.1177/0741088314556598.

Bakke, A. (2019). Writing for patients on the participatory web: Heuristics for purpose-driven personas. *IEEE Transactions on Professional Communication*, *62*(4), 318–333. https://doi.org/10.1109/TPC.2019.2946999.

Barton, E. (2005). Introduction to the special issue: The discourses of medicine. *Journal of Business and Technical Communication*, *19*(3), 245–248. https://doi.org/10.1177/1050651905275636.

Barton, E. & Eggly, S. (2009). Ethical or unethical persuasion? The rhetoric of offers to participate in clinical trials. *Written Communication*, *26*(3), 295–319. https://doi.org/10.1177/0741088309336936.

Barton, E., Thominet, L., Boeder, R. & Primeau, S. (2018). Do community members have an effective voice in the ethical deliberation of a behavioral institutional review board? *Journal of Business and Technical Communication*, *32*(2), 154–197. https://doi.org/10.1177/1050651917746460.

Bazerman, C. (1988). *Shaping written knowledge: The genre and activity of the experimental article in science* (Vol. 356). University of Wisconsin Press.

Bellwoar, H. (2012). Everyday matters: Reception and use as productive design of health-related texts. *Technical Communication Quarterly*, *21*(4), 325–345. https://doi.org/10.1080/10572252.2012.702533.

Berkenkotter, C. & Hanganu-Bresch, C. (2011). Occult genres and the certification of madness in a 19th-century lunatic asylum. *Written Communication*, *28*(2), 220–250. https://doi.org/10.1177/0741088311401557.
Bivens, K. M. (2017). Rhetorically listening for microwithdrawals of consent in research practice. In Lisa Melonçon and J. Blake Scott (Eds.), *Methodologies for the rhetoric of health & medicine* (pp. 138–156). Routledge. https://doi.org/10.4324/9781315303758-8.
Bloom-Pojar, R. (2018). *Translanguaging outside the academy: Negotiating rhetoric and healthcare in the Spanish Caribbean*. National Council of Teachers of English.
Brasseur, L. E. & Thompson, T. L. (1995). Gendered ideologies: Cultural and social contexts for illustrated medical manuals in Renaissance England. *IEEE Transactions on Professional Communication*, *38*(4), 204–215. https://doi.org/10.1109/47.475592.
Breuch, L.-A. K., Bakke, A., Thomas-Pollei, K., Mackey, L. E. & Weinert, C. (2016). Toward audience involvement: Extending audiences of written physician notes in a hospital setting. *Written Communication*, *33*(4), 418–451. https://doi.org/10.1177/0741088316668517.
Britt, R. K. & Hatten, K. N. (2016). The development and validation of the eHealth Competency Scale: A measurement of self-efficacy, knowledge, usage, and motivation. *Technical Communication Quarterly*, *25*(2), 137–150. https://doi.org/10.1080/10572252.2016.1149621.
Burleson, D. (2014). Communication challenges in the hospital setting: A comparative case study of hospitalists' and patients' perceptions. *Journal of Business and Technical Communication*, *28*(2), 187–221. https://doi.org/10.1177/1050651913513901.
Campeau, K. L. (2019). Vaccine barriers, vaccine refusals: Situated vaccine decision-making in the wake of the 2017 Minnesota measles outbreak. *Rhetoric of Health & Medicine*, *2*(2), 176–207. https://doi.org/10.5744/rhm.2019.1007.
Cedillo, C. V. (2020). Disabled and undocumented: In/visability at the borders of presence, disclosure, and nation. *Rhetoric Society Quarterly*, *50*(3), 203–211. https://doi.org/10.1080/02773945.2020.1752131.
Condit, C. M. (1990). *Decoding abortion rhetoric: Communicating social change*. University of Illinois Press.
Condit, C. M. (2018). Rhetoricians on human remaking and the project of genomics. *Rhetoric of Health & Medicine*, *1*(1), 19–36. https://doi.org/10.5744/rhm.2018.1007.
Connor, J. J. (1993). Medical text and historical context: Research issues and methods in history and technical communication. *Journal of Technical Writing and Communication*, *23*(3), 211–232. https://doi.org/10.2190/0P4Q-07X0-R2EV-WRD2.
De Hertogh, L. B. (2018). Feminist digital research methodology for rhetoricians of health and medicine. *Journal of Business and Technical Communication*, *32*(4), 480–503. https://doi.org/10.1177/1050651918780188.
Derkatch, C. (2016). *Bounding biomedicine: Evidence and rhetoric in the new science of alternative medicine*. University of Chicago Press. https://doi.org/10.7208/chicago/9780226345987.001.0001.
Ding, H. (2009). Rhetorics of alternative media in an emerging epidemic: SARS, censorship, and extra-institutional risk communication. *Technical Communication Quarterly*, *18*(4), 327–350. https://doi.org/10.1080/10572250903149548.
Ding, H. (2020). Crowdsourcing, social media, and intercultural communication about Zika: Use contextualized research to bridge the digital divide in global health inter-

vention. *Journal of Technical Writing and Communication*, *50*(2), 141–166. https://doi.org/10.1177/0047281620906127.
Eberhard, J. M. (2012). An annotated bibliography of literature on the rhetoric of health and medicine. *Present Tense*, *2*(2), 1–49.
Edenfield, A. C. (2019). Queering consent: Design and sexual consent messaging. *Communication Design Quarterly Review*, *7*(2), 50–63. https://doi.org/10.1145/3358931.3358938.
Edenfield, A. C., Colton, J. S. & Holmes, S. (2019). Always already geopolitical: Trans health care and global tactical technical communication. *Journal of Technical Writing and Communication*, *49*(4), 433–457. https://doi.org/10.1177/0047281619871211.
Edenfield, A. C., Holmes, S. & Colton, J. S. (2019). Queering tactical technical communication: DIY HRT. *Technical Communication Quarterly*, *28*(3), 177–191. https://doi.org/10.1080/10572252.2019.1607906.
Fountain, T. K. (2014). *Rhetoric in the flesh: Trained vision, technical expertise, and the gross anatomy lab*. Routledge.
Freeman, K. S. & Spyridakis, J. H. (2009). Effect of contact information on the credibility of online health information. *IEEE Transactions on Professional Communication*, *52*(2), 152–166. https://doi.org/10.1109/TPC.2009.2017992.
Gigante, M. E. (2018). Argumentation by self-model: Missing methods and opportunities in the personal narratives of popular health coaches. *Journal of Technical Writing and Communication*, *48*(3), 259–280. https://doi.org/10.1177/0047281617696984.
Gonzales, L., Bloom-Pojar, R., Perez, G., Leger, A., Sanchez, C., Rafaiel, R., Anyijong, J. Y., Raab, M., Mulac, B. & Brown, J. (2018). A dialogue with medical interpreters about rhetoric, culture, and language. *Rhetoric of Health & Medicine*, *1*(1), 193–212. https://doi.org/10.5744/rhm.2018.1002.
Gouge, C. C. (2017). Improving patient discharge communication. *Journal of Technical Writing and Communication*, *47*(4), 419–439. https://doi.org/10.1177/0047281616646749.
Graham, S. S. (2009). Agency and the rhetoric of medicine: Biomedical brain scans and the ontology of fibromyalgia. *Technical Communication Quarterly*, *18*(4), 376–404. https://doi.org/10.1080/10572250903149555.
Graham, S. S. & Herndl, C. (2013). Multiple ontologies in pain management: Toward a postplural rhetoric of science. *Technical Communication Quarterly*, *22*(2), 103–125. https://doi.org/10.1080/10572252.2013.733674.
Haas, C. (2009). Writing and medicine [Special issue]. *Written Communication*, *26*(3).
Hagge, J. (1995). Early engineering writing textbooks and the anthropological complexity of disciplinary discourse. *Written Communication*, *12*(4), 439–491. https://doi.org/10.1177/0741088395012004003.
Happe, K. E., Johnson, J. & Levina, M. (Eds.). (2018). *Biocitizenship: The politics of bodies, governance, and power* (Vol. 19). NYU Press.
Harris, B. L. (1991). Corpus Hippocraticum: One of Western civilization's earliest technical documents. *Technical Communication*, *38*(4), 598–599.
Heifferon, B. (2005). *Writing in the health professions*. Longman Publishing Group.
Heifferon, B. & Brown, S. (Ed.). (2000). Medical rhetoric [Special issue]. *Technical Communication Quarterly*, *9*(3).
Hernández, L. H. & Dean, M. (2020). "I felt very discounted": Negotiation of Caucasian and Hispanic/Latina women's bodily ownership and expertise in patient-provider interactions. In. E. Frost & M. Eble (Eds.), *Interrogating gendered pathologies* (pp. 101–120). Utah State University Press.

Jaeger, W. (1957). Aristotle's use of medicine as model of method in his ethics. *The Journal of Hellenic Studies*, *77*(1), 54–61. https://doi.org/10.2307/628634.

Kim, L., Young, A. J., Neimeyer, R. A., Baker, J. N. & Barfield, R. C. (2008). Keeping users at the center: Developing a multimedia interface for informed consent. *Technical Communication Quarterly*, *17*(3), 335–357. https://doi.org/10.1080/10572250802100451.

Kirkscey, R. (2019). Shifts and transpositions: An analysis of gateway documents for cancer genetic testing. *Rhetoric of Health & Medicine*, *2*(4), 384–414. https://doi.org/10.5744/rhm.2019.1018.

Koerber, A. & Graham, H. (2017). Theorizing the value of English proficiency in cross-cultural rhetorics of health and medicine: A qualitative study. *Journal of Business and Technical Communication*, *31*(1), 63–93. https://doi.org/10.1177/1050651916667533.

Koerber, A. & Still, B. (2008). Guest editors' introduction: Online health communication. *Technical Communication Quarterly*, *17*(3), 259–263. https://doi.org/10.1080/10572250802100329.

Lawrence, H. Y. (2020). *Vaccine rhetorics*. The Ohio State University Press.

Lindsley, T. (2015). Legitimizing the wound: Mapping the military's diagnostic discourse of traumatic brain injury. *Technical Communication Quarterly*, *24*(3), 235–257. https://doi.org/10.1080/10572252.2015.1044120.

Liz, J. (2020). Pathologizing Black female bodies: The construction of difference in contemporary breast cancer research. In. E. Frost & M. Eble (Eds.), *Interrogating gendered pathologies* (pp. 223–238). Utah State University Press.

Longo, B., Weinert, C. & Fountain, T. K. (2007). Implementation of medical research findings through insulin protocols: Initial findings from an ongoing study of document design and visual display. *Journal of Technical Writing and Communication*, *37*(4), 435–452. https://doi.org/10.2190/V986-K02V-519T-721J.

Lupton, D. (2002). Foucault and the medicalisation critique. In A. Petersen & R. Bunton (Eds.), *Foucault, health, and medicine* (pp. 94–107). Routledge.

Lynch, J. A. (2009). Articulating scientific practice: Understanding Dean Hamer's "gay gene" study as overlapping material, social and rhetorical registers. *Quarterly Journal of Speech*, *95*(4), 435–456. https://doi.org/10.1080/00335630903296168.

Lynch, J. A. & Zoller, H. (2015). Recognizing differences and commonalities: The rhetoric of health and medicine and critical-interpretive health communication. *Communication Quarterly*, *63*(5), 498–503. https://doi.org/10.1080/01463373.2015.1103592.

Melonçon, L. K. (2017). Patient experience design: Expanding usability methodologies for healthcare. *Communication Design Quarterly Review*, *5*(2), 19–28. https://doi.org/10.1145/3131201.3131203.

Melonçon, L. & Frost, E. A. (2015). Special issue introduction: Charting an emerging field: The rhetorics of health and medicine and its importance in communication design. *Communication Design Quarterly Review*, *3*(4), 7–14. https://doi.org/10.1145/2826972.2826973.

Mirel, B., Barton, E. & Ackerman, M. (2008). Researching telemedicine: Capturing complex clinical interactions with a simple interface design. *Technical Communication Quarterly*, *17*(3), 358–378. https://doi.org/10.1080/10572250802100477.

Moeller, M. (2015). Pushing boundaries of normalcy: Employing critical disability studies in analyzing medical advocacy websites. *Communication Design Quarterly Review*, *2*(4), 52–80. https://doi.org/10.1145/2721874.2721877.

Moore, K., Jones, N., Cundiff, B. & Heilig, L. (2018). Contested sites of health risks: Using wearable technologies to intervene in racial oppression. *Communication Design Quarterly*, *5*(4), 52–60. https://doi.org/10.1145/3188387.3188392.

Opel, D. S. & Hart-Davidson, W. (2019). The primary care clinic as writing space. *Written Communication*, *36*(3), 348–378. https://doi.org/10.1177/0741088319839968.

Paradis, J. (2019). Bacon, Linnaeus, and Lavoisier: Early language reform in the sciences. In P. Anderson, J. Brockman, and C. Miller (Eds.). *New essays in technical and scientific communication* (pp. 200–224). Routledge. https://doi.org/10.4324/9781315224060-16.

Popham, S. L. (2014). Hybrid disciplinarity: Métis and ethos in juvenile mental health electronic records. *Journal of Technical Writing and Communication*, *44*(3), 329–344. https://doi.org/10.2190/TW.44.3.f.

Popham, S. & Graham, S. L. (2008). A structural analysis of coherence in electronic charts in juvenile mental health. *Technical Communication Quarterly*, *17*(2), 149–172. https://doi.org/10.1080/10572250801904622.

Rose, E. J., Racadio, R., Wong, K., Nguyen, S., Kim, J. & Zahler, A. (2017). Community-based user experience: Evaluating the usability of health insurance information with immigrant patients. *IEEE Transactions on Professional Communication*, *60*(2), 214–231. https://doi.org/10.1109/TPC.2017.2656698.

Roundtree, A. K. (2017). Social health content and activity on Facebook: A survey study. *Journal of Technical Writing and Communication*, *47*(3), 300–329. https://doi.org/10.1177/0047281616641925.

Sánchez, F. (2020). Distributed and mediated ethos in a mental health call center. *Rhetoric of Health & Medicine*, *3*(2), 133–162. https://doi.org/10.5744/rhm.2020.1009.

Schryer, C., McDougall, A., Tait, G. R. & Lingard, L. (2012). Creating discursive order at the end of life: The role of genres in palliative care settings. *Written Communication*, *29*(2), 111–141. https://doi.org/10.1177/0741088312439877.

Schuster, M. L., Russell, A. L. B., Bartels, D. M. & Kelly-Trombley, H. (2013). "Standing in Terri Schiavo's shoes": The role of genre in end-of-life decision making. *Technical Communication Quarterly*, *22*(3), 195–218. https://doi.org/10.1080/10572252.2013.760061.

Scott, J. Blake. (2002). The public policy debate over newborn HIV testing: A case study of the knowledge enthymeme. *Rhetoric Society Quarterly*, *32*(2), 57–83. https://doi.org/10.1080/02773940209391228.

Scott, J. Blake. (2014). Afterword: Elaborating health and medicine's publics. *Journal of Medical Humanities*, *35*(2), 229–235. https://doi.org/10.1007/s10912-014-9279-3.

Scott, J. Bracken. (2016). Boundary work and the construction of scientific authority in the vaccines-autism controversy. *Journal of Technical Writing and Communication*, *46*(1), 59–82. https://doi.org/10.1177/0047281615600638.

Segal, J. (2005). *Health and the rhetoric of medicine*. SIU Press.

Segal, J. Z. (2009). Internet health and the 21st-century patient: A rhetorical view. *Written Communication*, *26*(4), 351–369. https://doi.org/10.1177/0741088309342362.

Sidler, M. & Jones, N. (2008). Genetics interfaces: Representing science and enacting public discourse in online spaces. *Technical Communication Quarterly*, *18*(1), 28–48. https://doi.org/10.1080/10572250802437317.

Skinner, C. (2012). Incompatible rhetorical expectations: Julia W. Carpenter's medical society papers, 1895–1899. *Technical Communication Quarterly*, *21*(4), 307–324. https://doi.org/10.1080/10572252.2012.686847.

Solomon, M. (1985). The rhetoric of dehumanization: An analysis of medical reports of the Tuskegee syphilis project. *Western Journal of Speech Communication*, *49*(4), 233–247. https://doi.org/10.1080/10570318509374200.

Spafford, M. M., Schryer, C. F., Mian, M. & Lingard, L. (2006). Look who's talking: Teaching and learning using the genre of medical case presentations. *Journal of Business and Technical Communication*, *20*(2), 121–158. https://doi.org/10.1177/1050651905284396.

Spoel, P. (2008). Communicating values, valuing community through health-care websites: Midwifery's online ethos and public communication in Ontario. *Technical Communication Quarterly*, *17*(3), 264–288. https://doi.org/10.1080/10572250802100360.

Stone, M. S. (1997). In search of patient agency in the rhetoric of diabetes care. *Technical Communication Quarterly*, *6*(2), 201–217. https://doi.org/10.1207/s15427625tcq0602_5.

TallBear, K. (2013). *Native American DNA: Tribal belonging and the false promise of genetic science*. University of Minnesota Press. https://doi.org/10.5749/minnesota/9780816665853.001.0001.

Teston, C. (2016). Rhetoric, precarity, and mHealth technologies. *Rhetoric Society Quarterly*, *46*(3), 251–268. https://doi.org/10.1080/02773945.2016.1171694.

Teston, C. (2018). Pathologizing precarity. In W. S. Hesford, A. C. Licona & C. Teston (Eds.), *Precarious rhetorics* (pp. 276–297). The Ohio State University Press.

Varpio, L., Spafford, M. M., Schryer, C. F. & Lingard, L. (2007). Seeing and listening: A visual and social analysis of optometric record-keeping practices. *Journal of Business and Technical Communication*, *21*(4), 343–375. https://doi.org/10.1177/1050651907303991.

Walkup, K. L. & Cannon, P. (2018). Health ecologies in addiction treatment: Rhetoric of health and medicine and conceptualizing care. *Technical Communication Quarterly*, *27*(1), 108–120. https://doi.org/10.1080/10572252.2018.1401352.

Walton, R. & DeRenzi, B. (2009). Value-sensitive design and health care in Africa. *IEEE Transactions on Professional Communication*, *52*(4), 346–358. https://doi.org/10.1109/TPC.2009.2034075.

Washington, H. A. (2006). *Medical apartheid: The dark history of medical experimentation on Black Americans from colonial times to the present.* Doubleday Books.

Welhausen, C. A. (2015). Power and authority in disease maps: Visualizing medical cartography through yellow fever mapping. *Journal of Business and Technical Communication*, *29*(3), 257–283. https://doi.org/10.1177/1050651915573942.

Willerton, R. (2008). Writing toward readers' better health: A case study examining the development of online health information. *Technical Communication Quarterly*, *17*(3), 311–334. https://doi.org/10.1080/10572250802100428.

Wise, B. (2018). Fetal positions: Fetal visualization, public art, and abortion politics. *Rhetoric of Health & Medicine*, *1*(3), 296–322. https://doi.org/10.5744/rhm.2018.1015.

Yusuf, M. O. (2022). *Discourse, materiality, and the users of mobile health technologies: A Nigerian case study*. Doctoral dissertation, Michigan Technological University.

Zappen, J. P. (2004). Scientific rhetoric in the nineteenth and early twentieth centuries: Herbert Spencer, Thomas H. Huxley, and John Dewey. In C. Bazerman and J. Paradis (Eds.), *Textual dynamics of the professions: Historical and contemporary studies of writing in professional communities* (pp. 145–167). The WAC Clearinghouse. https://wac.colostate.edu/books/landmarks/textual-dynamics/ (Originally published 1991 by University of Wisconsin Press).

21. Multimodality

Dirk Remley
Kent State University

On its surface, *multimodality* has a relatively basic meaning within technical communication. To paraphrase the definition of the New London Group (1996), there are multiple modes by which to represent a message (print-linguistic, ***visual***, audio, gestural, spatial), and any two or more of these modes can be combined to form a multimodal representation (p. 60). Thus, on its most basic level, the term *multimodal* means any combination of modes of representation to create a single artifact, and multimodal artifacts are the norm rather than the exception in technical communication practice. An example of such an artifact would be a training or instructional video that combines visual (images of people, graphics, and/or objects), audio (a voice-over or someone speaking), and gestural and spatial (a person demonstrating how to perform a given task) modes of representation.

Multimodality's historical development relative to contextual uses within the field, even as technical communication developed as a recognized discipline, complicates its treatment as a keyword. The term has been used relatively commonly since the mid-1990s; however, the concepts associated with it go back to early studies in ***literacy*** and even earlier scholarship in ***rhetoric*** and semiotics. Further, the various ways it is studied evolve as new technologies emerge. Its connections to rhetoric, literacy, and media technologies shape and complicate its treatment as a keyword in technical communication scholarship, ***pedagogy*** and practice.

Though not termed as such, multimodality is treated within the classical rhetorical scholarship of Aristotle (1991) and Quintilian (1922) as part of delivery. They recognized the importance of using gesture in conjunction with oration in persuading an ***audience***. That is, gestures and facial expression are visual/nonverbal actions that carry meaning and can enhance or complement oral communication. Subsequent generations of rhetorical theorists continued to study delivery-related implications for messages as communication technologies from the printing press to hypertext offered new ways to present multiple modes (see McCorkle, 2012). Rhetoric, as an academic discipline, has longstanding links to the field of technical communication; thus, it is not surprising that technical communication theorists had also taken up issues of multimodality long before the coining of the term.

Initially, technical communication scholars identified the use of multiple modes of representation in technical documents without using the term *multimodal*. Many of these studies focused on workplace literacy practices—how professionals communicated with each other in workplace settings and elements of page ***design*** (see, for example, Doheny-Farina, 1988; Odell & Goswami, 1985).

DOI: https://doi.org/10.37514/TPC-B.2023.1923.2.21

Some reviewed how graphics were used in those documents. For example, in a review of technical communication practices, Mary Beth Debs (1988) noted that writers tend to supplement text with graphics, acknowledging that "pictures serve additive function" (p. 19). Manuals are a common artifact discussed within technical communication; these often combine print-linguistic text (words) and visual elements. For example, Davida Charney et al. (1988) illustrate how the "minimalist manual" should include illustrations of examples of tasks (pp. 70–72), and John Carroll et al. (1988) show a visually appealing page design as an attribute of the "minimalist manual" (pp. 82, 85). Stephen Bernhardt (1986) describes the visual rhetoric of headings, print quality, and white space within a print-text fact sheet pertaining to wetlands and designed for multiple audiences—legislators, teachers, students, and the general ***public*** (pp. 71–72).

Transitioning from purely document-related consideration of multiple modes of representation and related analyses, scholars began looking more closely at multimodal forms of communication connected to technical communication practices beyond print documents in the late 1980s and into the 1990s. Muriel Zimmerman and Hugh Marsh (1989), for example, studied how storyboarding facilitated ***proposal*** development within a particular company. Further, Carroll et al. (1988) considered how hands-on instruction may affect learning within workplace settings.

As mentioned previously, the New London Group (1996) was among the first set of scholars to formally recognize and define the term *multimodality*. As the number of digital composing technologies increased and became more widely used, scholars encouraged recognition of multiple forms of literacy within a growing set of tools to use for creating messages and encouraged pedagogy that included literacy with those tools and forms of representation.

The study of multimodal rhetoric evolved in the early 2000s, as scholars shifted their focus from examining how technical communicators presented ***information*** in multimodal ways to understanding how various modal combinations affected audiences' ability to understand a message relative to technical communication purposes. Linguists Gunther Kress and Theo Van Leeuwen (2001, 2006) attempted to develop a theory of semiotics that integrated terminology that could describe the various rhetorical dynamics at work in multimodal forms of communication, facilitating rhetorical analyses of multimodal artifacts. This "grammar" of visual design (Kress & Van Leeuwen, 2006) has been used by many within scholarship of multimodality; a cursory review of Google Scholar in May 2020 finds that this work has been cited over 14,800 times, indicating its value as a theory of analysis for multimodal messages.

Other lines of ***research*** have focused on the benefits of multimodal communication for teaching and learning. Roxana Moreno and Richard Mayer (2000) found that certain combinations of visuals and text information affect cognition, particularly related to learning, suggesting a relationship between modes used to communicate and their rhetorical impact. In an instructional context, combining visual and audio modes of representation is more powerful for accomplishing the instructional purpose than using only audio narration or visuals alone. Mayer (2001) summarized their multimodal principle with the statement that people learn better when pictures and words are integrated into an instructional message than when only words are used (p. 63). If a picture is provided, people can make a visual connection more readily. Mayer also asserts that it is vital to eliminate extraneous material—words, images, and sounds—from any multimedia message. Such irrelevant information "competes for cognitive resources in working memory," disrupting the learner's ability to organize and retain relevant information (p. 113).

Technical communication scholars have, also, examined the relationship between multimodal artifacts and cognition. Jonathan Buehl (2016) calls attention to theories from multiple cognitive scientists that link to multimodal theory as applied to scientific texts. Wolfgang Schnotz (2005) reviewed several studies pertaining to the influence that working memory has on learning with multimedia, and she develops a model of text/picture comprehension that considers working memory. Visual images that integrate text are easier to process because fewer processes of working memory are involved. According to Alan Baddeley's (1986) model of working memory, there is a phonological (auditory) channel and a "visuo-spatial" (visual) channel associated with short term memory. By facilitating use of both channels, people can better process information than they can when too much of one system is used. This helps technical communicators design products that balance elements affecting cognition, improving an audience's ability to understand the message.

James Paul Gee (2003) connected literacy theory to multimodal practice, identifying a marriage between the semiotic domain and situated practice (p. 26). Gee argued that, as part of audience consideration, it is important to understand modes in which trainees have learned previously. So, some studies have considered relationships between modal combinations relative to multiliteracies and technical communication rhetoric relative to development of instructional materials. Matt Morain and Jason Swarts (2012), for example, allude to using students' "digital literacy" to develop an understanding of how to assess and create YouTube videos for instructional purposes (p. 6).

As these studies occurred, advancing multimodal theory relative to technical communication, teachers began integrating multimodal concepts and approaches into their classroom practices. A body of work emerged from studying such instruction (e.g., A. Bourelle et al., 2015; T. Bourelle et al., 2017; Katz & Odell, 2012). These studies range from helping students understand the possible uses of different media to compose technical communication products to how one may apply criteria—old and new—in assessing multimodal products developed by students. In their introduction to a special issue of *Technical Communication Quarterly*, for example, Susan Katz and Lee Odell (2012) acknowledge that "Confronted with the full range of affordances of digital media, we need to achieve a level of clarity that will help students wisely use these affordances" (p. 2). Andrew Bourelle et al. (2015) describe ways teachers can help students apply the rhetorical canons of invention, arrangement, ***style***, delivery, and memory to composing in new digital media. Cheryl Ball (2012), Christa Teston et al. (2019), and Pamela Takayoshi and Cynthia Selfe (2009) describe factors to consider in assessing multimodal work, including students' self-reflection of why they chose to use certain media with which to compose a message and the media's abilities and limitations. This reflection can help one understand how to select composing media for future projects relative to information that should be included and how to best represent that information given access to multiple modes of representation.

As reflected in the historical development of its treatment, scholars shifted between labelling the use of multiple modal combinations as *multimodal* and *multimedia*. As indicated above, *multimodality* increased in use as a term with the rapid development of various technologies that facilitated integrating multiple forms of representation in them. Moreno and Mayer (2000) demonstrate this synonymous use while describing studies of participants reacting to messages that included text and images. While the majority of their analyses revolve around performance of subjects relative to modal combinations, they use the term *multimedia* throughout their work. In concluding their article, they write,

> To foster the process of integrating, multimedia presentations should present words and pictures using modalities that effectively use available visual and auditory working memory resources. The major advance in our research program is to identify techniques for presentation of verbal and visual information that minimizes working memory load and promotes meaningful learning. (n.p.)

Scholarly publications in technical communication theory and pedagogy illustrate the favoring of the term "multimodality" in academic settings (e.g., Armfield et al., 2011; A. Bourelle et al., 2015; T. Bourelle et al., 2017). Stephen Frailberg (2012) and Dirk Remley (2015, 2017) illustrate favoring "multimodality" in case studies of practices, using the term "multimodal" instead of "multimedia" throughout their works, even including the term in the title of their works.

S. Scott Graham and Brandon Whalen (2008) illustrate the conflation of the two terms relative to a case study of a web designer's practices. They state, "The possibility of plurality in descriptions of digital communication media and ***genres*** has helped to generate a broad host of heteroglossic and hybrid theories, as well as an assortment of multi-prefixed neologisms (multimedia and multimodality being the most prominent)" (pp. 66–67). Claire Lauer (2009) found that "multimedia" is used by some in academia and tends to be the preferred term in industrial contexts to describe the same artifact (p. 231). Consequently, she states that instructors and scholars need to use multimedia "as a gateway term" when interacting with practitioners (p. 225). It is interesting to note, relative to the

Graham and Whalen (2008) article, that Graham is a technical communication scholar, while Whalen is a practitioner.

Several scholars, including Rich Rice and Carol Clark Papper (2005), Lauer (2009), and Andy Lucking and Thies Pfeiffer (2012), differentiate the two terms, though. These scholars state that multimodality describes the sign systems used to make meaning, while multimedia pertains to the tools by which such artifacts are distributed. Lucking and Pfeiffer (2012) state that "multimodality in a message is perceived as integrating more than one sensory interface and is perceived as multimedia if the message is conveyed using more than one medium" (p. 593). For example, software that facilitates creation of video that includes audio is multimedia. The video product itself is considered a multimodal artifact by most involved with technical communication in academia. Some have assessed the effectiveness of various media available to present technical information multimodally (e.g., Tufte, 2003); however, these studies focus on the media's technical and design capabilities and limitations.

Evolving from the study of the effects certain modal combinations may have on cognition, more recently, scholarship has begun considering neuroscientific or biological analyses associated with multimodal artifacts and related effectiveness relative to rhetoric. For example, Dirk Remley (2015, 2017) examines how multimodal artifacts used in technical communication settings affect neural dynamics to influence meaning and response. Examples included in his analyses range from website design and public service announcements to nurse and pilot training. Such consideration helps to show the biology of cognition with multimodal products, or why certain multimodal combinations are effective for certain audiences, which can help technical communicators design better materials.

Additionally, with the proliferation of video-gaming as an industry and its related value in developing remote control tools and practical skills, technical communication scholars have been studying its multimodal designs and uses for classroom activities and uses in industry (see, for example, Cata, 2017; Cooke et al., 2020; McDaniel & Dear, 2016; Robinson, 2016; and Vie, 2008).

As a concept of communication, multimodality complemented traditional notions of writing and composing. As noted above, the integration of graphics into technical documents was generally regarded as valuable practice; so, initially, multimodality fit well into technical communication analyses and pedagogy. Today, it has grown into a valued concept in technical communication. To a certain extent, it competes with the term "multimedia" synonymously when used in industry by technical communication practitioners.

References

Anderson, P. (1984). What technical and scientific communicators do: A comprehensive model for developing academic programs. *IEEE Transactions on Education*, *27*, 160–166. https://doi.org/10.1109/TE.1984.4321691.

Aristotle. (1991). *The art of rhetoric* (H. C. Lawson-Tancred, Trans.). Penguin.

Armfield, D. M., Gurak, L. J., Kays, T. M. & Weinberg, J. (2011). *Technical communication education in a digital, visual world.* IEEE. https://doi.org/10.1109/IPCC.2012.6408637.

Baddeley, A. D. (1986). *Working memory.* Oxford University Press.

Ball, C. E. (2012). Assessing scholarly multimedia: A rhetorical genre studies approach. *Technical Communication Quarterly, 21*, 61–77. https://doi.org/10.1080/10572252.2012.626390.

Bernhardt, S. A. (1986). Seeing the text. *College Composition and Communication, 37*, 66–78. https://doi.org/10.2307/357383.

Bourelle, A., Bourelle, T. & Jones, N. (2015). Multimodality in the technical communication classroom: Viewing classical rhetoric through a 21st century lens. *Technical Communication Quarterly, 24*, 306–327. https://doi.org/10.1080/10572252.2015.1078847.

Bourelle, T., Bourelle, A., Spong, S. & Hendrickson, B. (2017). Assessing multimodal literacy in the online technical communication classroom. *Journal of Business and Technical Communication, 31*, 222–255. https://doi.org/10.1177/1050651916682288.

Buehl, J. (2016). Assembling arguments: Multimodal rhetoric & scientific discourse. University of South Carolina Press. https://doi.org/10.2307/j.ctv6wgfc3.

Carroll, J. M., Kerker, P. L. S., Ford, J. R. & Mazur, S. (1988). The minimal manual. In E. Doheny-Farina (Ed.), *Effective documentation: What we have learned from research* (pp. 73–102). MIT Press.

Cata, A. (2017). *Playing with usability: Why technical communicators should examine mobile games* (Publication No. 5395) [Doctoral dissertation, University of Central Florida]. Electronic Theses and Dissertations. https://stars.library.ucf.edu/etd/5395.

Charney, D. H., Reder, L.M. & Wells, G.W. (1988). Studies of elaboration in instructional texts. In E. Doheny-Farina (Ed.), *Effective documentation: What we have learned from research* (pp. 47–72). MIT Press.

Cooke, L., Dusenberry, L. & Robinson, J. (2020). Gaming design thinking: Wicked problems, sufficient solutions, and the possibility space of games. *Technical Communication Quarterly, 29*(4), 1–14. https://doi.org/10.1080/10572252.2020.1738555.

Debs, M. B. (1988). A history of advice: What experts have to tell us. In E. Doheny-Farina (Ed.), *Effective documentation: What we have learned from research* (pp. 11–24). MIT Press.

Doheny-Farina, E. (Ed.). (1988). *Effective documentation: What we have learned from research.* MIT Press.

Frailberg, S. (2012). Reassembling technical communication: A framework for studying multilingual and multimodal practices in global contexts. *Technical Communication Quarterly, 22*, 10–27. https://doi.org/10.1080/10572252.2013.735635.

Gee, J. P. (2003). *What video games have to teach us about learning and literacy.* Palgrave McMillan.

Graham, S. S. & Whalen, B. (2008). Mode, medium and genre: A case study of decisions in new-media design. *Journal of Business and Technical Communication, 22*, 65–91. https://doi.org/10.1177/1050651907307709.

Katz, S. M. & Odell, L. (2012). Making the implicit explicit in assessing multimodal composition: Continuing the conversation. *Technical Communication Quarterly, 21*, 1–5. https://doi.org/10.1080/10572252.2012.626700.

Kress, G. (2003). *Literacy in the new media age.* Routledge. https://doi.org/10.4324/9780203299234.

Kress, G. & Van Leeuwen, T. (2001). *Multimodal discourse: The modes and media of contemporary communication*. Arnold.

Kress, G. & Van Leeuwen, T. (2006). *Reading images: The grammar of visual design*. Routledge. (Original work published 1996)

Lauer, C. (2009). Contending with terms: "Multimodal" and "multimedia" in the academic and public spheres. *Computers and Composition*, *26*, 225–239. https://doi.org/10.1016/j.compcom.2009.09.001.

Lucking, A. & Pfeiffer, T. (2012). Framing multimodal technical communication. In A. Mehler & L. Romary (Eds.), *Handbook of technical communication* (pp. 591–644). Walter de Gruyter. https://doi.org/10.1515/9783110224948.591.

Mayer, R. E. (2001). *Multimedia learning*. Cambridge University Press.

McCorkle, B. (2012). *Rhetorical delivery as technological discourse: A cross-historical study*. Southern Illinois University Press.

McDaniel, R. & Dear, A. (2016). Developer discourse: Exploring technical communication practices within video game development. *Technical Communication Quarterly*, *25*(3), 155–166. https://doi.org/10.1080/10572252.2016.1180430.

Morain, M. & Swarts, J. (2012). Youtuorial: A framework for assessing online instructional video. *Technical Communication Quarterly*, *21*, 6–24. https://doi.org/10.1080/10572252.2012.626690.

Moreno, R. & Mayer, R. E. (2000). A learner-centered approach to multimedia explanations: Deriving instructional design principles from cognitive theory. *Interactive Multimedia Electronic Journal of Computer-Enhanced Learning*, *2*. http://imej.wfu.edu/articles/2000/2/05/index.asp.

New London Group. (1996). A pedagogy of multiliteracies: Designing social futures. *Harvard Educational Review*, *66*, 60–92. https://doi.org/10.17763/haer.66.1.17370n67v22j160u.

Odell, L. & Goswami, D. (Eds.). (1985). *Writing in nonacademic settings*. Guilford.

Quintilian. (1922). *Institutio oratoria* (H. E. Butler, Trans.). Loeb Classical Library.

Remley, D. (2015). *How the brain processes multimodal technical instructions*. Baywood. https://doi.org/10.4324/9781315231556.

Remley, D. (2017). *The neuroscience of multimodal persuasive messages: Persuading the brain*. Routledge. https://doi.org/10.4324/9781315206325.

Robinson, J. (2016). Look before you lead: Seeing virtual teams through the lens of games. *Technical Communication Quarterly*, *25*(3), 178–190. https://doi.org/10.1080/10572252.2016.1185159.

Rice, R. & Papper, C. C. (2005). Moving beyond text-only pedagogy: Oral, print and electronic media in technical communication assignments. In C. Lipson & M. Day (Eds.), *Technical communication and the World Wide Web* (pp. 295–304). Lawrence Erlbaum.

Schnotz, W. (2005). An integrated model of text and picture comprehension. In R. E. Mayer (Ed.), *The Cambridge handbook of multimedia learning* (pp. 49–60). Cambridge University Press. https://doi.org/10.1017/CBO9780511816819.005.

Takayoshi, P. & Selfe, C. (2007). Thinking about multimodality. In C. Selfe (Ed.), *Multimodal composition: Resources for teachers*. (pp. 1–12). Hampton Press.

Teston, C., Previte, B. & Hashlamon, Y. (2019). The grind of multimodal work in professional writing pedagogies. *Computers and Composition*, *52*, 195–209. https://doi.org/10.1016/j.compcom.2019.01.007.

Tufte, E. (2003). *The cognitive style of PowerPoint*. Graphics Press.

Vie, S. (2008). Tech writing, meet *Tomb Raider*: Video and computer games in the technical communication classroom. *E-Learning*, 5(2), 157–166. https://doi.org/10.2304/elea.2008.5.2.157.

Zimmerman, M. & Marsh, H. (1989). Storyboarding and industrial proposal: A case study of teaching and producing writing. In C. B. Matalene (Ed.), *Worlds of writing: Teaching and learning discourse communities of the work* (pp. 203–221). Random House.

22. Pedagogy

Tracy Bridgeford
University of Nebraska at Omaha

Merriam-Webster's online dictionary definition of *pedagogy* provides perhaps the best context for understanding this term in technical communication. That is, it is the "art, science, or profession of teaching." Classrooms, more than workspaces, connect scholars and practitioners of technical communication in ways that led James Dubinsky (2004) to describe the field as a "pedagogical discipline" (p. 3). Indeed, regardless of milieu, both academic and industrial professionals debate the purpose and content of technical communication curricula more than they do other contexts of action, perhaps because classrooms so readily blend the scholarly with the pragmatic. The tensions among these stakeholders have often defined and sometimes divided the community and its discourse, resulting in a dichotomous pedagogical corpus and lexicon. The field's exchanges on such topics might be broadly categorized as focusing either on practice and production or on conceptual frameworks and their implications. Although this characterization is not precisely chronological in its manifestation, it is true that much of the work prior to 1980 was more production-oriented than theoretical, and work after 1980 is increasingly complex in its scope, depth, and conceptual rigor.

Prior to the widespread adoption of desktop-publishing technologies in the early 1980s, technical writers (and thus technical communication classrooms) emphasized the construction of coherent documents that represent commonplace industrial ***genres*** (e.g., reports, instructions, and manuals) primarily through the crafting of stylistically clear, concise texts that privileged expert ***knowledge*** over reader needs. Dwight W. Stevenson's (1981) *Courses, Components, and Exercises in Technical Communication* captures the industrial practicality of this moment. The evolution of such scholarship resulted in the publication of collections such as Paul Anderson, R. John Brockman, and Carolyn Miller's (1983) *New Essays in Technical and Scientific Communication*, Lynn Beene and Peter White's (1988) *Solving Problems in Technical Writing*, Bertie E. Fearing and Keats Sparrow's (1989) *Technical Writing: Theory and Practice*, Carol M. Barnum and Saul Carliner's (1991) *Techniques for Technical Communicators*, and Thomas T. Barker's (1991) *Perspectives on Software Documentation*.

Focus on document production, including page layout and the ***visual*** elements of ***design***, increased throughout the 1980s. During the past 40 years, significant attention has been devoted to the intersection of technical communication pedagogy and ***information*** production technologies and strategies. Teachers of technical communication were challenged to transform classroom practices to include page design and image preparation in ways that established relationships

DOI: https://doi.org/10.37514/TPC-B.2023.1923.2.22

between text and visual aspects of ***documentation*** (Bernhardt, 1986; Kostelnick & Roberts, 1999; Kramer & Bernhardt, 1996; Moore & Fitz, 1993; Schriver, 1997). Teaching visual content discussions followed in the 2000s, with scholars describing image-oriented pedagogies that included visual thinking and information design (Brumberger, 2005, 2007) and the ***rhetorical-ethical*** issues that accompany such a focus (Barton & Barton, 1993; Dragga & Voss, 2001). This relationship led to visual design textbooks (Kostelnick & Roberts, 2011), pedagogical collections (Brumberger & Northcut, 2013), and instruction on formatting texts and creating information graphics (Dragga, 2001; Kitalong, 2018).

It was not until web browsers adopted visual layouts for hypertext in 1992 that hypertext design, markup languages, and specialized design software crept into standard pedagogical practice. Perhaps because this shift did not gather momentum until the mid- to late-1990s, much of the discussion of design, technologies, and strategies blends the act of creation with complementary issues and challenges. Collections by Patricia Sullivan and Jannie Dautermann (1996), Stuart A. Selber (1997), Carol Lipson and Michael Day (2005), and Rachel Spilka (2010) span a range of topical intersections, including visual ***literacy*** and information design, programmatic implementation of technologies, interaction and ***collaboration***, ethical and legal responsibilities, power, access, and identity. This rich and deep conversation has inspired the following conversations:

- Explorations of power and politics in hypertext (Johnson-Eilola, 1997)
- Copyright and fair use (Herrington, 1998, 2010)
- Ethical action (Salvo, 2002)
- Preparedness to teach in online pedagogical spaces (Cargile Cook & Grant-Davie, 2005, 2013; Melonçon, 2007)
- The implications of moving online (Gurak & Duin, 2004)
- ***Social media*** (Potts, 2014)
- Rhetoric and community in online spaces and documents (Howard, 1996; Porter, 1998; Pullman, 2016)
- Diverse topical and strategic literacies required of technical communicators (McCarthy et al., 2011; Selber, 2004)

In parallel developments, industrial practice also expanded the implementation of ***content management*** systems. With this decentralized, modular approach to developing and publishing content came multiple strategic emphases: single sourcing (Albers, 2003; Eble, 2003; Robidoux, 2008), ***structured*** authoring and information architecture (Evia et al., 2015; Salvo, 2004, 2010), and content strategy (Andersen, 2008, 2014; Clark, 2018; Evia, 2019; Getto et al., 2020; Hart-Davidson et al., 2007; Potts & Gonzales, 2020). In a most recent collection, *Teaching Content Management in Technical and Professional Communication*, Tracy Bridgeford (2020) addresses what she calls a "pedagogical exigency" by bringing together a variety of approaches for teaching the various areas and competencies associated with content management.

Meanwhile, the technical communication (TC) discipline also engaged in constructing more sophisticated frameworks for gathering technologically enabled design practices, resulting in the turn to experience architecture (Potts & Salvo, 2016). Experience architecture (XA) itself represents the confluence of a number of conversations in technical communication over the past 30 years. Not only does it draw upon the scholarly exchange about technologies and design strategies introduced previously, XA represents the culmination of work in usability studies (Chong, 2016; Salvo & Ren, 2007; Mirel & Spilka, 2002; Redish, 2011; Sauer, 2018), user-centered and participatory design (Johnson, 1998; Spinuzzi, 2005), accessible and inclusive design (Frascara, 2015; Oswal & Melonçon, 2014), ***user experience*** design (Geisler, 2016), ***intercultural communication*** pedagogies (St.Amant, 2018; Thatcher & St.Amant, 2011), and workplace roles (Batova & Andersen, 2017).

In addition to technical communication teachers' ever-present awareness of changing industrial needs and expectations, the developments in classroom content and practices highlighted so far have been complemented by a parallel evolution of the shaping of pedagogy through theoretical concepts. Pedagogical influences driven by conceptual "turns" (rhetorical, social, cultural, and ***social justice***) have both changed and challenged the discipline's pedagogical habits and practices by introducing new ways of thinking about technical communication, workplace and classroom spaces, and scholarly methodologies. These turns, in turn, awakened other ways of positioning technical communication, the technical communicator, and the technical communication student. Rhetoric empowered us to explore writing in action and how we attend to the ***style***, ***audience***, and purpose in document creation; cultural studies offered perspectives of cultural contexts in ways that helped us understand how communities work; and social theory helped us focus on language and how it shapes reality and social justice, demonstrating ways to bring out new paths, new practices, and destabilizations.

Carolyn Miller's (1979) landmark article, "A Humanistic Rationale for Technical Writing," is credited with sparking what has become acknowledged as technical communication's rhetorical turn. The rhetorical turn represents a move to relocate (or at least challenge) the epistemological framework of technical communication, reclaiming technical discourse from ***science*** and engineering (disciplines that had not yet begun to acknowledge the communal construction of knowledge). By engaging in a rhetorical examination of technical documents, authority, and ethical values, scholars recast scientific and technical knowledge (and with it writing) as negotiated, constructed, and therefore evolving. The rhetorical turn continues to thrive (Smith, 1997), and from it emerge foci such as the implications of civic engagement (Dubinsky & Carpenter, 2004; Huckins, 1997), ***public*** intellectualism and service learning (Bowden & Scott, 2003; Sapp & Crabtree, 2002), innovation and creativity (Bridgeford et al., 2004), and ***ethics*** (Dombrowski, 2000; Dragga, 1997; Katz, 1992, 1993; Sullivan, 1990).

Additionally, pedagogical discussions in the 1980s and into the 1990s deepened and complicated theory-practice collaborations. Fearing and Sparrow (1989) brought to the community a theory-practice focus that shaped the pedagogical approaches during this time, some of which still define classroom practices today, such as Carolyn Miller's (1989) definition of technical communication as conduct, which gave us a new understanding of what we teach and how we teach it. Katherine Staples and Cezar M. Ornatowski (1997) reflected their understanding of technical communication as "founded in theory and oriented toward practice" (p. xii). Dubinsky (2004) collected the major articles that identified the critical issues for the technical communication classroom in ways that encouraged reflection in practice. By complicating the theory-practice classroom, pedagogy became more compelling, enriching our repertoire.

Overlapping with the rhetorical turn, a prevailing theory in academic contexts—social construction—permeates all modern conversations about pedagogy. This theoretical perspective posits that social action is not an individual act; rather, it is a communal emphasis that grows out of the culture and language from which it originates. Influenced by a social theory perspective, technical communication scholars and teachers moved from thinking about pedagogy as a forms-based product approach to a socially constructed process approach through notions of ***knowledge*** and its construction, discourse conventions, ***collaboration***, and community (Blyler & Thralls, 1993; Thralls & Blyler, 1993). Most notable, Nancy Roundy Blyler and Charlotte Thrall's (1993) article "The Social Perspective and Pedagogy in Technical Communication," as well as their edited collection, *Professional Communication: The Social Perspective*, meaningfully outline the pedagogical tenets and approaches of social theory and pedagogy (social construction, community, ideology, and the paralogic hermeneutics). By refocusing pedagogy on the contexts and actions affecting technical communication, scholars helped students see communication as contextualized, affecting the ***style***, writing, ***editing***, and ***design*** of technical documentation and content. Scholars across the spectrum drew from the social perspective's theoretical reach, addressing notions of ideology, gender, culture, and politics. From approaches advocating ***feminism***, to diversity and inclusion, to social justice, and to globalization and intercultural perspectives of technical communication, social theory expanded the possibilities of technical communication pedagogy and its practice.

The cultural turn during the 1990s and 2000s moved the field to a poststructuralist stance, empowering scholars to look at pedagogy beyond the way language shapes action, considering constructions of knowledge and power and how they play out in institutional contexts. This led to deeper meditations about the purpose of technical communication pedagogy. Two articles in particular broadened our pedagogical scope: Through an articulation lens, Jennifer Slack et al. (1993) argued for positioning technical communicators more within a context of power and authority as authors, and Johndan Johnson-Eilola (1996) opened the door to considering the role of technical communicators as symbolic analysts. Cultural studies theory enabled scholars to consider institution, knowledge, legitimation, and power and their effect on the culture of technical communication. As cultural agents, institutions contribute to the genre and style conventions that reinforce cultural norms and practices (Longo, 1998, 2000; Miller, 1984; Spinuzzi, 2003). During the late 1990s, Bernadette Longo provided a "cultural studies" approach to teaching technical communication that supported the ways discourse contributed to institutional relationships. In 2006, J. Blake Scott and Bernadette Longo and colleagues published *Critical Power Tools: Technical Communication and Cultural Studies*, moving technical communication teachers and students "from cultural critique to ethical civic action" (p. 196). This approach to teaching technical communication is concerned with the actions of a virtuous, ethical student (and future professional) who considers the different ideologies, identities, and legitimations of knowledge when creating technical documentation.

The foreseeable future of pedagogy challenges us to demonstrate that we can remain human centered in the face of social change, asking anew "what it means to call our field 'humanistic'" (Jones, 2016, p. 345). Emerging designs shift pedagogy more consciously toward social justice approaches that aim to bring forth aspects of technical communication that have previously been less explicitly acknowledged. In this way, our historical narratives about pedagogy are "disrupted," which, in turn, allows us to resee them from different perspectives (Jones et al., 2016). Such disruption reveals issues relevant to pedagogy such as diversity (Jones et al., 2014; Savage & Mattson, 2011), race and ethnicity (Banks, 2010; Savage & Matveeva, 2011; Williams & Pimentel, 2012, 2014), ***translation*** and localization (Agboka, 2013; Maylath & St.Amant, 2019), decolonization of our pedagogies (Agboka, 2014; Haas, 2012), and narrative or storytelling as a pedagogical tool that helps students contribute to practice and build empathy (Jones & Walton, 2018; Moore, 2013)—all areas that influence what and how we teach technical communication. In a collection focused specifically on social justice pedagogies, Angela M. Haas and Michelle F. Eble (2018) broke significant ground by highlighting social justice with *Key Theoretical Frameworks: Teaching Technical Communication in the Twenty-First Century*, a collection that parallels nicely with Tracy Bridgeford's (2018) collection of the same year that describes theory-driven practical approaches. This pedagogical reach builds on all past turns in what Walton et al. (2019) call the social justice turn.

As this short ***history*** shows, technical communication has always had a dichotomous relationship with its pedagogical lexicon. We have always endeavored to both prepare students to perform well in the workplace and to question the status quo. This tension is what drives invention and innovation in pedagogy, moving us away from a focus on writing only (the product) to a perspective of writing in context (the communicative situation). During the last four decades, we have moved from rhetorical discussions about humanistic and ethical to critical and cultural studies and social justice approaches, remaining committed to teaching *craft* as it updates with each turn. But as much of this history is told in a semi-chronological way, the truth is that multiple turns happen in overlapping ways, influencing and impacting each other. Each turn shows, perhaps, a different face of our humanistic genealogy. The future of pedagogy challenges us to demonstrate that we can remain human centered in the face of social change.

References

Agboka, G. Y. (2013). Participatory localization: A social justice approach to navigating unenfranchised/disenfranchised culture sites. *Technical Communication Quarterly*, *22*(1), 28–49. https://doi.org/10.1080/10572252.2013.730966.

Agboka, G. Y. (2014). Decolonial methodologies: Social justice perspectives in intercultural technical communication research. *Journal of Technical Writing and Communication*, *44*(3), 297–327. https://doi.org/10.2190/TW.44.3.e.

Albers, M. J. (2003). Single sourcing and the technical communication career path. *Technical Communication*, *50*(3), 335–343.

Andersen, R. (2008). The rhetoric of enterprise content management (ECM): Confronting the assumptions during ECM adoption and transforming technical communication. *Technical Communication Quarterly*, 17(1), 61–87. https://doi.org/10.1080/10572250701588657.

Andersen, R. (2014). Rhetorical work in the age of content management: Implications for the field of technical communication. *Journal of Business and Technical Communication*, *28*(2), 115–157. https://doi.org/10.1177/1050651913513904.

Anderson, P. V., Brockmann, J. R. & Miller, C. M. (1983). *New essays in technical and scientific communication: Research, theory, practice*. Routledge.

Banks, A. J. (2006). *Race, rhetoric, and technology: Searching for higher ground*. Routledge. https://doi.org/10.4324/9781410617385.

Barker, T. T. (Ed.). (1991). *Perspectives on software documentation: Inquiries and innovations*. Baywood Publishing Company, Inc.

Barnum, C. M. & Carliner, S. (Eds.). (1991). *Techniques for technical communicators*. Macmillan Publishing Company.

Barton, B. F. & Barton, M. S. (1993). Modes of power in technical and professional visuals. *Journal of Business and Technical Communication*, *7*(1), 138–162. https://doi.org/10.1177/1050651993007001007.

Batova, T. & Andersen, R. (2017). A systematic literature review of changes in roles/skills in component content management environments and implications for education. *Technical Communication Quarterly*, *26*(2), 173–200. https://doi.org/10.1080/10572252.2017.1287958.

Bazerman, C. & Paradis, J. (1991). *Textual dynamics of the professions*. University of Wisconsin Press.

Beene, L. & White, P. (1988). *Solving problems in technical writing*. Oxford University Press.

Bernhardt, S. A. (1986). Seeing the text. *College Composition and Communication*, *37*, 66–78. https://doi.org/10.2307/357383.

Bitzer, L. (1968). The rhetorical situation. *Philosophy and Rhetoric*, *1*(1), 1–15.

Blyler, N. R. & Thralls, C. (Eds.). (1993). *Professional communication: The social perspective*. Sage Publications.

Bowden, M. & Scott, J. B. (2003). *Service-learning in technical and professional communication*. Longman.

Bridgeford, T. (Ed.). (2018). *Teaching professional and technical communication: A practicum in a book*. Utah State University Press and University of Colorado, Colorado Springs. https://doi.org/10.4324/9780429059612.

Bridgeford, T. (Ed.). (2020). *Teaching content management in technical and professional communication*. Taylor & Francis and Routledge. https://doi.org/10.4324/9780429059612.

Bridgeford, T., Kitalong, K. S. & Selfe, D. (Eds.). (2004). *Innovative approaches to teaching technical communication*. Utah State University Press. https://doi.org/10.2307/j.ctt46nzds.

Brumberger, E. (2005). Visual rhetoric in the curriculum: Pedagogy for a multimodal workplace. *Business Communication Quarterly*, *68*(3), 318–333. https://doi.org/10.1177/1080569905278863.

Brumberger, E. (2007). Making the strange familiar: A pedagogical exploration of visual thinking. *Journal of Business and Technical Communication*, *21*(4), 376–401. https://doi.org/10.1177/1050651907304021.

Brumberger, E. (2018). Designing teaching to teach design. In T. Bridgeford (Ed.), *Teaching professional and technical communication: A practicum in a book* (pp. 105–121). Utah State University Press.

Brumberger, E., Lauer, C. & Northcut, K. (2013). Technological literacy in the visual communication classroom: Reconciling principles and practice for the "Whole" communicator. *Programmatic Perspectives*, *5*(2), 3–28.

Brumberger, E. & Northcut, K. (2013). *Designing texts: Teaching visual communication*. Routledge.

Cargile Cook, K. C. & Grant-Davie, K. (Eds.). (2005). *Online education: Global questions, local answers*. Baywood Publishing Company.

Cargile Cook, K. C. & Grant-Davie, K. (Eds.). (2013). *Online education 2.0: Evolving, adapting, and reinventing online technical communication*. Baywood.

Chong, F. (2016). The pedagogy of usability: An analysis of technical communication textbooks, anthologies, and course syllabi and descriptions. *Technical Communication Quarterly*, *25*(1), 12–28. https://doi.org/10.1080/10572252.2016.1113073.

Clark, D. (2018). Teaching content strategy in professional and technical communication. In T. Bridgeford (Ed.), *Teaching professional and technical communication: A practicum in a book* (pp. 58–71). Utah State University Press.

DeVoss, D., Jasken, J. & Hayden, D. (2002). Teaching intracultural and intercultural communication: A critique and suggested method. *Journal of Business and Technical Communication*, *16*(1), 69–94. https://doi.org/10.1177/1050651902016001003.

Dombrowski, P. (1994). *Humanistic aspects of technical communication*. Baywood.

Dombrowski, P. (2000). *Ethics and technical communication*. Allyn & Bacon.

Dragga, S. (1997). A question of ethics: Lessons from technical communicators on the job. *Technical Communication Quarterly*, *6*(2), 161–178. https://doi.org/10.1207/s15427625tcq0602_3.

Dragga, S. & Voss, D. (2001). Cruel pies: The inhumanity of technical illustrations. *Technical Communication*, *48*(3), 265–274.

Dragga, S. (Ed.). (2001). Ethics in technical communication [Special issue]. *Technical Communication Quarterly*, *10*(3).

Dubinsky, J. M. (2004). *Teaching technical communication: Critical issues for the classroom*. Bedford/St. Martin's.

Dubinsky, J. M. & Carpenter, J. H. (2004). Civic engagement [Special issue]. *Technical Communication Quarterly, 13*(3).

Duin, A. H. & Hansen, C. J. (1996). *Nonacademic writing: Social theory and technology*. Lawrence Erlbaum Associates.

Eble, M. (2003). Content vs. product: The effects of single sourcing on the teaching of technical communication. *Technical Communication*, 50(3), 344–349.

Evia, C. (2019). *Creating intelligent content with lightweight DITA*. Routledge. https://doi.org/10.4324/9781351187510.

Evia, C., Sharp, M. & Pérez-Quiñones, M.A. (2015). Teaching structed authoring and DITA through rhetorical and computational thinking. *IEEE Transactions on Professional Communication*, *58*(3), 328–343.

Faigley, L. (1985). Competing theories of process: A critique and a proposal. In L. Odell & D. Goswami (Eds.), *Writing in nonacademic settings* (pp. 231–248). Guilford.

Fearing, B. E. & Sparrow, W. K. (Eds.). (1989). *Technical writing: Theory and practice*. Modern Language Association.

Frascara, J. (2015). Design as culture builder. *Visual Language*, *49*(1/2), 9–11.

Geisler, C. (2016). Opening: Toward an integrated approach. *Journal of Writing Research*, *7*(3), 417–424. https://doi.org/10.17239/jowr-2016.07.03.05.

Getto, G., Labriola, J. T. & Ruszkiewicz, S. (Eds.). (2020). *Content strategy in technical communication*. Routledge. https://doi.org/10.4324/9780429201141.

Gurak, L. & Duin, A. H. (2004). The impact of the internet and digital technologies on teaching and research in technical communication. *Technical Communication Quarterly*, *13*(2), 187–198. https://doi.org/10.1207/s15427625tcq1302_4.

Hall, D. R. (1976). The role of invention in technical writing. *Technical Writing Teacher*, *4*, 13–24.

Halloran, S. M. (1978). Technical writing and the rhetoric of science. *Journal of Technical Writing and Communication*, *8*(2), 77–88. https://doi.org/10.2190/RM3A-U8F4-MK32-4XHK.

Hass, A. M. (2012). Race, rhetoric, and technology: A case study of decolonial theory, methodology, and pedagogy. *Journal of Business and Technical Communication*, *26*(3), 277–310. https://doi.org/10.1177/1050651912439539.

Haas, A. & Eble, M. F. (Eds.). (2018). *Key theoretical frameworks: Teaching technical communication in the twenty-first century*. Utah State University Press.

Hart-Davidson, W. Bernhardt, G., McLeod, M., Martine, R. & Grabill, J. T. (2007). Coming to content management: Inventing infrastructure for organizational knowledge work. *Technical Communication Quarterly, 17*(1), 10–34. https://doi.org/10.1080/10572250701588608.

Herndl, C. (1993). Teaching discourse and reproducing culture: A critique of research and pedagogy in professional and non-academic writing. *College Composition and Communication*, *44*(3), 349–363. https://doi.org/10.2307/358988.

Herndl, C. G., Fennell, B. A. & Miller, C. R. (1991). Understanding failures in organizational discourse: The accident at Three Mile Island and the Shuttle Challenger Disaster. In C. Bazerman & J. Paradis (Eds.), *Textual dynamics of the professions: Historical and contemporary studies of writing in professional communities* (pp. 279–305). University of Wisconsin Press.

Herrington, T. (1998). The interdependency of fair use and the First Amendment. *Computers and Composition, 15*(2), 125–143. https://doi.org/10.1016/S8755-4615(98)90050-0.
Herrington, T. (2010). Crossing global boundaries: Beyond intercultural communication. *Journal of Business and Technical Communication, 24*(4), 516–539. https://doi.org/10.1177/1050651910371303.
Hoft, N. (1995a). A curriculum for the research and practice of international technical communication. *Technical Communication, 42*(4), 650–652.
Hoft, N. (1995b). *International technical communication*. Wiley.
Howard, T. (1996). Who owns electronic texts? In P. Sullivan & J. Dautermann (Eds.), *Electronic literacies in the workplace: Technologies of writing* (pp. 177–198). National Council of Teachers of English.
Howard, T. W. (2018). Teaching usability testing: Coding usability testing data. In T. Bridgeford (Ed.), *Teaching professional and technical communication: A practicum in a book* (pp. 176–202). Utah State University Press.
Huckins, T. N. (1997). Technical writing and community service. *Journal of Business and Technical Communication, 11*(1), 49–59. https://doi.org/10.1177/1050651997011001003.
Johnson, R. R. (1998). *User-centered technology*. State University of New York Press.
Johnson-Eilola, J. (1996). Relocating the value of work: Technical communication in a post-industrial age. *Technical Communication Quarterly, 5*(3), 245–270. https://doi.org/10.1207/s15427625tcq0503_1.
Johnson-Eilola, J. (1997). *Nostalgic angels: Rearticulating hypertext writing*. Ablex Publishing Company.
Jones, N. N. (2016). The technical communicator as advocate: Integrating a social justice approach in technical communication. *Journal of Technical Writing and Communication, 46*(3), 342–361. https://doi.org/10.1177/0047281616639472.
Jones, N. N., Moore, K. R. & Walton, R. (2016). Disrupting the past to disrupt the future: An antenarrative of technical communication. *Technical Communication Quarterly, 25*(4), 211–229. https://doi.org/10.1080/10572252.2016.1224655.
Jones, N. N., Savage, G. & Yu, H. (2014). Tracking our progress: Diversity in technical and professional communication programs. *Programmatic Perspectives, 6*(1), 132–152. https://cptsc.org/wp-content/uploads/2018/04/vol6-1.pdf.
Jones, N. N. & Walton, R. (2018). Using narratives to foster critical thinking about diversity and social justice. In A. Haas & M. Eble (Eds.), *Key theoretical frameworks for teaching technical communication in the 21st century* (pp. 241–267). Utah State University Press.
Katz, S. B. (1992). The ethic of expediency: Classical rhetoric, technology, and the Holocaust. *College English, 54*: 255–275.
Katz, S. B. (1993). Aristotle's Rhetoric, Hitler's Program, and the Ideological Problem of Praxis, Power, and Professional Discourse. Journal of Business and Technical Communication, 7(1): 37–62. https://doi.org/10.1177/1050651993007001003.
Kitalong, K. S. (2018). What do instructors need to know about teaching information graphics? A multiliteracies approach. In T. Bridgeford (Ed.), *Teaching professional and technical communication: A practicum in a book* (pp. 89–104). Utah State University Press.
Kostelnick, C. & Roberts, D. D. (2011). *Designing visual language: Strategies for professional communicators*. Longman.
Kramer, R. & Bernhardt, S. (1996). Teaching text design. *Technical Communication Quarterly, 5*(1), 35–60. https://doi.org/10.1207/s15427625tcq0501_3.

Lipson, C. & Day, M. (Eds.). (2005). *Technical communication and the World Wide Web.* Lawrence Erlbaum Associates. https://doi.org/10.4324/9781410612953.

Longo, B. (1998). An approach for applying cultural study theory to technical writing research. *Technical Communication Quarterly, 7*(1), 53–73. https://doi.org/10.1080/10572259809364617.

Longo, B. (2000). *Spurious coin: Science, management, and the history of technical writing in the 20th century.* State University of New York.

Maylath, B. & St.Amant, K. (Eds.). (2019). *Translation and localization: A guide for technical and professional communication.* Routledge. https://doi.org/10.4324/9780429453670.

McCarthy, J. E., Grabill, J. T., Hart-Davidson, W. & McLeod, M. (2011). Content management in the workplace: Community, context, and a new way to organize writing. *Journal of Business and Technical Communication, 25*(4), 367–395. https://doi.org/10.1177/1050651911410943.

Melonçon, L. (2007). Exploring electronic landscapes: Technical communication, online learning, and instructor preparedness. *Technical Communication Quarterly, 16*(1), 31–53. https://doi.org/10.1080/10572250709336576.

Miller, C. R. (1979). A humanistic rationale for technical writing. *College English, 40*(6), 610–617. https://doi.org/10.2307/375964.

Miller, C. R. (1984). Genre as social action. *Quarterly Journal of Speech, 70*, 151–167. https://doi.org/10.1080/00335638409383686.

Miller, C. R. (1989). What's practical about technical writing? In B. E. Fearing & W. K. Sparrow (Eds.), *Technical writing: Theory and practice* (pp. 14–24). Modern Language Association.

Mirel, B. & Spilka, R. (Eds.). (2002). *Reshaping technical communication: New directions and challenges for the 21st century.* Lawrence Erlbaum Associates. https://doi.org/10.4324/9781410603739.

Moore, K. R. (2013). Exposing the hidden relations: Storytelling, pedagogy, and the study of policy. *Journal of Technical Writing and Communication, 43*(1), 63–78. https://doi.org/10.2190/TW.43.1.d.

Moore, P. & Fitz, C. (1993). Using Gestalt theory to teach document design and graphics. *Technical Communication Quarterly, 2*(4), 389–410. https://doi.org/10.1080/10572259309364549.

O'Keefe, A. (1985). Teaching technical writing. In M. G. Moran & D. Journet (Eds.), *Research in technical communication: A bibliographic sourcebook* (pp. 85–113). Greenwood Press.

Oswal, S. K. & Melonçon, L. (2014). Paying attention to accessibility when designing online courses in technical and professional communication. *Journal of Business and Technical Communication, 28*(3) 271–300.

Pedagogy. (n.d.). In *Merriam-Webster's online dictionary.* Retrieved from: https://www.merriam-webster.com/dictionary/pedagogy.

Porter, J. E. (1998). *Rhetorical ethics and internetworked writing.* Ablex.

Potts, L. (2014). *Social media in disaster response: How experience architects can build for participation.* Routledge.

Potts, L. & Gonzales, L. (2020). Teaching content strategy in technical communication. In T. Bridgeford (Ed.), *Teaching content management in technical and professional communication* (pp. 59–71). Taylor & Francis. https://doi.org/10.4324/9780429059612-3.

Potts, L. & Salvo, M. (Eds.). (2016). *Rhetoric and experience architecture*. Parlor Press.

Pullman, G. (2016). *Writing online: Rhetoric for the digital age*. Hackett.

Pullman, G. & Gu, B. (Eds.). (2008). Guest editors' introduction: Rationalizing and rhetoricizing content management. *Technical Communication Quarterly*, *17*(1), 1–9. https://doi.org/10.1080/10572250701588558.

Pullman, G. & Gu, B. (Eds.). (2009). *Content management: Bridging the gap between theory and practice*. Baywood Publishing Company.

Redish, J. (1985). The plain English movement. In S. Greenbaum (Ed.), *The English language today* (pp. 125–138). Pergamon.

Redish, J. (2011). Overlap, influence, intertwining: The interplay of UX and technical communication. *Journal of Usability Studies*, *6*(3), 90–101.

Robidoux, C. (2008). Rhetorically structured content: Developing a collaborative single-sourcing curriculum. *Technical Communication Quarterly*, *17*(1), 110–143. https://doi.org/10.1080/10572250701595652.

Salvo, M. J. (2002). Critical engagement with technology in the computer classroom. *Technical Communication Quarterly*, *11*(3), 317–337. https://doi.org/10.1207/s15427625tcq1103_5.

Salvo, M. J. (2004). Rhetorical action in professional space: Information architecture as critical practice. *Journal of Business and Technical Communication*, *18*(1), 39–66. https://doi.org/10.1177/1050651903258129.

Salvo, M. J. (2009). Ethics of engagement: User-centered design and rhetorical methodology. *Technical Communication Quarterly*, *10*(3), 273–290. https://doi.org/10.1207/s15427625tcq1003_3.

Salvo, M. & Ren, J. (2007). Participatory assessment: Negotiating engagement in a technical communication program. *Technical Communication*, *54*(4), 424–439.

Salvo, M. J. & Rosinski, P. (2010). Information design: From authoring text to architecting virtual spaces. In R. Spilka (Ed.), *Digital literacy for technical communication: Twenty-first century theory and practice* (pp. 103–127). Routledge.

Salvo, M. J. & Zoetwewy, M. W. (2007). User-centered technology in participatory culture: Two decades "Beyond a Narrow Conception of Usability Testing." *IEEE Transactions on Professional Communication*, *50*(4), 320–332. https://doi.org/10.1109/TPC.2007.908730.

Sapp, D. A. & Crabtree, R. D. (2002). A laboratory in citizenship: Service learning in the technical communication classroom. *Technical Communication Quarterly*, *11*(4), 411–432. https://doi.org/10.1207/s15427625tcq1104_3.

Sauer, G. (2018). Applying usability and user experience within academic contexts: Why progress remains slow. *Technical Communication Quarterly*, *27*(4), 362–371. https://doi.org/10.1080/10572252.2018.1521637.

Savage, G. J. (1996). Redefining the responsibilities of teachers and the social position of the technical communicator. *Technical Communication Quarterly*, *5*(3), 309–328. https://doi.org/10.1207/s15427625tcq0503_4.

Savage, G. & Mattson, K. (2011) Perceptions of racial and ethnic diversity in technical communication programs. *Programmatic Perspectives*, *3*(1), 5–57.

Savage, G. & Matveeva, N. (2011). Toward racial and ethnic diversity in technical communication programs: A study of technical communication in Historically Black Colleges and Universities and Tribal Colleges and Universities in the United States. *Programmatic Perspectives*, *3*(1): 58–85.

Schriver, K. (1991). Plain language for expert or lay audiences: Designing text using protocol-aided revision. Technical Report No. 46. Center for the Study of Writing: 148–172.

Schriver, K. (1997). *Dynamics in document design*. Wiley.

Schriver, K. (2000). Readability formulas in the new millennium: What's the use? *AM Journal of Computer Documentation*, *24*(3), 138–140. https://doi.org/10.1145/344599.344638.

Scott, J. B., Longo, B. & Willis, K. V. (2006). *Critical power tools: Technical communication and cultural studies*. State University of New York Press.

Selber, S. (1997). *Computers and technical communication: Pedagogical and programmatic perspectives*. Ablex Publishing Corporation.

Selber, S. (2004). *Multiliteracies for a digital age*. Southern Illinois University Press.

Selzer, J. (1983). What constitutes a "readable" style? In P. V., Anderson, R. J. Brockmann & C. R. Miller (Eds.), *New essays in teaching and scientific communication*. Baywood.

Slack, J., Miller, D. J. & Doak, J. (1993). The technical communicator as author: Meaning, power, authority. *Journal of Business and Technical Communication*, *7*(1), 12–36. https://doi.org/10.1177/1050651993007001002.

Smith, M. O. (1997). Intertextual connections to "A Humanistic Rationale for Technical Writing." *Journal of Business and Technical Communication*, *11*(2), 192–222. https://doi.org/10.1177/1050651997011002003.

Spilka, R. (2010). *Digital literacy for technical communication: 21st century theory and practice*. Routledge. https://doi.org/10.4324/9780203866115.

Spinuzzi, C. (2003). *Tracing genres through organizations: A sociocultural approach to information design*. MIT Press. https://doi.org/10.7551/mitpress/6875.001.0001.

Spinuzzi, C. (2005). The methodology of participatory design. *Technical Communication*, *52*(2), 163–174.

Spinuzzi, C. (2007). Technical communication in the age of distributed work. *Technical Communication Quarterly*, *16*(3), 265–277. https://doi.org/10.1080/10572250701290998.

St.Amant, K. (2018). Teaching international and intercultural technical communication: A comparative online credibility approach. In T. Bridgeford (Ed.), *Teaching professional and technical communication: A practicum in a book* (pp. 222–236). Utah State University Press.

St.Amant, K. (2020). Teaching content management for global and cross-cultural contexts. In T. Bridgeford (Ed.), *Teaching content management in technical and professional communication* (pp. 195–212). Routledge. https://doi.org/10.4324/9780429059612-11.

Staples, K. & Ornatowski, C. (1997). *Foundations for teaching technical communication: Theory, practice, and program designs*. Ablex Publishing Corporation.

Stevenson, D. W. (1981). *Courses, Components, and Exercises in Technical Communication*. National Council for Teachers of English.

Sullivan, D. (1990). Political-ethical implications of defining technical communication as a practice. *Journal of Advanced Composition*, *10*(2), 375–386.

Sullivan, P. & Dautermann, J. (Eds.). (1996). *Electronic literacies in the workplace: Technologies of writing*. National Council of Teachers of English.

Thatcher, B. & St.Amant, K. (2011). *Teaching intercultural rhetoric and technical communication: Theories, curriculum, pedagogies, and practices*. Baywood Publishing.

Thralls, C. & Blyler, N. R. (Eds.). (1993). The social perspective and pedagogy in technical communication. [Special issue]. *Technical Communication Quarterly, 2*(3). https://doi.org/10.1080/10572259309364540.

Walton, R., Moore, K. & Jones, N. (2019). *Technical communication and the social justice turn*. Routledge.

Williams, M. F. & Pimentel, O. (2012). Introduction: Race, ethnicity, and technical communication. *Journal of Business and Technical Communication*, *26*(3), 271–276. https://doi.org/10.1177/1050651912439535.

Williams, M. F. & Pimentel, O. (Eds.). (2014). *Communicating race, ethnicity, and identity in technical communication*. Baywood.

23. Plain Language

Russell Willerton
Po'okela Solutions, LLC

The plainness of a text, like beauty, is in the eye of the beholder. One who says to a mechanic, "Please tell me in plain English why the car won't run," hopes the mechanic will use familiar words and clear examples instead of insider jargon and complex explanations. The mechanic, in this case, will have to watch the questioner for signs that the answer is clear and understood. People in government, health and medicine, and legal services started to realize decades ago that their constituents could not understand letters, brochures, and policy documents written in dense jargon and presented with poor page ***design***. Over the past several decades, plain language has become an approach focused on helping non-expert readers—citizens, consumers, medical patients, and others going about their lives—understand and act upon important documents they receive. Plain-language texts may be recognized by their surface features, but the plain-language approach goes deeper than the surface level. For decades, technical communicators have advocated for audiences by applying plain-language principles and testing documents with readers.

Most who work in plain language today would take a descriptive approach to defining the term rather than a prescriptive (or proscriptive) one. That is, they identify traits that make language plainer rather than setting requirements for what a plain passage should or should not contain. These traits include using familiar vocabulary instead of complex jargon, writing shorter sentences instead of longer ones, writing with clear subjects and active verbs, and using section headings and white space to make reading easier (see a summary in Kimble, 2012, pp. 5–10). The Center for Plain Language, in defining plain language, focuses on the reception of a document by its ***audience***:

> A communication is in plain language if its wording, structure, and design are so clear that the intended readers can easily find what they need, understand what they find, and use that ***information***. The definition of "plain" depends on the audience. What is plain for one audience may not be plain at all for another audience. (Center for Plain Language, 2023a)

To understand what plain language means, it is important to know how the term has been used over time. The term *plain English* preceded *plain language* as applied to creating readable, usable documents. Currently, initiatives for *plain English* often appear more frequently in the UK and in some Commonwealth countries, while *plain language* is often used in the US, Canada, and other countries.

DOI: https://doi.org/10.37514/TPC-B.2023.1923.2.23

A text's plainness comes from the ***style*** the writer used to write it. Style is one of the canons of classical ***rhetoric***, and debates over which style is appropriate for particular situations go back many centuries. Edward P.J. Corbett and Robert J. Connors (1999) note that rhetoricians identified three fundamental levels of style: the *low* or *plain* style, the *middle* or *forcible* style, and the *high* or *florid* style. Quintilian, say Corbett and Connors, wrote that plain style was best for instructing audiences, middle for moving them, and high for charming them (1999, p. 21).

The advocacy for plain style has a long ***history***. Tom McArthur (1991) points out that the Host in Chaucer's *Canterbury Tales* calls on the (educated) Clerke of Oxenford to speak plainly to reach the pilgrims in the group. Authors of technical books in English in the 16th century used plain style, but such books got less attention than traditional literary genres (Tebeaux, 1997). The first reference to "plain English" as a style choice may be from Robert Cawdrey in the 17th century. Cawdrey's *Table Alphabeticall* of 1604, the first known English dictionary, was written for women, who had much less access to education and less familiarity with Latinate terms (McArthur, 1991, p. 13). Denise Tillery (2005) writes that several advocated for plain style in ***science*** writing in the 17th century, including Francis Bacon, Margaret Cavendish, and Jane Sharp.

In the 1940s and 1950s, parallel developments in the US and the UK led people to reconsider how governments and large organizations should communicate with their constituents. Advocates for change acknowledged that citizens and even employees within governmental organizations struggle to respond appropriately when they do not understand official, bureaucratic language. Karen Schriver (2017) provides a detailed and authoritative account of plain language activities in the US between 1940 and 2015. Schriver shows how, over several decades, successive efforts by government employees, academic researchers, Congress, and plain-language advocates in industry led bureaucracies to communicate to their constituents more effectively.

Over the years, multiple organizations have been formed to advocate for plain language. Organizations including Clarity International and Plain Language Association International connect people around the world who share the goal of communicating clearly with a variety of audiences. Both organizations sponsor conferences, and Clarity International also publishes its own journal. The Center for Plain Language is a U.S. nonprofit that advocates for plain language in government and industry. As part of its ***public*** outreach, the Center for Plain Language issues an annual report card to assess how well federal agencies follow the Plain Writing Act of 2010.

Plain language is an active area of professional activity that continues to grow and develop. Academics in professional and technical communication and other fields continue to ***research*** the history, impacts, and best practices for plain language (e.g., Matveeva et al., 2017). A set of four key terms beginning with "p" provides a way to navigate that research. Plain language is manifest in *products*, in *process*, in *principle*, for a *payoff*.

As Schriver (2017) notes, plain language is manifest in a *product* (a document of some kind). The product is plain enough if it meets the needs of its audience through its language and design. Surface measures of plainness, such as syllable counts and sentence lengths, frequently correlate with audience judgments about a document's helpfulness and usefulness. Insider jargon, which frequently is complex and unfamiliar, often adds a layer of complexity that editors strive to remove. Shorter words and shorter sentences are frequently easier for audiences to understand. In some cases, an organization like an insurance company or a state health agency might require documents for consumers to reflect certain readability scores or grade-level scores (see "Style" in this volume for more information). That said, plain-language practitioners have stated for decades that surface measurements of readability are not enough to ensure that a document is usable by its intended audience. And yet, practitioners of plain language frequently use some measure to assess the surface characteristics of their documents. Informal, unsystematic "eyeball tests" of document readability are not sufficient.

Design choices are also part of a plain-language practitioner's toolset. By skillfully using white space, typography choices, bulleted or numbered lists, and other ***visual*** cues, writers and designers can complement written content to create a document that is plain and easy to use by the audience (Garner, 2013).

Schriver (2017) also writes that plain language is manifest through *process*. The process of creating effective documents in plain language is just as complex and iterative as that of creating complicated documents for expert users. Testing draft documents with members of the target audience has long been part of plain-language practice; if the target audience cannot use a document effectively, it is not plain enough. Janice C. (Ginny) Redish (2000) emphasizes that an effective process is critical for producing plain-language documents that work for users; merely following "a few guidelines for sentences and words" is not enough (p. 165). Willerton (2015) provides profiles of organizations that create documents in plain language. Reinforcing Redish's point, these profiles show that effective plain-language processes are clear, they allow for iteration and recursion (i.e., incorporating feedback from audience members and experts), and they are focused on serving the audience.

Over several decades, it has been clear that a *principle* of serving public audiences ethically is central to plain language work. "Service is inherent in the mission and components of the plain-language movement" (Matveeva et al., 2017, p. 337). Many plain-language advocates have come from government agencies, where documents with unclear language can prevent citizens from receiving services or benefits. Schriver's (2017) history shows many instances in which government workers realized plain language serves constituents better. In particular, as Willerton (2015) writes, plain language helps readers navigate BUROC situations—situations that are bureaucratic (involving large, complex organizations and policies), unfamiliar (faced infrequently), rights-oriented (related to rights held as consumers and even as citizens), and critical (with significant consequences). A court trial, for example, is certainly a BUROC situation. Legal documents in plain language, such as jury instructions and court rules, can reduce confusion and help lawyers, jurors, and judges to focus on facts of particular cases (e.g., Kimble, 2012).

In the UK in the late 1970s, Martin Cutts and others campaigning for plain English in government shredded government forms in a public protest in London's Parliament Square (Cutts, 2009, p. xv) to show that unclear documents are bad for citizens. Organizations like Healthwise (Willerton, 2015) and Health Literacy Media (Health Literacy Media, 2023) create information in plain language to support health ***literacy***. Documents in plain language can help people with lower health literacy skills to learn more and to make better decisions around ***health*** and ***medicine***. Iva W. Cheung (2017) shows how people in underrepresented and marginalized groups, people who deal with negative stereotypes about themselves, and people with disabilities often face persistent stresses that negatively affect their cognitive load. Cheung argues that communicators have an ethical imperative to use plain language to reduce the cognitive load and promote ***social justice*** for oppressed people who need information to make important decisions (p. 454). In one example, Aisha T. Langford and colleagues (2020) describe how they used plain language to develop a decision support tool for Black and Hispanic audiences to use in considering whether to participate in cancer clinical trials (CCTs). CCTs are an activity in which members of nonwhite racial groups are often underrepresented. Langford et al. used interviews with cancer patients, a survey of cancer patients, and usability testing to develop a web-based tool in English and Spanish that Black and Hispanic patients viewed favorably. The principle of helping audiences to learn, to use benefits owed them, and to make informed decisions about their lives is central to plain language.

This principle of serving public audiences separates plain language work from other for-profit applications of technical communication and information design. At times, however, boundaries may blur. Federal legislation such as the Truth in Lending Act and the Fair Credit Reporting Act has required certain consumer contracts to have information in plain language. Some for-profit businesses do use plain language in their communication. For example, insurance companies sometimes enter and win in the Center for Plain Language (2023b) ClearMark Awards for effective plain language documents. The U.S. Securities and Exchange Commission also has rules requiring investment companies to use plain language when writing certain prospectus documents (Securities and Exchange Commission, 2023).

Finally, plain language documents frequently provide better results than those written in bureaucratese; plain language *pays off*. Joseph Kimble's (2012) book *Writing for Dollars, Writing to Please* provides vignettes of 23 instances in which plain language documents saved organizations time and money, along with 27 vignettes about studies showing that audiences prefer plain-language documents over less-plain counterparts. Later research (Campbell et al., 2017; Trudeau & Cawthorne, 2017) reinforces these studies, showing that working professionals frequently prefer documents in plain language over those in more complex language. Advocates for plain legal language, including Kimble (2012, 2017) and Bryan Garner (2013), oppose the idea that writing in legalese helps lawyers protect themselves and their ***profession***. Instead, Kimble, Garner, and others say that clear language creates trust between clients and counsel, and that clear legal documents are better than convoluted ones.

There are many examples of plain-language documents that embody plain language as product, process, principle, and payoff. One example is the Field Guides for Ensuring Voter Intent (Center for Civic Design, 2023), which are written for local elections officials. This project was first led by Dana Chisnell, a fellow of the Society for Technical Communication, an expert and author on usability testing, and a concerned citizen-turned-consultant for local elections issues. Chisnell was drawn to election operations after 2000, in which the winning electoral votes from Florida were awarded to George W. Bush after a contentious recount and numerous problems from the "butterfly" ballot design that confused some voters. Chisnell spent several years learning how elections are run and found that information that could help local elections officials had been collected and published, but it was written for academic audiences. Chisnell and her team distilled some of this information into a set of small booklets, each of which fits in a shirt pocket when printed (Willerton, 2015). Chisnell's team produced documents written in plain language and presented on clean, orderly pages. The processes were iterative and audience-focused. The principle of serving the audience—who is serving citizens by administering the voting process—is evident. The payoff from these documents (downloadable from the Center for Civic Design) comes through ballots that are more usable, poll workers who are trained better, voter education guides that are more effective, and local elections websites that tell voters what they need to know.

Redish (2000) notes that over time, document design, plain language, and information design have been used to describe the same core activity. While more than one label may apply in a given situation, plain language stands apart from others with its four P's—particularly the *principle* of serving an audience using an organization's information to accomplish critical tasks. With this emphasis on principle, plain language provides a framework suitable for responding to the social justice turn in technical communication (e.g., Walton et al., 2019). Plain language is an approach that technical communicators can use to create effective documents while meeting audience needs ethically and respectfully.

References

Campbell, K. S., Amare, N., Kane, E., Manning, A. D. & Naidoo, J. S. (2017). Plain-style preferences of US professionals. *IEEE Transactions on Professional Communication*, 60(4), 401–411. https://doi.org/10.1109/TPC.2017.2759621.

Center for Civic Design. (2023). *Field guides to ensuring voter intent*. https://civicdesign.org/fieldguides/.

Center for Plain Language. (2023a). About. https://centerforplainlanguage.org/about/.

Center for Plain Language. (2023b). ClearMark Awards. https://centerforplainlanguage.org/awards/clearmark/.

Cheung, I. W. (2017). Plain language to minimize cognitive load: A social justice perspective. *IEEE Transactions on Professional Communication*, *60*(4), 448–457. https://doi.org/10.1109/TPC.2017.2759639.

Corbett, E. P. J. & Connors, R. J. (1999). Classical rhetoric for the modern student (4th ed.). Oxford University Press.

Cutts, M. (2009). *Oxford guide to plain English* (3rd ed.). Oxford University Press.

Garner, B. A. (2013). *Legal writing in plain English* (2nd ed.). University of Chicago Press. https://doi.org/10.7208/chicago/9780226031392.001.0001.

Health Literacy Media. (2023). What we do. https://www.healthliteracy.media/whatwedo.

Kimble, J. (2012). *Writing for dollars, writing to please: The case for plain language in business, government, and law*. Carolina Academic Press.

Kimble, J. (2017). *Seeing through legalese: More essays on plain language*. Carolina Academic Press.

Langford, A. T., Hawley, S. T., Stableford, S., Studts, J. L. & Byrne, M. M. (2020). Development of a plain language decision support tool for cancer clinical trials: Blending health literacy, academic research, and minority patient perspectives. *Journal of Cancer Education*, *35*(3), 454–461. https://doi.org/10.1007/s13187-019-1482-5.

Matveeva, N., Moosally, M. & Willerton, R. (2017). Plain language in the twenty-first century: Introduction to the special issue on plain language. *IEEE Transactions on Professional Communication*, *60*(4), 336–342. https://doi.org/10.1109/TPC.2017.2759619.

McArthur, T. (1991). The pedigree of plain English. *English Today*, *7*(3), 13–19. https://doi.org/10.1017/S0266078400005642.

Redish, J. C. (2000). What is information design? *Technical Communication*, *47*(2), 163–166.

Schriver, K. (2017). Plain language in the US gains momentum: 1940–2015. *IEEE Transactions on Professional Communication*, 60(4), 343–383. https://doi.org/10.1109/TPC.2017.2765118.

Tebeaux, E. (1997). *The emergence of a tradition: Technical writing in the English Renaissance, 1475–1640*. Baywood.

Tillery, D. (2005). The plain style in the seventeenth century: Gender and the history of scientific discourse. *Journal of Technical Writing and Communication*, *35*(3), 273–289. https://doi.org/10.2190/MRQQ-K2U6-LTQU-0X56.

Trudeau, C. R. & Cawthorne, C. (2017). The public speaks, again: An international study of legal communication. *University of Arkansas at Little Rock Law Review*, *40*(2), 249–282.

U.S. Securities and Exchange Commission. (2023, April 10). Plain writing initiative. https://www.sec.gov/plainwriting.

Walton, R., Moore, K. & Jones, N. (2019). *Technical communication after the social justice turn: Building coalitions for action.* Routledge. https://doi.org/10.4324/9780429198748.

Willerton, R. (2015). *Plain language and ethical action: A dialogic approach to technical content in the twenty-first century*. Routledge. https://doi.org/10.4324/9781315796956.

24. Profession

Gerald Savage
Illinois State University

Profession and professionalization are concepts that engage scholars and practitioners in the technical communication (TC) field, sometimes explicitly (e.g., through a study of "the profession of technical communication" or through seeking professional certification in TC), but perhaps more often implicitly (i.e., through the daily work of teaching, ***researching***, or producing TC).

The *Oxford English Dictionary* indicates that *profession* has had a number of meanings, depending upon the nation using it, but its root meaning is a "declaration," as in declaration of religious faith, a vow upon entering a religious order, or a declaration of property or person as for a public registry (Oxford University Press, n.d.). By the 15th century CE, *profession* could also have the sense of a professional occupation: "An occupation in which a professed ***knowledge*** of some subject, field, or science is applied; a vocation or career, especially one that involves prolonged training and a formal qualification" (Oxford University Press, n.d.). In the 16th century, the term could also refer to any occupation, ranging from skilled trades to thievery (Oxford University Press, n.d.), although as recent as 1711, the clergy, law, and medicine remained the three traditional professions (Addison, 1891, p. 78).

In the field of TC, *profession* and *professionalization* are terms that can signify key debates or ongoing tensions. For some, the idea of profession is a settled issue: Technical communication simply and obviously is a bona fide profession and is regularly referred to as a profession. For others, profession is a goal the field is still working toward through a process called professionalization—a process that is envisioned or described in various ways upon which all discussants may not agree.

Those who view TC as a profession can point to the facts that it provides regular employment, that it is what practitioners say they do for a living, that it is a term of occupational identity, and (though by no means necessarily) that it may be a formal job title. It is a term that encompasses all of the tasks they perform in their work. That means it isn't something they may do as part of a job that has a different title, jobs like engineer, manager, physician—jobs where the work requires reports, ***proposals***, instructions, and the like but only as aspects of their primary duties.

The existence of national professional organizations for technical communicators is further testament to the professionalization of the field. The largest organization, the Society for Technical Communication has more than 6,000 members according to their most recently available report (Society for Technical Communication, 2014).

Indeed, to some, that technical communication is a profession seemed to be settled over 40 years ago with Robert Connors' (1982) widely cited study "The

DOI: https://doi.org/10.37514/TPC-B.2023.1923.2.24

Rise of Technical Writing Instruction in America." Although the study's focus is on the emergence of "technical writing" as a teaching specialization, Connors asserts that, following World War II, the practice of technical writing became "a job in itself" (p. 341) and "the profession of technical and scientific writing grew and matured" in the 1950s (p. 342).

However, Connors' view has been considerably complicated by later studies from the 1990s to the early 2000s (Carliner, 1994; Hayhoe, 1997; Kynell, 1996; Kynell-Hunt & Savage, 2003, 2004; Pinelli & Barclay, 1992; Savage, 1999). Edward Malone (2011), notably, recognizes that professionalization was a deliberate, intentional movement beginning in 1953. Malone finds that professionalization was, and perhaps remains, an ongoing struggle involving establishing professional organizations, defining a body of knowledge specific to the field, codifying ethical standards, developing venues for certifying practitioners, gaining legal recognition for the profession, and establishing accreditation for academic programs (p. 287). These are essentially the criteria defined by Abraham Flexner (1915) more than a century ago.

Malone (2011) documents the work of professionalization leaders of the 1950s who recognized that the field of TC lacked most of the characteristics of a profession at that time. Clearly, these early leaders were not viewing profession in Connors' terms, as simply "a job in itself." Some of them found guidance in theories of profession and professionalization, including Flexner's work. Using those criteria to identify a starting point for professionalization, they formed the first professional organizations for technical communicators, The Association of Technical Writers and Editors in New York, The Society of Technical Writers in Boston, and a year later, the Technical Publishing Society in Los Angeles. These organizations eventually merged to form a truly national organization, the Society of Technical Writers and Publishers (STWP) in 1960 (Malone, 2011, p. 289). STWP changed its name to the Society for Technical Communication (STC) in 1971 and ultimately became the largest professional organization in the field (Society for Technical Communication, n.d.a). Malone (2015a, 2015b) also has called attention to significant roles of women in organizing these professional associations.

A body of knowledge for technical communication has been discussed and debated several times since the 1950s. In 1957, Israel Sweet, a higher education administrator, argued that identifying a body of knowledge was, by nature, the responsibility of academics. This would require ***research*** on multiple fronts, a process that did not actually begin in any concerted way for another 30 years. Although technical communication teachers were being recruited shortly after the war to meet growing industry demands for technical writers (Connors, 1982; Kynell, 1996; Whitburn, 2009), and although some of those teachers began publishing articles and textbooks almost immediately, their publications focused on teaching practices—essentially teaching tips for freshly appointed and inexperienced college instructors (Cunningham & Harris, 1994; Cunningham & Hertz, 1970; Kynell & Tebeaux, 2009). It was not until the 1980s that a need for serious research—theoretical and empirical—was realized, and not, initially, with much enthusiasm. In fact, the primary motivation was the desire and the necessity for recognition and status of technical writing teachers who were marginalized in traditional English departments. In practical terms, this meant tenure, promotion, and increased salaries. It did not bear much fruit until the 1990s as technical communication scholars made the transition from writing about teaching based on their own experience to learning how to apply or develop theory to address the teaching and practice of workplace writing and to conduct actual research in such matters.

Not until the early 2000s did a body of knowledge become a mission in the field, spurred by a growing concern for certification of technical communication practitioners. Certification, or "recognition or validation by a professional organization (including a college or university) or agency that an *individual* possesses the qualifications for engaging in a specific profession" (Turner & Rainey, 2004, p. 234), is an essential condition for market closure, but it was impossible to develop ***assessment*** criteria and certification standards without a "a codified body of knowledge as the basis for certification" (Rainey et al., 2005, p. 335). A project was organized by STC in 2007 to accomplish this goal. It was called the "Technical Communication Body of Knowledge (TCBOK) initiative" project, with a "task force" of professors from four universities and two industry representatives (Coppola, 2010, pp. 11–13). The task force quickly decided that the body of knowledge and certification were separate concerns and focused their work on TCBOK alone. Over the next two or three years, a web portal was made available to the public. It is operational today at https://www.tcbok.org/, although it is not represented as complete or definitive of the technical communication field or as a basis for certification, even for STC's certificate program, which was established in 2011. Instead, according to the certificate program website, "the body of knowledge STC is using . . . is Johnson-Sheehan's textbook *Technical Communication Today*" (Society for Technical Communication, n.d.b). As of June 2020, a total of 340 certificates had been awarded by STC.

A concern for ***ethics*** is characteristic of established professions. This is sometimes expressed in terms of altruism—a primary commitment to exercising one's professional knowledge and practice not only for the good of clients (or audiences and users in the case of technical communication) but equally, for the good of society at large (Anteby et al., 2016; Evetts, 2006; Noordegraaf, 2015; Saks, 2011; Sciulli, 2005). Flexner (1915) may have been the first to state this idea, and his may be the most eloquent:

> Devotion to well-doing is thus more and more likely to become an accepted mark of professional activity; and as this development proceeds, the pecuniary interest of the individual practitioner of a given profession is apt to yield gradually before an increasing realization of responsibility to a larger end. (p. 581)

Most established professions have, in fact, devised a code of ethics specific to the nature of their practice. In the technical communication field, there are presently three statements of ethical principles: the ATTW Code of Ethics, the IEEE

Code of Conduct, and STC's Ethical Principles, all of which are available on the organizations' websites. These three codes vow loyalty to professional colleagues, clients, and audiences, as well as to the laws of the land. However, none of them call for altruism in the sense of selfless service to others. STC comes closest in explicitly recognizing an ethical responsibility to "respect cultural variety and other aspects of diversity in our clients, employers, development teams, and audiences." This could count as an important step toward Flexner's (1915) ideal of a maturing profession "increasing realization of responsibility to a larger end" (p. 581).

The value of professional status, particularly in the traditional sense most studies have assumed, has not been universally accepted in our field. This is evident in the ways that women, LGBTQ people, and people of marginalized races and ethnicities have been (when not simply excluded) sidelined, exoticized, or closeted in technical and professional communication roles just as they have been in the larger society. Natasha N. Jones et al. (2016) point out that the 1990s feminist movement in technical communication, which they suggest may have begun the ***social justice*** awakening in the field, called attention to the patriarchal values and assumptions about technical communication as a profession. Feminist scholars showed that women had been active, and sometimes leading, scholars and practitioners for well over a century, although with little or no recognition.

With the rise of social justice awareness, several new research methodologies have been making their way not only into research design but also into approaches to teaching and practice. Participatory action methodologies, feminist theories—including Black feminist theories—and queer theory have influenced not only how technical communication is practiced but in what domains of culture and society our field is or should be working. Decolonial methodologies are being applied to expose the ways that colonial and neo-colonial ideologies and practices have shaped and sustained traditional institutions and structures of power, including traditional ways technical communication has been practiced (Agboka, 2014; Haas, 2012).

Probably the first organized effort to address social justice in technical communication began in 2003 with the formation of the Council for Programs in Technical and Scientific Communication's (CPTSC) Diversity Committee (Selfe, 2004), which began by questioning the lack of racial and ethnic diversity and calling attention to the racist assumptions and structures in our academic programs and workplaces. Today, we are seeing traditional notions of profession being challenged in multiple ways. Discourse and ***genre*** conventions are being critiqued by scholars like Laura Gonzales (2018) and Cruz Medina (2014). Alternative ***rhetorics***, including nonverbal rhetorics, have been identified by Matthew B. Cox (2018), Temptaous T. Mckoy (2019), Cecilia Shelton (2020), and Brittany Hull et al. (2020).

It is often presumed that professional status confers an exalted identity upon those who have such status. As Jennifer Daryl Slack and J. Macgregor Wise (2005) argue, "In general, identity affects how a person is placed in culture: how important they are, how they are treated, and what possibilities are open to them" (p. 149). Professional identity is supposedly earned by the acquisition of expertise certified by academic degrees, professional organizations, and/or licensure. However, one's placement in culture is also conferred by factors in which a person may have no choice, including race, gender, and other, usually intersectional, forms of embodiment. As Hull et al. (2020) argue,

> Because minority bodies are always, already under scrutiny and subject to explanation and qualification, they are often conditioned to be aware of and responsive to the presumed standards of professionalism just to survive. . . . Black women embody dual identities and the pressure to conform to spaces where they were not welcome historically must be negotiated almost every day. (p. 7)

Cecilia Shelton (2020) brings a Black feminist perspective to bear upon the traditional concept of expertise, challenging "the violence of expertise that feigns an apolitical neutrality in service of the status quo" (p. 28).

Our scholarship, teaching, and practice, as Miriam F. Williams (2014) has emphasized, must recognize "those communicative practices used to negatively impact historically marginalized groups and identify new practices that can be used to encourage cultural competence within institutions and communities" (pp. 1–2.) Rebecca Walton (2016) takes up this charge in calling for "embracing human dignity and human rights as the first principle of communication and the foundational value of the TPC [technical and professional communication] field" (p. 402).

The turn to social justice may well be on the way to redefining technical communication as a profession uniquely qualified to guide those who seek its services in designing professional communications that are just, liberating, and accessible to all stakeholders, including those—humans and nonhumans—whose only stake is in the effects and consequences of the rhetorical and material actions that technical communication facilitates.

References

Addison, J. (1891). *The spectator* (No. 21). George Routledge & Sons. https://www.gutenberg.org/files/12030/12030-h/SV1/Spectator1.html#section21.

Agboka, G. Y. (2014). Decolonial methodologies: Social justice perspectives in intercultural technical communication research. *Journal of Technical Writing and Communication*, *44*(3), 297–327. https://doi.org/10.2190/TW.44.3.e.

Anteby, M., Chan, C. K. & Dibenigno, J. (2016). Three lenses on occupations and professions in organizations. *The Academy of Management Annals*, *10*(1), 183–244. https://doi.org/10.5465/19416520.2016.1120962.

Association of Teachers of Technical Writing. (n.d.). *Code of ethics*. https://attw.org/about-attw/code-of-ethics/.

Carliner, S. (1994). A call to research. *Technical Communication*, *41*(4), 615–619.

Connors, R. J. (1982). The rise of technical writing instruction in America. *Journal of Technical Writing and Communication*, *12*(4), 329–352. https://doi.org/10.1177/004728168201200406.

Coppola, N. W. (2010). The technical communication body of knowledge initiative: An academic-practitioner partnership. *Technical Communication, 57*(1), 11–25.

Cox, Matthew B. (2018). Working closets: Mapping queer professional discourses and why professional communication studies need queer rhetorics. *Journal of Business and Technical Communication, 33 (1),* 1–25. https://doi.org/10.1177/1050651918798691.

Cunningham, D. H. & Harris, J. G. (1994). Undergraduate technical and professional writing programs: A question of status. *Journal of Technical Writing and Communication, 24*(2), 127–137. https://doi.org/10.2190/TA1Y-72AH-05YM-UKEY.

Cunningham, D. H. & Hertz, V. (1970). An annotated bibliography on the teaching of technical writing. *College Composition and Communication, 21*(2), 177–186.

Evetts, J. (2006). Trust and professionalism: Challenges and occupational changes. *Current Sociology, 54*(4), 515–531. https://doi.org/10.1177/0011392106065083.

Flexner, A. (1915). Is social work a profession? In *Proceedings of the 42nd Annual Session of the National Conference of Charities and Correction, Baltimore, MD.* Hildmann.

Gonzales, L. (2018). *Sites of translation: What multilinguals can teach us about digital writing and rhetoric.* University of Michigan Press.

Haas, A. M. (2012). Race, rhetoric, and technology: A case study of decolonial technical communication theory, methodology, and pedagogy. *Journal of Business and Technical Communication, 26*(3), 277–310. https://doi.org/10.1177/1050651912439539.

Hayhoe, G. F. (1997). What research do we need, and why should practitioners care? *Technical Communication, 44*(1), 19–21.

Hull, B., Shelton, C. D. & Mckoy, T. (2020). Dressed but not tryin' to impress: Black women deconstructing "professional" dress. *Journal of Multimodal Rhetorics, 3*(2), 7–20.

IEEE, *IEEE code of conduct* (2014). https://www.ieee.org/content/dam/ieee-org/ieee/web/org/about/ieee_code_of_conduct.pdf.

Jones, N. N., Moore, K. R. & Walton, R. (2016). Disrupting the past to disrupt the future: An antenarrative of technical communication. *Technical Communication Quarterly, 25*(4), 211–229. https://doi.org/10.1080/10572252.2016.1224655.

Kynell, T. C. (1996). *Writing in a milieu of utility: The move to technical communication in American engineering programs 1850–1950.* Ablex.

Kynell-Hunt, T. & Savage, G. J. (Eds.). (2003). *Power and legitimacy in technical communication: The historical and contemporary struggle for professional status.* Baywood.

Kynell, T. & Savage, G. J. (Eds.). (2004). *Power and legitimacy in technical communication: Strategies for professional status.* Baywood.

Kynell, T. & Tebeaux, G. J. (2009). The Association of Teachers of Technical Communication: The emergence of professional identity. *Technical Communication Quarterly 18(2),* 107–141. 10.1080/10572250802688000.

Malone, E. A. (2011). The first wave (1953–1961) of the professionalization movement in technical communication. *Technical Communication, 58*(4), 285–306. https://doi.org/10.1177/0047281615569480.

Malone, E. A. (2015a). Eleanor McElwee and the formation of IEEE PCS. *Journal of Technical Writing and Communication, 45*(2), 104–133. https://doi.org/10.1080/10572252.2015.1001291.

Malone, E. A. (2015b). Women organizers of the first professional associations in technical communication. *Technical Communication Quarterly, 24*(2), 121–146.

Mckoy, T. T. (2019). *Y'all call it technical and professional communication, we call it# ForTheCulture: The use of amplification rhetorics in Black communities and their implications for technical and professional communication studies* [Doctoral dissertation, East Carolina University]. http://hdl.handle.net/10342/7421.

Medina, C. (2014). Tweeting collaborative identity: Race, ICTs, and performing *Latinadad*. In M. F. Williams & O. Pimentel (Eds.), *Communicating race, ethnicity, and identity in technical communication* (pp. 63–86). Baywood.

Noordegraaf, M. (2015). Hybrid professionalism and beyond: (New) forms of public professionalism in changing organizational and societal contexts. *Journal of Professions and Organization*, *2*, 187–206. https://doi.org/10.1093/jpo/jov002.

Oxford University Press. (n.d.). Profession. In *Oxford English Dictionary*. Retrieved March 1, 2023, from https://www.oed.com.

Pinelli, T. E. & Barclay, R. O. (1992). Research in technical communication: Perspectives and thoughts on the process. *Technical Communication*, *39*(4), 526–532.

Rainey, K. T., Turner, R. K. & Dayton, D. (2005). Do curricula correspond to managerial expectations? Core competencies for technical communicators. *Technical Communication*, *52*(3), 323–352.

Saks, M. (2011). Defining a profession: The role of knowledge and expertise. *Professions & Professionalism*, *2*(1), 1–10. https://doi.org/10.7577/pp.v2i1.151.

Savage, G. J. (1999). The process and prospects for professionalizing technical communication. *Journal of Technical Writing and Communication*, *29*(4), 355–381. https://doi.org/10.2190/7GFX-A5PC-5P7R-9LHX.

Sciulli, D. (2005). Continental sociology of professions today: Conceptual contributions. *Current Sociology*, *53*(6), 915–942. https://doi.org/10.1177/0011392105057155.

Selfe, C. L. (2004). *CPTSC Committee for Diversity report*. Council for Programs in Technical and Scientific Communication.

Shelton, C. (2020). Shifting out of neutral: Centering difference, bias, and social justice in a business writing course. *Technical Communication Quarterly*, *29*(1), 18–32. https://doi.org/10.1080/10572252.2019.1640287.

Slack, J. D. & Wise, J. M. (2005). *Culture and technology: A primer.* Peter Lang.

Society for Technical Communication. (n.d.a). *About STC*. https://www.stc.org/about-stc/.

Society for Technical Communication. (n.d.b). *Certified professional technical communicator (CPTC) practitioner*. https://www.stc.org/certification/.

Society for Technical Communication. (n.d.c). *Ethical principles*. https://www.stc.org/about-stc/ethical-principles/.

Turner, R. K. & Rainey, K. T. (2004). Certification in technical communication. *Technical Communication Quarterly*, *13*(2), 211–234. https://doi.org/10.1207/s15427625tcq1302_6.

Walton, R. (2016). Supporting human dignity and human rights: A call to adopt the first principle of human-centered design. *Journal of Technical Writing and Communication*, *46*(4), 402–426. https://doi.org/10.1177/0047281616653496.

Walton, R., Moore, K. R. & Jones, N. N. (2019). *Technical communication after the social justice turn: Building coalitions for action*. Routledge. https://doi.org/10.4324/9780429198748.

Whitburn, M. D. (2009). The first weeklong technical writers' institute and its impact. *Journal of Business and Technical Communication*, *23*(4), 428–447. https://doi.org/10.1177/1050651909338801.

Williams, M. F. (2014). Introduction. In M. F. Williams & O. Pimentel (Eds.), *Communicating race, ethnicity, and identity in technical communication* (pp. 1–4). Baywood.

25. Project Management

Benjamin Lauren
University of Miami

The word "project" was borrowed from Latin in the mid-14th century, and in English "project" came to mean "a plan, draft, scheme, or table of something; a tabulated statement; a design or pattern according to which something is made" (Oxford University Press, n.d.). This definition can be broadly understood as a way to create forward momentum of some kind of initiative. With this definition, juxtaposed with the concept of "management," or what the *Oxford English Dictionary* defines as "the application of skill or care in the manipulation, use, treatment, or control (of a thing or person), or in the conduct of something" (Oxford University Press, n.d.), a fuller picture emerges of project management as a plan, or plans, of work patterns that are controlled by some kind of individual or force. This historical conceptualization holds relatively true for technical communicators today.

The earliest forms of project management shaped the growth and safety of societies by managing the process and workforce responsible for manufacturing dwellings and monuments; creating regular access to water, including irrigation; developing systems for waste; and building farms, roads, tools, and other elements of everyday life. The people responsible for managing these projects ranged across occupations such as architects, builders, blacksmiths, farmers, and even artists.

The roots of contemporary project management practices can be traced to at least three historical moments in western culture. The first was the rise of the railroad; the second is mass manufacturing of automobiles; and the third is World War II. In *Control Through Communication*, JoAnna Yates (1993) described how the development of the American railroad system necessitated a system for sharing ***information*** in the form of reports and other documents across distances, particularly through the rise of corporations that required redundancies in order to operate effectively. To do so, corporations had to develop systems highly reliant on the technologies available to them at the time. Therein, Yates discussed an emphasis on the constraints and affordances of ***technology***, and its impact on how projects were controlled and information was coordinated across distributed teams of people.

The scientific management principles offered by Taylorism were adopted and prized by automaker Henry Ford, who had an important influence on concepts still discussed in project management scholarship today: efficiency (i.e., how quickly individuals or teams can produce and coordinate quality output) and productivity (how much quality output individuals or teams can contribute during a defined period of time). In particular, Ford focused a great deal on the efficiency of line

DOI: https://doi.org/10.37514/TPC-B.2023.1923.2.25

workers, going so far as to base employee wages on meeting or exceeding quotas. The culture of efficiency and its emphasis on productivity is nearly a ubiquitous concern of many workplaces that employ technical communicators. Technical communication scholars like Erin Frost (2016) and Joanna Schreiber (2017) argue that we must be critical of philosophies like efficiency and productivity as they directly influence how we position and manage teams of people in the workplace. Rather, Frost and Schreiber alike suggest a realignment with the efficiency paradigm that emphasizes inclusion and reimagining what it means to engage, or what can be broadly understood as motivating and inviting participation.

Scientific management also influenced the United States military in World War II, which really worked to professionalize project management as praxis, or as a series of dependable practices for planning work. Mirroring the development of technical communication as a field, project management found its footing as a ***profession*** during World War II. The U.S. Navy is widely credited for what was called the PERT program (Program Evaluation and Review Technique), which was developed well into the 1970s. Up until this time, Gantt charts were one of the more universally adopted planning tools used by project managers, but the PERT program built on scientific management principles to improve methods of estimating time to complete projects, hoping to establish best practices for planning, scheduling, and coordinating teams of people.

The time estimation concepts that were described in the PERT program are often still used today, including critical path analysis (i.e., the longest amount of possible time a project could take) or lead time (i.e., how much time is needed from the completion of one project to the beginning of another one). The PERT principles were used to plan projects like space exploration or even to manage the military policies implemented during the Cold War. Technical communicators involved in product ***documentation***, drafting, and the development of instructional materials often worked on teams using PERT principles. Later, as ***knowledge*** work began to focus on software development, additional coordinating tools were developed to help visualize planning principles forwarded by PERT. For instance, ***visual*** planning techniques like a Kanban board visualize how work is coordinated. Technical communication scholars have also created visualizations to coordinate information across teams, such as Clay Spinuzzi et al. (2006), who created a system for visualizing the activity streams of projects.

A through-line can be traced from the PERT method to lean manufacturing, which was made popular by Toyota in the 1980s. Lean manufacturing aimed to improve efficiency of teams by removing bureaucracy and empowering managers to make budgetary decisions. Lean approaches also began to be adopted into corporate environments during the same period of time. As Nikil Saval (2014) described, the conceptualization of doing more with less and empowering managers to make their own financial decisions as embedded into an organizational structure proved attractive to corporate management structures in the 1980s, especially because the political times called for high levels of intrapreneurial activities (i.e., an entrepreneurial mindset used inside an organization) to achieve higher levels of productivity.

Technical communicators had important influence during the 1970s and 1980s on both work patterns of developing texts and project management methods. As Ginny Redish (2010) showed, there is a long ***history*** of technical communicators reimagining approaches for ***designing*** documents and making products more usable all the way back to the 1970s. The usability testing protocols offered by practitioner-scholars like Karen Schriver, JoAnn Hackos, and Stephanie Rosenbaum not only influenced the focus of the work of technical communicators, but also how the work was managed and coordinated across groups of people. For example, the document cycling and publication processes of instructional materials required the development of new technologies, systems, and a strategy for involving a range of stakeholders in those processes. Furthermore, the study of effective document design principles and ***user experience*** were well established by several technical communication scholars in the 1980s and 1990s (e.g., Hackos & Redish, 1998; Schriver, 1996).

By the 1990s, as more technology corporations began to manufacture products (like software) that relied on computing technologies, lean approaches to managing projects were abandoned by many software engineers. Rather than engineering heavy machinery, such as cars or refrigerators, software engineers were suddenly writers in that they were authoring code, and drawing from iterative approaches to do so. A building, once constructed, cannot easily be changed. Software, once programmed, could easily be changed, and continuous updates of software became a feature of the product rather than a bug. Development teams were more often interdisciplinary, staffed by people with flexible skillsets who understood both the technical requirements of a system and user needs. As a result, computer engineers could no longer rely on processes meant for manufacturing, and software development teams began to develop flexible processes and procedures, such as Extreme Programming, which served as a predecessor for the flexibility of Agile development processes.

Already, technical communicators had been regularly working with subject-matter experts, such as engineers, to write technical documentation for products as a way to help users operate computing systems with ease. Software development processes positioned some technical communicators as usability specialists as well as communication designers. As a result, technical communicators became familiar with iterative forms of development and flexible project management processes and procedures (see Dicks, 2004; Hackos, 2007).

In 2001, the Agile Manifesto was published online, which became one of the most disruptive and important moments in project management history since the scientific management processes developed prior to World War II and the subsequent advancements of the PERT program. The Agile Manifesto squarely rejected previous ways of thinking about project management developed for manufacturing activities, decentralizing the role from a particular individual to a group of people working collectively. As a result, a range of communication practices were developed to support Agile's main tenets of supporting individuals, creating flexibility, ***collaborating*** with customers, and developing working software. The facilitation of this work created new roles for project managers under the term Scrum Master or Agile Coach. The large difference between the traditional project manager and a Scrum Master was significant, as Scrum Masters were considered experts on Agile practices, whereas certified project managers might have expertise across several domains. While the intellectual shift toward Agile practices is largely traced to this time period (and those who signed the Agile Manifesto), scholarship in technical communication demonstrates that several practitioners were also advocating for what can be described as "agile" practices in the early 1980s (see Redish, 2010).

With the rise of ***content management***, and later, content strategy (Anderson & Batova, 2015), project management in technical communication became more focused on delivering and managing the content organizations shared about their products. Project management as a means of managing texts, people, and projects did not disappear; rather, it continued to evolve with technology and technical communication as a profession. Rather than publishing booklets teaching individuals how to use a product, technical communicators helped to design products that are to be intuitive on their own. As such, many technical communicators today are also involved in content creation that supports a range of activities—from promoting and advertising content to helping customers understand the features of a product. Managing the delivery of this content became a key way technical communicators acted as project managers in an Agile workplace (see Hart-Davidson et al., 2007). Agile and lean development strategies were adapted to work alongside content management and strategy techniques (see Lauren, 2018), and digital governance frameworks developed for organizations to manage their footprint and messaging in a digital world (Welchman, 2015). Digital governance work made clear that organizations and institutions would need a specialist, or team of specialists, to manage their online content, but in a way that involved a variety of stakeholders. In other words, project managers needed to develop skills of involving people in complex processes to create alignment across organizations.

In 2020, the Dice Second Quarter Jobs Report showed that project management skills are the second most desirable trait for new job seekers, but how these skills are utilized depends quite a bit on the organizational structure and its general focus. Whether managing people, texts, or projects, many technical communicators will find that experience with project management is not only foundational to their success, but also a central organizing feature of knowledge work today.

Given the near ubiquitous need for project management skills and experience in the professional lives of technical communicators, instructors have developed coursework to teach students how to manage information and communication design work. One popular approach was offered by Rebecca Pope-Ruark (2012), who taught English students the Scrum framework to manage group projects.

Several other examples exist as well, including frameworks for managing community engagement work ethically and effectively (Gonzales & Turner, 2019). As well, scholarship on technical communication as a field frequently surfaces trends related to project management as a practice, such as James Dubinsky's (2015) discussion of facilitation as an important part of the technical communicator's skillset. No doubt, project management will remain an important element of how to develop, manage, and strategize communication work.

References

Andersen, R. & Batova, T. (2015). The current state of component content management: An integrative literature review. *IEEE Transactions on Professional Communication*, *58*(3), 247–270. https://doi.org/10.1109/TPC.2016.2516619.

Dicks, S. (2004). *Management principles and practices for technical communicators*. Longman.

Dubinsky, J. (2015). Products and processes: Transition from "product documentation to . . . integrated technical content." *Technical Communication*, *62*(2), 118–134.

Frost, E. A. (2016). Apparent feminism as a methodology for technical communication and rhetoric. *Journal of Business and Technical Communication*, *30*(1), 3–28. https://doi.org/10.1177/1050651915602295.

Gonzales, L. & Turner, H. N. (2019). Challenges and insights for fostering academic-industry collaborations in UX. In *Proceedings of the 37th ACM International Conference on the Design of Communication (SIGDOC '19)* (Article 21, pp. 1–6). Association for Computing Machinery. . https://doi.org/10.1145/3328020.3353921.

Hackos, J. T. (2007). *Information development: Managing your documentation projects, portfolio, and people*. Wiley. (Original work published 1997)

Hackos, J. T. & Redish, J. C. (1998). *User and task analysis for interface design*. Jon Wiley and Sons.

Hart-Davidson, W., Bernhardt, G., McLeod, M., Rife, M. & Grabill, J. T. (2007). Coming to content management: Inventing infrastructure for organizational knowledge work. *Technical Communication Quarterly*, *17*(1), 10–34. https://doi.org/10.1080/10572250701588608.

Lauren, B. (2018). *Communicating project management: A participatory rhetoric for development teams*. Routledge. https://doi.org/10.4324/9781315171418.

Oxford University Press. (n.d.). Management. In *Oxford English Dictionary*. Retrieved April 11, 2023, from https://www.oed.com.

Oxford University Press. (n.d.). Project. In *Oxford English Dictionary*. Retrieved April 11, 2023, from https://www.oed.com.

Pope-Ruark, R. (2012). We Scrum every day: Using Scrum project management framework for group projects. *College Teaching*, *60*(4), 164–169. https://doi.org/10.1080/87567555.2012.669425.

Redish, J. (2010). Technical communication and usability: Intertwined strands and mutual influences. *IEEE Transactions on Professional Communication*, *53*(3), 191–201. https://doi.org/10.1109/TPC.2010.2052861.

Saval, N. (2014). *Cubed: A secret history of the workplace*. Doubleday.

Schreiber, J. (2017). Toward a critical alignment with efficiency philosophies. *Technical Communication*, *64*(1), 27–37.

Schriver, K. (1996). *Dynamics in document design: Creating text for readers*. Wiley.

Spinuzzi, C., Hart-Davidson, W. & Zachry, M. (2006). Chains and ecologies: Methodological notes toward a communicative-mediational model of technologically mediated writing. In *Proceedings of the 24th Annual International Conference on Design of Communication, SIGDOC 2006* (pp. 43–50). ACM. https://doi.org/10.1145/1166324.1166336.

Welchman, L. (2015). *Managing chaos: Digital governance by design*. Rosenfeld Media.

Yates, J. & American Council of Learned Societies. (1989). *Control through communication: The rise of system in American management*. Johns Hopkins University Press.

26. Proposal

Richard Johnson-Sheehan
Purdue University

An internet search for the word *proposal* will bring up several common usages, the two most prominent being marriage proposals and self-described "modest proposals"—though most of the proposals in this second category fail to understand that Jonathan Swift's classic satire was ironic and not meant to be modest. The third most popular usage of the word *proposal* is the one that most interests us in technical and professional communication: a document that defines a problem or opportunity and then presents a plan or method for solving or taking advantage of that problem. A proposal in this sense is a ***genre*** used to get things done in workplaces and civic life (Johnson-Sheehan, 2008).

The proposal is one of the oldest and most powerful genres in technical and professional communication. The ability to write persuasive proposals to clients, customers, and funding sources can make or break a high-tech business or organization (Sant, 2012). Whether someone works in an engineering firm, a scientific laboratory, or a nonprofit organization, they will need to write persuasive proposals as part of their career.

The English word *proposal* derives from the word *propos*, which probably arrived in the British Isles in 1066 with Norman invaders. In Middle French, the term *proposer* meant "to intend, purpose," according to the *Oxford English Dictionary* (Oxford University Press). This French word was derived from the Latin word *propositum*, which means a "plan, intention, design" (Provost, 1961). Thus, the root word for proposal is *posit*, which means "position, posture, situation." The prefix *pro-* adds a sense of direction or support, and the suffix *-al* means "related to" or "the kind of."

As shown in the *Oxford English Dictionary*, the English meaning of the word *proposal*, as we know it now, probably originated in the mid-14th century, when the usage of the word *propos* narrowed to mean "purpose." Not long afterward, the word *proposal* became common in the English language, perhaps due to the rise of English commerce and industry.

Today, proposals are usually categorized according to how they were initiated. A proposal can be either *external* or *internal*. An external proposal is written from one company or organization to another, usually to pitch a product or service. An internal proposal is written to be used within a company or organization, usually to present new ideas for products, services, or processes.

Proposals can also be categorized by who initiates them. A proposal can be either *solicited* (requested by the customer or client) or *unsolicited* (initiated by the provider without being requested by the customer or client). These categories

DOI: https://doi.org/10.37514/TPC-B.2023.1923.2.26

overlap with the external/internal distinction. For example, an external solicited proposal is one that has been requested by a customer or client from another company or organization. An external solicited proposal usually begins when the customer or client sends out an advertisement called a request for proposals (RFP) that describes the desired product or service. A typical RFP will summarize the current problem, state the project objectives, explain the scope of the project, provide an overview of the company or organization, and specify expectations and deliverables (Hamper & Baugh, 2011, p. 56). The RFP will also include information about submission deadlines, assessment procedures, points of contact, and formatting. Depending on the industry, RFPs can also be called a request for bids (RFB), call for proposals (CFP), request for application (RFA), information for bid (IFB), call for quotes (CFQ), or advertisement for bids (AFB). Each of these types of RFPs will signal the specific kinds of information that the customer or client is seeking in the proposal.

An internal solicited proposal, meanwhile, is usually one that was requested by a supervisor or management within the writer's company or organization. In our increasingly entrepreneurial workplaces (or "in-trepreneurial" workplaces), it is becoming common for management to solicit proposals from their divisions or teams. This process puts these divisions and teams into competition with each other, urging them to compete for the company's limited pool of resources.

An external unsolicited proposal is typically a sales device through which salespeople at one company reach out to another company, introducing themselves and pitching their products and services (Sant, 2012). Consultants use unsolicited external proposals to make clients or customers aware of solutions to problems that they aren't sure how to handle or may even not know exist.

An internal unsolicited proposal might be written by employees to their managers, making suggestions for changes to products, services, or corporate operating procedures. Usually, internal unsolicited proposals come about because a person or team identifies a persistent problem and decides to offer a plan to management for solving that problem.

Recently, pre-proposals (also known as letters of intent in nonprofit settings) have become more common as a way to streamline the proposal process (Markin, 2015). A pre-proposal can be as short as two pages long, allowing the proposing company or organization to describe the project or service in general terms. Then, the customer or client will invite a limited number of providers to submit full proposals. The pre-proposal process is advantageous for both providers and their customers or clients. By asking for pre-proposals rather than full proposals, the customers or clients can limit the final bidding to providers who seem to best understand what is needed and have the ability to provide it. Pre-proposals also allow customers and clients to give providers feedback that helps them craft better full proposals. Providers, meanwhile, save time because they don't need to write full proposals for all the RFPs that interest them. Instead, they only write full proposals for companies or organizations that have already reviewed and responded favorably to the ideas in their pre-proposals.

Another recent change, especially in this entrepreneurial age, is a shift to less formal and briefer proposals (Copel Communications, 2016). Just as an entrepreneur might wear a hoodie and jeans to pitch a new startup, the tone of these proposals can be intentionally informal and personal. Nevertheless, these "informal" proposals, just like those entrepreneurs in hoodies, are very serious and the stakes can be high. The informal tone is intended to put the readers at ease, and the shorter length is designed to encourage them to actually look over the proposal. Typically, if the customer or client expresses interest, an informal proposal will then be revised into a much longer formal proposal. The formal proposal, with its cover page, table of contents, abstract, appendixes, and itemized budget, becomes the *de facto* contract that spells out the formal offer.

Proposals come in many forms and sizes, which reflects the highly flexible nature of this genre (Northcut et al., 2009). Like most documents, a proposal typically has an introduction, body, and conclusion. The body of a proposal can be arranged into a variety of ***structures***, but it will usually make five major moves that often take the form of separate sections: (1) a *background* or *narrative* that explains the problem or opportunity by describing its causes and effects; (2) a list of *objectives* or *aims*, which are the goals any plan would need to achieve to solve the problem or take advantage of the opportunity; (3) a *project plan* or *methods* that describes how those objectives would be achieved; (4) the *qualifications* of the people who would do the work; and (5) the *costs* and *benefits*, which attempt to persuade the readers that the deliverables of the project would be worth the price (Johnson-Sheehan, 2008). Each of these sections plays a unique role.

Background or Narrative. After the introduction, a proposal will typically include an analysis of the customer's or client's problem or opportunity by identifying its causes and the likely effects. When experienced proposal writers draft this section, they usually spend a great amount of time asking two questions: 1) "What exactly is causing the problem our customer or client is trying to solve?" and 2) "What has changed recently to create this problem?" (Johnson-Sheehan, 2008). Of course—to use consultant-speak—a problem is always an opportunity in disguise. That is true, but using the word "problem" adds a sense of urgency in this section that holds the customers' or clients' attention (Miner & Ball, 2019, p. 91).

The word "narrative" is a tip-off to how proposal writers often approach the writing of this section. They will tell a story that identifies the main characters (protagonists and antagonists) as well as the events, causes, and effects that brought those characters to the current problem state. Experienced proposal writers will use narrative techniques, such as setting the scene, using rising action, and describing a climax, to explain how the problem emerged and how the problem will affect the customer or client in the future.

Statement of Objectives or Aims. Stating the objectives or aims is typically a major pivot point at which a proposal transitions from describing the problem (looking backward) to presenting a plan for solving that problem (looking forward). The objectives or aims are designed to focus the readers' attention, revealing the goals that need to be achieved to solve the problem. Typically, in a solicited proposal, these objectives or aims will be aligned with, but not duplicate, the evaluation criteria named in the RFP.

Project Plan or Methods. After stating its objectives, a proposal typically includes a step-by-step description of the work, which is called the "Project Plan" or "Methods." While generating the content of this section, proposal writers will often ask themselves, "What are the three to seven major steps required to achieve the objectives or aims?" Then, for each of those major steps, they will come up with three to seven minor steps needed to achieve the major step.

Proposal writers often use the following three questions to fill out the content for each major and minor step:

- "How will we complete each step?" Identify each major step and then describe the minor steps needed to complete that major step.
- "Why are these steps needed?" After stating each major step and its minor steps, spend a little time, perhaps a sentence or two, explaining why that step would be handled that way.
- "What are the deliverables or outcomes of each major step?" After describing each major step, explain what will be finished (products, services, reports, data sets, software, etc.) by the end of the step. Specifically, mention things that will be delivered (i.e., deliverables) to the customer or client (Johnson-Sheehan, 2008).

This How-Why-What pattern can be very persuasive to the readers because they see *how* the project will be completed step-by-step, *why* each step is needed, and *what* kinds of deliverables they will receive.

Qualifications. Even the best plan or methods won't work if the right people aren't in place to implement it. Most proposals will include a "Qualifications" section that describes the provider's management and labor, facilities and equipment, and prior experiences with similar projects. Individual qualifications can be expressed through biographical statements, resumes, curriculum vitas, or other genres that summarize the skills and experiences of the project team.

Costs and Benefits. Usually, a summary of the costs and benefits concludes the proposal by trying to persuade the readers that the benefits of saying "yes" to the proposal are worth the costs. The overall price of the project (the costs) is stated in a straightforward and unapologetic way. Proposal writers may include the bottom-line figure, a small table that categorizes the major costs, or a fully itemized budget.

After the costs are stated, many proposal writers will summarize the three to seven major benefits of the project. These benefits had been previously identified as the deliverables within the description of the project plan. Here at the end of the proposal, they are reframed as benefits and used to balance the costs. This allows the readers to do a quick cost-benefits analysis as they finish reading the proposal. The key move in this concluding section is to convey to the readers, "Here's what you get" for your money. The customer or client needs to be persuaded that the plan will solve their problem and that they will receive substantial benefits after the problem is solved.

Table 26.1 shows how each of these sections of a proposal is used in various disciplines. As shown in the table, the genre itself is similar across fields, but the genre's flexibility allows it to be used in many different ways.

Always remember that proposals are unabashedly persuasive in nature, and both sides know it. The readers are fully aware that something is being pitched to them. The proposal writers know their job is to use persuasion to sell the readers a solution. Proposal writers do this by showing the customers or clients that they understand the problem, that they have a reasonable plan for solving that problem, that they have the right people to do the work, and that the benefits of doing the project clearly outweigh the costs.

Table 26.1 Similarities in Proposals Across Various Fields.

	Business Proposals	**Engineering Proposals**	**Science Proposals**	**Nonprofit Proposals**
Background or Narrative	Service the client may not know they need	Problem with a manufacturing process	Literature review that highlights a gap in research	Problem in the community that needs to be solved
Objectives or Aims	List of the customer's needs	Criteria for determining a successful change to process	Version of funding agency's criteria for obtaining funding	Version of a funding source's evaluation criteria
Project Plan or Methods	Step-by-step description of service and how it would work for the customer	Step-by-step description of how the change would be implemented	Step-by-step description of the research methodology	Step-by-step description of the program to address a problem
Qualifications	Company backgrounder	Bios of the engineering and design team	Bios and CVs of the team of scientists and facilities	Bios and resumes of the nonprofit's administrators
Costs and Benefits	Estimate of the costs and benefits of the new service	Estimate of the costs and benefits of a new process	Significance and impact of the research	Project costs and impact on the community

The proposal, as one of the core genres in technical and professional communication, will be around as long as people have new ideas. In the future, proposals will continue evolving to match the speed and fluidity of today's networked and global workplace. They will likely become briefer, more visual, and more interactive, taking the form of slide decks, poster canvases, and ***multimodal*** presentations. In new forms, these brief, visual, and interactive proposals will still make the same major moves as traditional written proposals, but they will be designed for clients and customers who want to see more, read less, and be entertained.

References

Copel Communications. (2016, September 6). *Formal vs. informal proposals: Which nets more?* https://www.copelcommunications.com/blog/formal-vs-informal-proposals-which-nets-more.

Hamper, R. & Baugh, S. (2011). *Handbook for writing proposals.* McGraw-Hill.

Johnson-Sheehan, R. (2008). *Writing proposals* (2nd ed.). Allyn and Bacon; Longman.

Markin, K. (2015, January 21). Writing a preproposal: Leave them wanting more. *The Chronicle of Higher Education*, 62. https://www.chronicle.com/article/writing-a-preproposal-leave-them-wanting-more/.

McMurrey, D. & Arnett, J. (2019). 02.04: Proposals *Open technical communication*, 5. https://digitalcommons.kennesaw.edu/opentc/5.

Miner, J. & Ball, K. (2019). *Proposal planning and writing.* ABC-CLIO.

Northcut, K. M., Crow, M. L. & Mormile, M. (2009). Proposal writing from three perspectives: Technical communication, engineering, and science. In *2009 IEEE International Professional Communication Conference.* IEEE. https://doi.org/10.1109/IPCC.2009.5208695.

Oxford University Press. (n.d). Proposal. In *Oxford English Dictionary*. Retrieved May 5, 2023, from https://www.oed.com.

Provost, A. J. (1961). *Junior classic Latin dictionary: Latin-English and English-Latin.* Follett.

Sant, T. (2012). *Persuasive business proposals* (3rd ed.). Amacon.

27. Public

Kristen R. Moore
University at Buffalo

Many approaches to technical communication (TC) locate the field's work in corporate, industrial, and scientific workspaces—not the public sphere (Rude, 2008). And yet both the cultural turn (Scott et al., 2006) and the ***social justice*** turn (Haas & Eble, 2018; Walton et al., 2019) provide examples of the role of technical communication outside of these more traditional spaces. From public policy to ***health and medical communication*** to community-based literacies, technical communication often serves the public. And yet "the public" is not such a straightforward audience or set of users as one might hope.

Outside of TC, the term *public* has been well-theorized by scholars in communication (e.g., Asen, 2000; Goodnight, 2012), philosophy (e.g., Fraser, 1990; Habermas, 1962/1991), and ***rhetoric*** and writing (e.g., Flower, 2008; Long, 2008; Rice, 2012), among others (notably, Warner, 2005 in literary studies). In TC, however, the theoretical takeaways often fade into the background of practice and application. As such, this entry provides an overview of the theoretical debates by organizing them into four key (if false) dichotomies that affect TC (see Table 27.1): public vs. private, the public vs. publics, public vs. counterpublic, and public vs. community. Rather than linger in the theoretical, this entry focuses on the ways each of these dichotomies affects the practice of technical communication in industry, ***pedagogy***, and sites of ***research***.

Public vs. Private. If we consult Jürgen Habermas (1962/1991), the public realm of authority, the public sphere (of cafes, politics, and the market), and the private realm (of the house and civil work spaces) exist as separate spheres (p. 30). Habermas writes that the public sphere includes "a realm of our social life in which something approaching public opinion can be formed," noting that "citizens behave as a public body when they confer in an unrestricted faction . . . about matters of general interest" (Habermas, qtd. in Hauser, 1999). As the site of communication, the public sphere provides freedom from the restrictions of either the private sphere or the state. Yet theoretical critiques of Habermas argue that the differences between public and private are not so clear (e.g., Berlant and Warner, 1998).

Technical communication often exists in corporate, scientific, or government spaces; these spaces seem to live in a gray area between Habermas' public and private, entering into a discourse community that seems to be neither general enough to be considered "public" or intimate enough to be considered "private." This dichotomy serves as the foundation for how TC has traditionally been conceptualized. Yet, countless examples draw our attention to the limits of conceptualizing public as separate from private. Katherine Durack (1997), for example,

DOI: https://doi.org/10.37514/TPC-B.2023.1923.2.27

highlighted how technologies used in the domestic (or private sphere) have been ignored by technical communicators because they are gendered. From a public vs. private perspective, however, the gendered nature of the ***technology*** she discusses is also wrapped up in the gendered nature of the home as a private sphere.

Many forms of technical communication mediate between the traditional public and private spheres, collapsing the dichotomy on itself. For example, the adjustable mortgage rate documents that Natasha Jones and Miriam Williams (2017) analyze are designed to be used by all members of the public to enable work in the public sphere. Yet, as they discuss, language use in these documents deeply affects and reflects the private realm: where someone lives and how they can make personal decisions about where to make a home. Similarly, medical/health communication and policies are often written to articulate policies for the public but ultimately affect activities and decisions that often occur in what might be called the private sphere (the body). Medical and health-related technical communication contexts, from DIY hormone replacement instructions (Edenfield et al., 2019) to fertility tracking apps (Novotny & Hutchinson, 2019) to HIV testing (Scott, 2003), provide examples of technical communication that defy the public vs. private dichotomy.

Table 27.1. Theoretical and Practical Takeaways That Emerge from Four Key (False) Dichotomies for Understanding the Complexity of the Term *Public*

Dichotomy	Theoretical Takeaway	Practical Takeaway
Public vs. Private	The distinction between public and private is murky at best.	TC often mediates between the traditional public and private spaces.
The Public vs. Publics	No single, homogenous public exists.	When moving into the public sphere, our audiences must be broken down into stakeholders, users, and localized contingents.
Public vs. Counterpublic	People traditionally excluded from the public sphere—marginalized groups or oppressed groups—often constitute their own groups in opposition to the dominant public sphere.	When considering information products of all kinds, the conceptualization of a public must account for the fact that not all groups have historically been considered central, important, or worthy of our attention.
Public vs. Community	Publics have been theorized as gatherings of strangers without shared interests or common goals; communities, on the other hand, provide loci for shared decision-making and values.	When considering public-facing technical communication, understanding community-driven values encourages a kind of localization that helps TC address injustice and solve discrete problems.

The Public vs. Publics. Early articulations of the public sphere articulated it as a location occupied by a singular entity: *the* public. Political discourse occurs in public, of course, but it is also marketed to "the public." Despite the supposed neutrality of technical communication, scholars like Steven B. Katz (1992) and Dale Sullivan (1990) demonstrate that TC is political and addresses public problems; as such, the public is an audience for TC. But it's complex and in no way homogenous. Indeed, as Robert Asen (2000) argues, "A single, overarching public sphere ignores or denies social complexity insofar as it invokes a notion of publicity as contemporaneous face-to-face encounters among all citizens potentially affected by issues under consideration." This complexity has driven most theorists to articulate that multiple publics exist in any communication scenario.

The dichotomy between "the public" and "publics" has importance because technical communicators are often challenged to create a single technical document (a webpage, a policy, an instruction manual) that works for "the public." Susan Youngblood (2012), for example, demonstrates the complexities of developing emergency-planning websites for "the public" where ***information*** products must meet the demands of a number of stakeholders. In these cases, "the public" remains ambiguous at best and might best be described as "anyone who reads the document."

Technical communicators have handled the need to communicate with "the public" through a range of best practices, most notably ***accessibility*** standards and ***plain language***. Accessibility standards and ***user experience*** testing, for example, allow for ***designers*** to ensure that even if and as "the public" is conceptualized as homogeneous, public-facing information products have a base-level of accessibility for a wide range of users. Plain language standards also provide a foundation for addressing "the public" in its diversity and difference by simplifying language for the widest range of readers and users. Yet even with these strategies, the problems facing technical communicators writing for "the public" are many: Different users will use the document or technology in different ways (Johnson, 1998). In other words, there is never really just one public; rather, there are many publics who "gather" around the same document or technology for different purposes. As a result, technical communicators navigate public-facing projects using user-centered approaches, breaking down "the public" into stakeholder and user groups whenever possible (Acharya, 2017; Zoetewey & Staggers, 2004).

Public vs. Counterpublic. Perhaps the most important result of the public(s) conversation is the acknowledgement that some publics exist in contradistinction to what might be called *the* public—those at the margin (minoritized groups and individuals) versus those in the center (typically those who most closely resemble Audre Lorde's mythical norm: straight, white, male, Christian, and middle class). For example, in *Technical Communication after the Social Justice Turn*, Rebecca Walton, Natasha Jones, and I describe the ways able-bodied users are often *de facto*, leaving those with disabilities at the margin (Walton et al., 2019). This example demonstrates the need not only to articulate that there are multiple publics but also that those publics are unequally positioned to navigate political and institutional authorities. The concept of counterpublics (Warner, 2005) offers an import frame for understanding these inequities. As Michael Warner (2005) observes, "Some publics . . . are more likely than others to stand in for *the* public, to frame their address as the universal discussion of the people" (p. 117).

Counterpublics are "not merely a subset of the public"; instead, they are defined in contradistinction to the dominant or mainstream (Warner, 2005, p. 118). Subordinated by the dominant public, counterpublics (including women, workers, and people of color, among others) have "'no arenas for deliberation among themselves about their needs, objectives, and strategies'" (Fraser, qtd. in Warner, p. 118). In her articulation of "the" Black Public Sphere, Catherine Squires (2002) takes this further, arguing not only that counterpublics exist, but that they sometimes operate differently in order to thrive or survive. In the wake of political inequity, then, counterpublics develop as resistant, oppositional, or contrary to the dominant public.

The implications of this dichotomy have caused a tectonic shift in the field of TC. It is not enough to acknowledge that there are multiple publics; instead, technical communicators must understand the way that power and oppression imbue the public sphere. W. Michelle Simmons (2007) provides a foundational example of this as she articulates the ways TC practices needed to shift in order to ethically and justly accommodate those with less power in an environmental case. The role of systemic oppression has become prominent in the field's social justice turn, emphasizing the need for technical communicators to consider *counter*publics. Emma Rose and Rebecca Walton (2015), for example, articulated the ways particular users of public-transit systems (homeless bus riders) are often vulnerable to (and under-consulted on) system changes. Similarly, Lucía Durá and colleagues (2019) revealed the way Latinx migrants have limited support to navigate end-of-life contexts in the United States.

Public vs. Community. In the *Journal of Business and Technical Communication (JBTC)* special issue on business and technical communication in the public sphere, a number of articles address the impact TC can have in the public sphere and "convey a quiet optimism about the possibilities of using and improving texts for solving problems in the public sphere" (Rude, 2008). The first article in the issue begins "In a community we call Harbor . . ." and then describes "finding a way to work effectively with communities marked by severe distrust and broken relationships" (Blythe et al., 2008, p. 279). This linguistic move provides insight into a final proposed dichotomy: public vs. community.

TC has, as demonstrated in this entry, engaged with the public sphere in many ways, but often, there is slippage between public and community work. For example, the work of Dura et al. (2019) mentioned above arguably focuses on counterpublics, but the authors describe their project as a form of community-based user experience (UX). What do we get from *community* that we don't otherwise get from *publics*?

Warner (2005) describes a public as a collection of strangers; he argues that publics are formed through the circulation of documents. The public or publics cannot be known because they aren't stable and cannot be pre-determined. Communities, on the other hand, are intimate collections of individuals. When Stuart Blythe and colleagues (2008) describe Harbor as a community, it is because the group is a known entity, an emplaced and connected group of individuals. Community, in other words, focuses on connection and what is shared among individuals. Walton and colleagues (2015) demonstrate as much when they discuss their ***research*** in Rwanda. Focusing on the community, their research emerged as messy, deeply contextualized, and fundamentally collaborative. The focus on community provided these two research groups with an ability to engage with members who have specific needs, individualized stories, and culturally specific ***knowledge***. When technical communicators write for a collection of strangers (or a public), our orientation towards those individuals may become distanced, neutral, and objective; this neutrality, as Cecilia Shelton (2020) argues, can do harm. Shifting to a community-based framework may be one strategy for critically engaging those who have traditionally been excluded from "the public sphere," that is, the counterpublics.

The various dichotomies about "public" don't hold together under scrutiny and do not create easily defined categories or labels, yet they offer productive tensions to consider the way the concept of the public has and continues to affect TC practice.

References

Acharya, K. R. (2017). User value and usability in technical communication: A value-proposition design model. *Communication Design Quarterly Review*, *4*(3), 26–34. https://doi.org/10.1145/3071078.3071083.

Asen, R. (2000). Seeking the "counter" in counterpublics. *Communication Theory*, *10*(4), 424–446. https://doi.org/10.1111/j.1468-2885.2000.tb00201.x.

Berlant, L. & Warner, M. (1998). Sex in public. *Critical inquiry*, *24*(2), 547–566.

Blythe, S., Grabill, J. T. & Riley, K. (2008). Action research and wicked environmental problems: Exploring appropriate roles for researchers in professional communication. *Journal of Business and Technical Communication*, *22*(3), 272–298. https://doi.org/10.1177/1050651908315973.

Durá, L., Gonzáles, L. & Solis, G. (2019, October). Creating a bilingual, localized glossary for end-of-life-decision-making in borderland communities. In *Proceedings of the 37th ACM International Conference on the Design of Communication* (pp. 1–5). https://doi.org/10.1145/3328020.3353940.

Durack, K. (1997). Gender, technology and the history of technical communication. *Technical Communication Quarterly*, *6*(3), 249–260. https://doi.org/10.1207/s15427625tcq0603_2.

Edenfield, A. C., Holmes, S. & Colton, J. S. (2019). Queering tactical technical communication: DIY HRT. *Technical Communication Quarterly*, *28*(3), 177–191. https://doi.org/10.1080/10572252.2019.1607906.

Flower, L. (2008). *Community literacy and the rhetoric of public engagement.* Southern Illinois University Press.

Fraser, N. (1990). Rethinking the public sphere: A contribution to the critique of actually existing democracy. *Social Text*, (25/26), 56–80. https://doi.org/10.2307/466240.

Goodnight, G. T. (2012). The personal, technical, and public spheres: A note on 21st century critical communication inquiry. *Argumentation and Advocacy*, *48*(4), 258–267. https://doi.org/10.1080/00028533.2012.11821776.

Haas, A. M. & Eble, M. F. (Eds.). (2018). *Key theoretical frameworks: Teaching technical communication in the twenty-first century*. Utah State University Press.

Habermas, J. (1991). *The structural transformation of the public sphere: An inquiry into a category of bourgeois society*. MIT press. (Original work published 1962)

Hauser, G. A. (1999). *Vernacular voices: The rhetoric of publics and public spheres*. University of South Carolina Press.

Johnson, R. R. (1998). *User-centered technology: A rhetorical theory for computers and other mundane artifacts*. SUNY press.

Jones, N. N., Moore, K. R. & Walton, R. (2016). Disrupting the past to disrupt the future: An antenarrative of technical communication. *Technical Communication Quarterly*, *25*(4), 211–229. https://doi.org/10.1080/10572252.2016.1224655.

Jones, N. N. & Williams, M. F. (2017). The social justice impact of plain language: A critical approach to plain-language analysis. *IEEE Transactions on Professional Communication*, *60*(4), 412–429. https://doi.org/10.1109/TPC.2017.2762964.

Katz, S. B. (1992). The ethic of expediency: Classical rhetoric, technology, and the Holocaust. *College English*, *54*(3), 255–275. https://doi.org/10.2307/378062.

Long, E. (2008). *Community literacy and the rhetoric of local publics*. Parlor Press.

Lorde, A. (1984). *Sister outsider: Essays and speeches*. Crossing Press.

Novotny, M. & Hutchinson, L. (2019). Data our bodies tell: Towards critical feminist action in fertility and period tracking applications. *Technical Communication Quarterly*, *28*(4), 332–360. https://doi.org/10.1080/10572252.2019.1607907.

Rice, J. (2012). *Distant publics: Development rhetoric and the subject of crisis*. University of Pittsburgh Press. https://doi.org/10.2307/j.ctt5vkftk.

Rose, E. J. & Walton, R. (2015, July). Factors to actors: Implications of posthumanism for social justice work. In *Proceedings of the 33rd Annual International Conference on the Design of Communication* (pp. 1–10). https://doi.org/10.1145/2775441.2775464.

Rude, C. D. (2008). Introduction to the special issue on business and technical communication in the public sphere: Learning to have impact. *Journal of Business and Technical Communication*, *22*(3), 267–271. https://doi.org/10.1177/1050651908315949.

Scott, J. B. (2003). *Risky rhetoric: AIDS and the cultural practices of HIV testing*. Southern Illinois University Press.

Scott, J. B., Longo, B. & Wills, K. V. (2006). *Critical power tools*. SUNY Press.

Shelton, C. (2020). Shifting out of neutral: Centering difference, bias, and social justice in a business writing course. *Technical Communication Quarterly*, *29*(1), 18–32. https://doi.org/10.1080/10572252.2019.1640287.

Simmons, W. M. (2007). *Participation and power: Civic discourse in environmental policy decisions*. SUNY Press.

Squires, C. R. (2002). Rethinking the Black public sphere: An alternative vocabulary for multiple public spheres. *Communication Theory*, *12*(4), 446–468. https://doi.org/10.1111/j.1468-2885.2002.tb00278.x.

Sullivan, D. L. (1990). Political-ethical implications of defining technical communication as a practice. *Journal of Advanced Composition*, *10*(2), 375–386.

Walton, R., Moore, K. & Jones, N. (2019). *Technical communication after the social justice turn: Building coalitions for action*. Routledge. https://doi.org/10.4324/9780429198748.

Walton, R., Zraly, M. & Mugengana, J. P. (2015). Values and validity: Navigating messiness in a community-based research project in Rwanda. *Technical Communication Quarterly*, *24*(1), 45–69. https://doi.org/10.1080/10572252.2015.975962.

Warner, M. (2005). *Publics and counterpublics*. Zone books.

Youngblood, S. A. (2012). Balancing the rhetorical tension between right to know and security in risk communication: Ambiguity and avoidance. *Journal of Business and Technical Communication*, *26*(1), 35–64. https://doi.org/10.1177/1050651911421123.

Zoetewey, M. E. & Staggers, J. (2004). Teaching the Air Midwest case: A stakeholder approach to deliberative technical rhetoric. *IEEE Transactions on Professional Communication*, *47*(4), 233–243. https://doi.org/10.1109/TPC.2004.837969.

28. Research

Chris Lam
University of North Texas

Research is a nebulous term that can mean many different things to many different people. For some, research is equated with lengthy manuscripts as if the *output* of the research is the research itself. For others, research is conflated with the act of data collection and/or data analysis. In this keyword essay, I will examine how both of these definitions are incomplete in and of themselves. It is helpful, though, to first begin with a simple definition of the word *research* and then unpack and contextualize how this definition applies specifically to technical communication research. According to the *Oxford English Dictionary* (Oxford University Press, n.d.), *research* was first used as a verb in the late 16th century and derives from two morphemes (*re + search*). "Re" as a prefix is defined "with the general sense of 'back' or 'again'," and "search" is defined as the "examination or scrutiny for the purpose of finding a person or thing." While there are two primary definitions of research in the *Oxford English Dictionary*, the second is most relevant to academic research and to this essay:

> Research: Systematic investigation or inquiry aimed at contributing to knowledge of a theory, topic, etc., by careful consideration, observation, or study of a subject. In later use also: original critical or scientific investigation carried out under the auspices of an academic or other institution.

This definition, while only 43 words, provides much descriptive detail about research. It 1) qualifies research (*systematic*), 2) describes the act of research (*investigation or inquiry)*, 3) provides motive (*aimed at contributing to knowledge of a theory*), and 4) describes the methods in which research can be accomplished (*by careful consideration, observation, or study of a subject).*

To begin, it is important to clarify the distinction between product (the tangible output of research) and process (the act of doing research). For technical communication researchers, this distinction has significant ramifications because it can reveal competing values. For instance, in institutional contexts that more closely align with the social sciences, peer-reviewed journal articles are the gold standard. On the other hand, for technical communication faculty in humanistic departments, value may be more highly placed on scholarly monographs. In addition to differences in product, technical communicators have also historically diverged on both approaches and methods to research due to the diverse research training backgrounds in which technical communicators find themselves, which include ***rhetoric*** and composition, communication studies, human factors, and

DOI: https://doi.org/10.37514/TPC-B.2023.1923.2.28

linguistics (St.Amant & Melonçon, 2016). Regardless of background, a shared understanding that research involves *both* process *and* product and an acknowledgment that diversity exists within both of those categories are important starting points to understanding research within the context of technical communication. This essay will contextualize research within technical communication by outlining approaches, methods, and motives for research in the field.

There have been two primary approaches to research in technical communication as outlined in Davida Charney's foundational 1996 essay, "Empiricism is Not a Four-Letter Word." In her article, she clearly delineates two major schools of thought surrounding approaches to research in technical communication. On one hand, Charney describes a group of scholars who champion subjectivist methods (largely equated with qualitative methods). Subjectivists have been historically critical of objectivist methods, particularly in their ties to "patriarchal institutions of power" (Lay, 1991), no doubt inspired by Carolyn Miller's (1979) landmark work "A Humanistic Rationale for Technical Writing." On the other hand, there is an objectivist camp of scholars who argue that empirical approaches to research are essential to ***knowledge*** building. Nancy Coppola and Norbert Elliot (2005) similarly draw the distinction between big ***science*** and bricolage. Charney (1996) concludes her essay by asserting that "over-reliance on qualitative studies and repeated disparagement of objective methods is creating a serious imbalance in studies of technical and professional writing" (p. 590). She goes on to argue that "the numerous socially-situated ethnographies and case studies, excellent though each may be, cannot by themselves sufficiently extend and refine our methods and our knowledge base" (p. 590). Though Charney's essay was published in 1996, recent scholarship in technical communication suggests that there remains an over-reliance on subjectivist methods. For example, in a 2017 study, Chris Lam and Ryan Boettger examined 117 articles over a five-year period (2012–2016) and found a vast majority using subjectivist methods. Charney's allusion to knowledge gets at the third part of the *Oxford English Dictionary* definition of research: motive. As defined, the motivation of research is to contribute to knowledge of a theory. But, as Charney argues, if there is an overreliance on a particular approach to research, a knowledge base cannot be fully realized. The debate between objectivist and subjectivist methods was/is not only about methods themselves. Like the *Oxford English Dictionary* definition, it is merely one part of what makes research *research*. What Charney and others are arguing is that, while methods are important, the qualification, action, and motive of technical communication research are equally important.

While there are two primary approaches to research in technical communication, there are also foundational methods utilized by technical communicators. Research methods garner a lot of debate, but they are merely a means to an end. They act as a tool that allows researchers to answer research questions. According to George Hayhoe and Pam Estes Brewer (2020), technical communication has relied on five major methodological traditions: quantitative, qualitative, critical theory, literature review, and mixed methods. While this is true, it may be more helpful to understand methods within the context of technical communication by viewing methods through the lens of the data source or object of study. Most prominently, technical communicators have been interested in studying written texts. To study written texts, a variety of methodological traditions have been employed by technical communicators, including rhetorical analysis, discourse analysis, and content analysis. In her seminal work on integrating a ***social justice*** approach to technical communication, Natasha N. Jones (2016) further advocates for historical and archival research of texts that utilizes decolonial approaches. Also recently, innovative ***visual*** methods (McNely, 2013) and methods associated with big data (Graham et al., 2015) have been used to examine a variety of texts. Technical communicators have also studied people including practicing technical communicators, students, and faculty. Technical communicators have used methods including surveys, interviews, focus groups, diary studies, and participatory research to study people. Finally, technical communicators study contexts in which people interact with ***technology***. Methods like card sorting, participant observation, usability, and contextual inquiry have been used to examine these interactions. As McNely et al. (2015) put it, "technical communication's methodological and theoretical pluralism reveals the rich and diverse tapestry of opportunities for research and practice" (p. 6).

A final area that warrants discussion is debates surrounding the motive and purpose of research in technical communication. Simply put, why should we do research in the first place? What is the end goal of that research? If research is meant to contribute to a body of knowledge, what then is the role of researcher in facilitating the application of knowledge into practice? Certainly, there is much room for varying opinions, but an examination of the field's five major journals (*Technical Communication*, *Technical Communication Quarterly*, *Journal of Technical Writing and Communication*, *Journal of Business and Technical Communication*, and *IEEE Transactions on Professional Communication*) reveals varying publication practices in regards to knowledge application. For example, *Technical Communication* and *IEEE Transactions on Professional Communication* both require a "practitioner takeaways" section in their research reports. This is a clear signal that these publication venues value applied research and are trying to explicitly draw connections between academia and industry. While much of the motivation behind technical communication research has historically centered on "pragmatic topics," Jones et al. (2016) argue for research that is also motivated by ***feminism***, race and ethnicity, community engagement, and ***accessibility***, among other important areas for research. While motivations behind technical communication research are diverse, they are also often marred by the competing academic motivation of earning tenure and promotion. That is, it has also been argued that publication venues in technical communication "function as repositories for tenure and promotion materials" (Boettger & Friess, 2016, p. 322). When motivations for research become confounded by pressures to publish (i.e., the publish or perish paradigm), researchers may find themselves at odds with an original intent to put knowledge into practice. This can be seen in the research questions we choose to pursue and research topics we choose to explore. There is wide consensus in the field that there remains a divide between academics and practitioners and that research plays a vital role in bridging that divide (Melonçon & St.Amant, 2018. That is, if researchers attempt to answer questions that are relevant to practitioners, research output would necessarily be applied in practice. However, there is no clear consensus around what these fieldwide research questions ought to be or what topics are worth pursuing. Carolyn Rude (2009) attempted to address this lack of consensus by helpfully delineating fieldwide research questions. She outlined four major areas for research including disciplinarity, ***pedagogy***, practice, and social change (Rude, 2009). While these categories for research questions are clear in theory, recent research has found that there is still much misalignment between the questions academics pose and their relevance to practice. In studying the research topics of technical communicators over a 30-year period, Ryan Boettger and Erin Friess (2016) found little change over time. They argue that this, on one hand, could indicate "solidification of the core attributes of the field" (Boettger & Friess, 2016, p. 321). However, on the other hand, they argue "the amount of defined differences within our forums when compared to the size of our field could be symptomatic of the field's identified fragmentation" (p. 321).

While it can be tempting to delineate technical communication's diversity of approaches, methods, and motives to research as mutually exclusive and competing, examining the impact of such diversion requires more nuance. Charney herself never argued one approach at the exclusion of the other. Part of this necessary nuance around research in technical communication must focus on addressing problematic research practices within the field. Recent scholarship about research in technical communication has pointed to a lack in systematic and rigorous research, the very first qualification of research in *Oxford English Dictionary's* definition. In an article written in 2004, Ann Blakeslee and Rachel Spilka describe the state of technical communication research (Blakeslee & Spilka, 2004). A recurring problem in technical communication research is that "research in our field is too often predetermined to fulfill theoretical models rather than being used to challenge or build onto such models" (Blakeslee & Spilka, 2004, p. 76). It is the academic equivalent of proof-texting and rarely utilizes a systematic approach to research. Blakeslee and Spilka also discuss methods and accurately describe the field's plurality of methods as an asset, rather than a drawback. Rather than highlighting divisions between objectivist and subjectivist approaches to research, they highlight the necessity for both in advancement of knowledge. While advocating both approaches, they do point out a lack of awareness of methodological alignment to research questions. Specifically, they write, "Charney questions whether we have a good enough sense of which methods are helpful for which questions, and she proposes that we strive to do a better job, overall, of matching methods to questions" (Blakeslee & Spilka, 2004, p. 80). Lisa

Melonçon and Kirk St.Amant (2018) echo this point as they advocate for more sustainable research in technical communication that explicitly connects the dots between research questions, data collected and analyzed, and implications of the research in the reporting of research. The lack of systematic research is also discussed by S. Scott Graham (2017) when he describes much foundational knowledge in technical communication to be built upon lore, rather than systematic, empirical research. A common call for addressing this problem is a commitment to systematic and extensive training in methods, regardless of which approach researchers favor (Blakeslee & Spilka, 2004). Training in methods has also been addressed by many others in the field (Campbell, 2000; Boettger & Lam, 2013).

There is no clear answer to what research questions and topics should be emphasized in modern technical communication scholarship. But, to conclude this essay on research, it is essential to point out that a shared understanding of research, as defined in this essay, is one step in a potential path forward. That is, if the field can agree that research is 1) systematic, 2) investigative, 3) aimed at contributing to a body of knowledge, and 4) requires some method of investigation, research may be, as Melonçon and St.Amant (2018) put it, sustainable.

References

Blakeslee, A. & Spilka, R. (2004). The state of research in technical communication. *Technical Communication Quarterly*, *13*(1), 73–92. https://doi.org/10.1207/S15427625TCQ1301_8.

Boettger, R. K. & Friess, E. (2016). Academics are from Mars, practitioners are from Venus: Analyzing content alignment within technical communication forums. *Technical Communication*, *63*(4), 314–327.

Boettger, R. K. & Lam, C. (2013). An overview of experimental and quasi-experimental research in technical communication journals (1992–2011). *IEEE Transactions on Professional Communication*, *56*(4), 272–293. https://doi.org/10.1109/TPC.2013.2287570.

Campbell, K. S. (2000). Research methods course work for students specializing in business and technical communication. *Journal of Business and Technical Communication*, *14*(2), 223–241. https://doi.org/10.1177/105065190001400203.

Charney, D. (1996). Empiricism is not a four-letter word. *College Composition and Communication*, *47*(4), 567–593. https://doi.org/10.2307/358602.

Coppola, N. W. & Elliot, N. (2005). Big Science or bricolage: An alternative model for research in technical communication. *IEEE Transactions on Professional Communication*, *48*(3), 261–268. https://doi.org/10.1109/TPC.2005.853932.

Graham, S. S. (2017). Data and lore in technical communication research: Guest editorial. *Communication Design Quarterly Review*, *5*(1), 8–25. https://doi.org/10.1080/10572252.2015.975955.

Graham, S. S., Kim, S. Y., Devasto, D. M. & Keith, W. (2015). Statistical genre analysis: Toward big data methodologies in technical communication. *Technical Communication Quarterly*, *24*(1), 70–104. https://doi.org/10.1080/10572252.2015.975955.

Hayhoe, G. & Brewer, P. E. (2020). *A research primer for technical communication: Methods, exemplars, and analyses*. Routledge. https://doi.org/10.4324/9781003080688.

Jones, N. (2016). The technical communicator as advocate: Integrating a social justice approach in technical communication. *Journal of Technical Writing and Communication, 46*(3), 342–361. https://doi.org/10.1177/0047281616639472.

Jones, N. N., Moore, K. R. & Walton, R. (2016). Disrupting the past to disrupt the future: An antenarrative of technical communication. *Technical Communication Quarterly, 25*(4), 211–229. https://doi.org/10.1080/10572252.2016.1224655.

Lam, C. & Boettger, R. (2017). An overview of research methods in technical communication journals (2012–2016). In *IEEE International Professional Communication Conference*. IEEE. https://doi.org/10.1109/IPCC.2017.8013953.

Lay, M. M. (1991). Feminist theory and the redefinition of technical communication. *Journal of Business and Technical Communication, 5*(4), 348–370. https://doi.org/10.1177/1050651991005004002.

McNely, B. J. (2013). Visual research methods and communication design. In *SIGDOC '13: Proceedings of the 31st ACM International Conference on Design of Communication* (pp. 123–132). Association for Computing Machinery. https://doi.org/10.1145/2507065.2507073.

McNely, B., Spinuzzi, C. & Teston, C. (2015). Contemporary research methodologies in technical communication. *Technical Communication Quarterly, 24*(1), 1–13. https://doi.org/10.1080/10572252.2015.975958.

Melonçon, L. & St.Amant, K. (2018). Empirical research in technical and professional communication: A 5-year examination of research methods and a call for research sustainability. *Journal of Technical Writing & Communication, 49*(2), 128–155. https://doi.org/10.1177/0047281618764611.

Miller, C. R. (1979). A humanistic rationale for technical writing. *College English, 40*(6), 610–617. https://doi.org/10.2307/375964.

Oxford University Press. (n.d.). Research. In *Oxford English Dictionary*. Retrieved June 8, 2020, from https://www.oed.com/.

Rude, C. D. (2009). Mapping the research questions in technical communication. *Journal of Business and Technical Communication, 23*(2), 174–215. https://doi.org/10.1177/1050651908329562.

St.Amant, K. & Melonçon, L. (2016). Addressing the incommensurable: A research-based perspective for considering issues of power and legitimacy in the field. *Journal of Technical Writing and Communication, 46*(3), 267–283. https://doi.org/10.1177/0047281616639476.

29. Rhetoric

James E. Porter
Miami University

Very simply, *rhetoric* is the art of effective communication—in a wide variety of situations, from technical reports, web videos, ***social media*** postings, scholarly articles, proposals, and memos written at work, to everyday oral and written interactions among colleagues, friends, and family.

But rhetoric takes in more than spoken and written words. It includes all forms of symbolic interaction used to express, instruct, persuade, build relationships, and delight, including images, ***data visualizations***, bodily gestures, facial expressions, tattoos, mathematical expressions, music, movies, a thumbs up emoji at a Zoom meeting, a #BlackLivesMatter sign displayed at a ***public*** march, an Aztec codex pictogram (Baca, 2009), a quilt containing coded instructions to guide slaves to freedom (Banks, 2006), and other multimedia and nonverbal forms of expression. How a parent speaks to a child—both what they say and how they say it—that's rhetoric, too, or even just smiling at the child to express love. We practice rhetoric all the time, whenever we interact with others, even if we do not always label it *rhetoric*.

Rhetoric is also a formal academic field of study and of teaching—a humanistic, university-level discipline where scholars evaluate and critique communication practices and build theories, conduct ***research***, and recommend best practices for effective communication. At the university, rhetoric scholars are typically housed in departments of writing and rhetoric, communication, media studies, English, and/or technical/professional communication. But rhetoric as an applied field of practice extends across all university disciplines—business, engineering, ***science***, nursing, psychology, mathematics, computer ***technology***, graphic ***design***, music, education, etc.—since all academic disciplines form their ***knowledges***, necessarily, through writing and communication practices.

Rhetoric has long been closely linked with technical (and scientific and professional) communication, as evidenced by the considerable body of scholarship and ***research*** that builds upon and develops this connection and by the number of graduate and undergraduate degree programs whose identities link these two areas. Rhetoric provides the vital historical and theoretical grounding for technical/professional communication—that is, the operative principles that help us understand how to communicate effectively in professional contexts.

The definition that rhetoric is the art of effective communication sounds simple, but it begs a lot of questions and hides numerous complexities and several long-standing historical arguments. In fact, there are many competing definitions of rhetoric (Burton, 2016; Eidenmuller, 2020; Smit, 1997)—and many different

DOI: https://doi.org/10.37514/TPC-B.2023.1923.2.29

views of the scope and usefulness of rhetoric, even within the field of technical/professional communication.

There are two main competing views of rhetoric: a robust historical and scholarly one, but also a more pejorative, public usage that sees rhetoric as ***style*** in the superficial sense, as artificial ornamentation, verbal flourish, and bombast; rhetoric is dressing up ideas to make them seem more persuasive. The artificial ornamentation has the potential to be harmful, if it distracts, distorts, misleads, or skews the truth to achieve persuasive effect. In the public realm, the term *rhetoric* is almost always used in a disparaging way to refer to the lies or distortions of others. It is seen as the opposite of clarity, facts, reality, truth (Porter, 2020).

The more accurate historical view sees rhetoric as a noble art of truthful and ***ethical*** communication aimed not at deceiving an ***audience*** in order to persuade but rather at engaging audiences in order to teach them or interact with them cooperatively to address social needs and problems. Rhetoric is the necessary means by which we interact productively, cooperatively, collaboratively—in order to avoid conflict, promote positive relations, and achieve our goals. Rhetoric is inherently good, in other words—though of course it can be practiced badly.

Etymologically, *rhetoric* is a Greek (Attic) term: *Rhētorikē* is the art of speaking. *Rhētōr* refers to the speaker, orator, artist of discourse, or teacher of speaking. Roman rhetoricians sometimes referred to the art as *rhetorica*, using the Greek, or the Latin *oratoria* (MacDonald, 2014). Rhetoric theory certainly existed before and beyond the Greeks—different rhetorical concepts from other locations and ancient cultures (Lipson & Binkley, 2004)—but the term *rhetoric* itself comes from the ancient Greeks.

In the Mediterranean tradition, rhetoric emerged as a formal area of study in the 5th century BCE Athens, in the treatises of the Sophistic rhetoricians and in the schools of Isocrates, Plato, and Aristotle. The classical Greek, and then, later, Roman, rhetoricians recognized rhetoric as being its own distinct realm of knowledge important to the functioning of the *polis*, the Greek city state of Athens, and the republic of Rome. Rhetoric was the means by which civic life happened—at least in a democracy that permitted different voices to be heard. (Though not all voices were heard—not the voices of women or slaves.) The realm of rhetoric, according to Aristotle (*Rhetoric*, Book 1.3), was political speeches in the Athenian Assembly (deliberative), legal arguments (forensic), and speeches of praise (or blame) at ceremonial events (epideictic). In short, rhetoric was synonymous with public oratory. Rhetoric was also closely aligned with persuasion, as Aristotle defined rhetoric as "the faculty of observing in any given case the available means of persuasion" (*Rhetoric*, Book 1.2). As writing technologies improved and became more widely available (paper, stylus, ink), writing, too, became part of rhetoric.

The negative view of rhetoric in the Western tradition comes from Plato, specifically from his dialogue *Gorgias* (380 BCE). In *Gorgias*, Plato seems to dismiss rhetoric as "flattery . . . cookery . . . counterfeit," as largely a false art of placating or manipulating audiences. And yet in a later dialogue, *Phaedrus* (370

BCE), Plato acknowledges that, if used properly, rhetoric can move us toward the truth—if the rhetor possesses true knowledge and is motivated ultimately toward achieving good for others.

The Roman rhetorician Cicero had a broad view of the art: "The greatest orator is the one whose speech instructs, delights, and moves the spirit of the audience. To instruct is an obligation, to give pleasure a free gift, to move them is required" (*De Optimo*, I.3–4). Here, Cicero identifies rhetoric as having multiple purposes, with instruction as key—that is, to teach, instruct, inform is a requirement for rhetoric. That obligation has always been a strong purpose in technical/professional communication, and perhaps the primary one: reporting ***information*** in a way that instructs and helps audiences understand and use technology.

Quintilian's definition of rhetoric, from *Institutio Oratoria* (96 CE), even more strongly links rhetoric to ethical obligation, and particularly to the ***ethics*** of character. He defines rhetoric as "the art of speaking well" (2.14.5) or "a good man speaking well" (12.1.1). That definition insists that the rhetor must, first, be a virtuous person—*vir bonus*—or else they will not have the rhetorical credibility (*ethos*) to compel an audience. The good rhetor speaks with knowledge and expertise, and that expertise is very much guided by their public position, by their commitment to the pursuit of truth and knowledge, and by their obligation to the *polis*.

In other words, all acts of rhetoric should produce value, achieve some positive result for somebody—with the ultimate goal being the good of the *polis*, the republic, the state, and the citizens within it (Porter, 2020). Technical/professional communication has long defined its rhetorical mission as helping the reader or end user—in using clear and concise language, in designing usable documents, in creating accurate and valuable data visualizations, in conducting valid usability studies as a means of creating usable/useful interfaces, etc. These are ethical obligations to audience implicit in the rhetorical practices that define technical/professional communication.

Historically, rhetoric has had a queasy relationship with science—which led to disputes in the 20th century about the relevance of rhetoric to technical and scientific communication: i.e., about whether rhetoric was a helpful theoretical framework for the field. That debate has been settled now—yes, it is highly relevant and helpful—but it was not a given at first.

The European Enlightenment philosopher scientists of the 17th and 18th centuries saw rhetoric as antithetical to science. The Royal Society of London, founded in 1660, provides plentiful examples of hostility to rhetoric, seeing it as standing for unnecessary ornamentation, elaborate expression, and metaphoric bombast. Thomas Sprat, one of the founders of the society, referred to rhetoric as "this vicious abundance of Phrase, this trick of Metaphors, this volubility of Tongue, which makes so great a noise in the world" (1667, p. 111). The Royal Society certainly contributed to enshrining the degraded notion of rhetoric as false, as trickery, as ornamentation, and as a means of hiding the truth rather than revealing it.

According to Carolyn Miller (1979), this tension between science and rhetoric pertains to the positivism that science often promotes: "Science has to do with observation and logic, the only ways we have of approaching external, absolute reality. Rhetoric has to do with symbols and emotions, the stuff of uncertain, incomplete appearances" (p. 611). Because rhetoric deals in uncertainties, ambiguities, complexities, and probabilities—rather than certainties—it seems opposed to science.

However, the communication of scientists requires rhetorical knowledge (Gross, 1990)—e.g., about how to assemble data, organize it, design charts and graphs, and express conclusions clearly. Science relies on logic, reasoning, facts, and analysis, which is the rhetorical realm of *logos*—one of the three key persuasive appeals Aristotle emphasizes. In other words, science is not opposed to rhetoric; it *needs* rhetoric in order to develop and communicate scientific knowledge.

Historically, rhetoric has always had to adapt to change—to technological changes in communication media certainly, but not only those. How will rhetoric continue to adapt to meet the changing needs of society and recent developments in technology? Two key developments are the emergence of *cultural rhetorics* and *machine writing/rhetoric*, both of which fall under the heading of *posthumanist rhetorics* (Sackey et al., 2019)—i.e., rhetoric theories that challenge traditional humanistic assumptions about the nature of human communication.

For many years, scholars in rhetoric, technology, and technical/professional communication have argued the need to treat matters of race, ethnicity, gender, sexuality, ability/disability, and culture broadly understood as central to the field. The traditional inclination to treat these concerns as neutral," as monolithic, or, worse, as extraneous or irrelevant to considerations of technology and technical communication, needs to end (Cobos et al., 2018; Haas, 2012). Cultural concerns, especially the recognition of diversity as well as the acknowledgement of inequity in power relations (e.g., colonialism), are essential to the *techne* of rhetoric.

For technical communication, such a concern would mean, for example, viewing the Flint Water Crisis of 2014 as not simply a neutral technological failure but also as a failure of social relations involving race, socioeconomic status, power, inequity, and politics (see Sackey et al., 2019). Writing a technical report in this context without acknowledging how a white political power structure operated to deny, neglect, and ignore the material needs of the Black community is to instrumentalize the technology by removing the human element. It is, in short, to miss the point altogether. Technology, or technological communication, cannot overlook or neglect the broader social context and the material conditions of the human experience, the human suffering, the Black bodies, many of them children, that are the core of this rhetorical context. Similarly, cultural factors are important in the design of technology, as effective design needs to consider the diversity of users and the varying expectations, attitudes, and abilities that different users are likely to bring to technology use (Sun, 2006, 2012).

Technical/professional communication needs to prepare for the day when writing and communication will be produced mostly by machines, with humans functioning more in the role of editorial oversight. Artificial intelligence (AI) writing systems are already doing writing tasks previously done by humans—not just ***editing*** and simple text processing, but actual full text composition. AI writing agents transcribe meetings and produce minutes (Voicea's Eva), write emails to set up appointments (x.ai's Amy), and communicate via text chat with customers (customer service bots). AI systems publish news stories (the *Washington Post*'s Heliograf), create financial reports (Narrative Science's Quill), produce marketing copy (Persado), (co)write emails (Google Compose), and even produce entire documents from simple prompts (ChatGPT). Quite simply, we are already immersed in AI-created professional communications (McKee & Porter, 2020, 2021). Increasingly, technical com municators will be expected to collaborate/co-write with machines.

Rhetoric must always reinvent itself for new times, adapting to new media, new technologies, and changing social attitudes about what is appropriate, just, fair, logical, and factual. Nonetheless, the fundamental definition remains unchanged: *Rhetoric is the art of effective communication*—learning it, practicing it, teaching it—in whatever time and place and cultural moment we are in, with whatever communication technologies we are using.

References

Aristotle. (n.d.). *Rhetoric* (W. R. Roberts, Trans.). Internet Classics Archive. http://classics.mit.edu/Aristotle/rhetoric.html.

Baca, D. (2009). The Chicano Codex: Writing against historical and pedagogical colonization. *College English*, *71*(6), 564–583.

Banks, A. (2006). *Race, rhetoric, and technology: Searching for higher ground*. Erlbaum https://doi.org/10.4324/9781410617385.

Burton, G. O. (2016). *Rhetoric*. Silvae Rhetoricae. http://rhetoric.byu.edu/Encompassing%20Terms/rhetoric.htm.

Cicero. (n.d.). *De optimo genere oratorum*. Vicifons. https://la.wikisource.org/wiki/De_optimo_genere_oratorum.

Cobos, C., Rios, G. R., Sackey, D. J., Sano-Franchini, J. & Haas, A. M. (2018). Interfacing cultural rhetorics: A history and a call. *Rhetoric Review*, *37*(2), 139–154. https://doi.org/10.1080/07350198.2018.1424470.

Eidenmuller, M. E. (2020). *Scholarly definitions of rhetoric*. American Rhetoric. https://www.americanrhetoric.com/rhetoricdefinitions.htm.

Gross, A. G. (1990). *The rhetoric of science*. Harvard University Press.

Haas, A. M. (2012). Race, rhetoric, and technology: A case study of decolonial technical communication theory, methodology, and pedagogy. *Journal of Business and Technical Communication*, *26*(3), 277–310. https://doi.org/10.1177/1050651912439539.

Lipson, C. S. & Binkley, R. A. (Eds.). (2004). *Rhetoric before and beyond the Greeks*. SUNY Press.

MacDonald, M J. (2014). Glossary of Greek and Latin rhetorical terms. In *The Oxford handbook of rhetorical studies*. Oxford Handbooks Online. https://www.oxfordhandbooks.com/view/10.1093/oxfordhb/9780199731596.001.0001/oxfordhb-9780199731596-miscMatter-9.

McKee, H. A. & Porter, J. E. (2020). Ethics for AI writing: The importance of rhetorical context. In *Proceedings of 2020 AAAI/ACM Conference on AI, Ethics, and Society (AIES'20)*. https://doi.org/10.1145/3375627.3375811.

McKee, H. A. & Porter, J. E. (2021). Intertext: Writing machines and rhetoric. In A. H. Duin & I. Pedersen (Eds.), *Writing futures: Collaborative, algorithmic, autonomous*. (pp. 27–52). Springer.

Miller, C. R. (1979). A humanistic rationale for technical writing. *College English*, *40*(6), 610–617. https://doi.org/10.2307/375964.

Plato. (1871). *Phaedrus* (B. Jowett, Trans.). Internet Classics Archive. http://classics.mit.edu/Plato/phaedrus.html.

Plato. (1892). *Gorgias* (B. Jowett, Trans.). Internet Classics Archive. http://classics.mit.edu/Plato/gorgias.html.

Porter, J. E. (2020). Recovering a good rhetoric: Rhetoric as *techne* and *praxis*. In J. Duffy & L. Agnew (Eds.), *Rewriting Plato's legacy: Ethics, rhetoric, and writing studies* (pp. 15–36). Utah State University Press.

Quintilian. (1920–1922). *Institutes of oratory* (*Institutio oratoria*) (H. E. Butler, Trans.). Loeb Classical Library. https://penelope.uchicago.edu/Thayer/E/Roman/Texts/Quintilian/Institutio_Oratoria/home.html.

Sackey, D. J., Boyle, C., Xiong, M., Rios, G., Arola, K. & Barnett, S. (2019). Perspectives on cultural and posthumanist rhetorics. *Rhetoric Review*, *38*(4), 375–401. https://doi.org/10.1080/07350198.2019.1654760.

Smit, D. W. (1997). The uses of defining rhetoric. *Rhetoric Society Quarterly*, *27*(2), 39–50. https://doi.org/10.1080/02773949709391092.

Sprat, T. (1667). *History of the Royal Society of London, for the Improving of Natural Knowledge*. http://name.umdl.umich.edu/A61158.0001.001.

Sun, H. (2006). The triumph of users: Achieving cultural usability goals with user localization. *Technical Communication Quarterly*, *15*(4), 457–481. https://doi.org/10.1207/s15427625tcq1504_3.

Sun, H. (2012). *Cross-cultural technology design: Creating culture-sensitive technology for local users*. Oxford University Press. https://doi.org/10.1093/acprof:oso/9780199744763.001.0001.

30. Risk Communication

J. Blake Scott
University of Central Florida

The Western concept of risk is "a relatively novel phenomenon, seeping into European languages in the last 400 years," writes Gabe Mythen (2004), though there is no clear consensus about the term's etymology (p. 13). Among other meanings, scholars have traced *risk* to the Arabic word *risq* associated with wealth and fortune and to the Latin word *riscus* as referencing a slippery place or a steep rock or cliff that sailors must look out for in uncharted waters. Over the past few centuries, risk was increasingly quantified to measure possible outcomes in areas such as insurance and finance, where it was tied to probability more than uncertainty (Mythen, 2004, p. 13). Although technical and professional communication (TPC) scholars have continued to explore and dimensionalize the relationship between risk and uncertainty (Sauer, 2002; Walsh & Walker, 2016), this distinction has become blurred in the common contemporary understanding of risks as anticipated and uncertain dangers or threats.

By the late 20th century, "the term 'risk' obtained a pervasive and even intrusive presence in almost all institutionalized discursive fields in modern western societies" (van Loon, 2002, p. 5). A range of institutional efforts—such as government agencies, laws and regulations, and consulting firms—have been formed to predict, prepare for, and manage risks, particularly environmental, ***public*** health, and medical ones. Such efforts generated the modern field of risk analysis, which Alonzo Plough and Sheldon Krimsky (1987) described as "concerned primarily with predicting or quantifying the risks of 'scientifically identified hazards'" (p. 5). They added that risk analysis and management, as informed by decision ***science*** (developed in World War II), faced the challenge of connecting "the *assessment of risk*" to "political decisions concerning the *types, levels, and distribution* of risk [and resources to address it] acceptable to a society" (Plough & Krimsky, 1987, p. 5).

Although risk communication has been a more prevalent thread of ***research*** in communication studies (including health communication) and cognitive psychology (Reamer, 2015), it has become "an increasingly important aspect of the work of both technical experts and professional communicators" (Waddell, 1995, p. 1). We can track our field's engagement with risk communication along a general trajectory that moves from more *narrowly technical*, to ***rhetorical** and social*, and then to *cultural, material, and political attention* to risk communication, and that expands our notions of technical risk communicators' roles and responsibilities.

Risk communication was born from the need to convey to the public and other stakeholders levels of risks and their significance, and to gain cooperation with "decisions, actions or policies aimed at managing or controlling such risks"

DOI: https://doi.org/10.37514/TPC-B.2023.1923.2.30

(see the definition by Covello et al., 1988, p. 112). It was also born from the growing recognition of a disconnect between expert and public conceptions of risk, and a growing distrust in risk management authorities, exigencies that several scholars have linked to the environmental advocacy movement that began in the 1960s (see Grabill & Simmons, 1998). Steven B. Katz and Carolyn R. Miller (1996) argued that the initial goal of risk communication was "'correcting' the public's 'risk perceptions' so they would better match the 'risk analyses' made by experts" (p. 116). This goal has been critiqued by risk communication scholars, including those in TPC, as grounded in a technocratic model characterized by an overvaluation of expert risk determination and assessment, the one-way transmittal of ***information*** from expert to public, and an assumption that public questioning of expert risk information is grounded in irrationality and must be corrected (see Katz & Miller, 1996; Rowan, 1994; Waddell, 1995).

Starting in the 1980s, strictly technocratic approaches to risk assessment and communication gave way to a broader engagement of psychological and social considerations for bridging the expert-public divide, as evidenced by discussions of trust, motivations, values, and experiences, and by research on risk perception and the social amplification of risk (McComas, 2006; Powell & Leiss, 1997). This shift was accompanied by a recognition of risk as socially and rhetorically constructed (see Field-Springer & Striley, 2014; Hilyard, 2014), and by more social and participatory models of risk communication. In concert with this shift, Plough and Krimsky (1987) advocated for a sociocultural definition of risk that more expansively accounts for communication "from any source to any recipient" (p. 7) and broader considerations of risk understanding and acceptability (p. 6). In her call for a rhetorical model of risk communication, Katherine E. Rowan (1994) pointed the way for technical and professional communicators to consider the challenges of persuasion and participation, including around the cultivation of credibility (p. 403).

Extending the social turn of risk communication, TPC scholars and rhetoricians have further conceptualized risk as rhetorically and socially constructed and risk communication as necessitating the fuller involvement of those affected by risk management decisions. A number of such scholars, some of whom also identify as rhetoricians of science, ***technology***, and medicine, have focused on risk communication in case studies of specific, time-bound risk crises and controversies (Reamer, 2015, p. 350; see also Jensen, 2015), while others have sought to expand this purview to longitudinal studies of changing risk communication strategies (Reamer, 2015) or to public-relations-oriented risk communication by researchers (Giles, 2010).

Some studies of specific crises have offered retroactive analysis of internal communication failures leading up to a particular crisis (e.g., Dombrowski, 1991; Herndl et al., 1991; Winsor, 1988). Because of its focus, this work overlaps with the area of emergency management and ***crisis communication***. Other studies have examined TPC involved in the more public engagement of risk around environmental, health, or other controversies, focusing on the social-rhetorical dimensions of the communication between experts and publics or area communities, and pointing to ways to improve the communication processes, texts, and spaces involved (Katz & Miller, 1996; Nagelhout et al., 2009; Stratman et al., 1995). As Ed Nagelhout and colleagues (2009) noted, TPC scholars have increasingly argued "that decisions about risk should be the shared responsibility of all stakeholders" (p. 229). In his discussions of environmental communication efforts about sustainable development in the Great Lakes ecosystem, Craig Waddell (1995) called for a multi-directional "social constructionist model" in which "*all participants* also communicate, appeal to, and engage values, beliefs, and emotions" in making policy decisions (p. 207; see also Katz & Miller, 1996).

Jeffrey T. Grabill and W. Michele Simmons (1998) went further, critiquing the limitations of "negotiated" approaches that have responded to the technocratic limiting of public input, arguing that they idealize public participation without addressing challenges to shared decision-making, including asymmetrical power relations and a limited conception of stakeholders (p. 430). They called their alternative model a "critical rhetoric of risk communication," arguing for the public's involvement from the beginning of risk definition and assessment and thereby collapsing the distinction between risk assessment/analysis and risk communication (p. 417). Simmons (2007) extended this work in her case analyses of environmental policymaking, arguing that risk management institutions typically separate public participation from actual policy formation. Reminding us that citizens also have expertise, Simmons advocated for more fully participatory processes distinguished by shared decision-making (rather than, say, "strategic or "pseudoparticipatory" approaches that are still expert-driven) by offering flexible heuristics for assessing citizen roles and identifying "spaces and moments" for impactful contributions (p. 133).

Discussions of more participatory models of risk communication also have suggested more expansive roles for technical and professional communicators. Departing from Barbara Mirel (1994), Grabill and Simmons (1998) argued that such communicators should do more than disseminate or mediate risk assessment, but rather are "uniquely qualified" to participate in risk assessment and related communication and policymaking, through the construction of user ***knowledge*** (e.g., usability research), and through the facilitation of public involvement and action (pp. 434–435). Although technical communicators might face challenges in facilitating stakeholder input (see Youngblood, 2012), Grabill and Simmons called on technical communicators to be symbolic analysts and user/public advocates who move "between ranges and varieties of experts and nonexperts" (p. 434.). Simmons (2007) added that technical communication specialists can help citizens and citizen groups build technical capacity for information sharing and policymaking involvement, including in both institutional and extra-institutional contexts. Huiling Ding (2009) later critiqued some more participatory models and roles as overly idealistic and Western-centric, noting that they assume "that technical communicators play key roles in risk communication processes" (p. 331) and that they overlook "larger power issues such as national/regional protectionism, corporate interests, and systematic governmental censorship" (p. 332).

In addition to more social and rhetorical models of risk communication, technical and professional communicators have turned to its cultural, material, and political dimensions. Some have examined these dimensions in specific workplace contexts fraught with risk. Beverly Sauer's (2002) work on risk communication in hazardous mining environments is noteworthy for its nuanced, contextualized analysis of how miners manage the "dynamic uncertainty" of their environments and the multiple levels and types of institutional and cultural knowledges at play. In discussing ways to improve technical risk communication in such contexts, including for ***visual*** representations and embodied forms of training, Sauer resisted an easy separation between risk analysis, risk communication, and user uptake and negotiation. In another study of safety communication, in this case for Latino construction workers, Carlos Evia and Ashley Patriarca (2012) argued that additional considerations of language and other differences among stakeholders are needed to develop more responsively ***designed*** and culturally attuned forms of communication.

Other scholars have examined cultural and material dimensions of stakeholder-driven risk assessment and decision-making in ***medical/health communication*** contexts, aiming to empower patients, health consumers, and health publics. For example, Candice Welhausen (2017) examined consumers' localized, "do-it-yourself" (DIY) risk assessment through disease-tracking apps such as "Flu Near You." Lora (Arduser) Anderson (2017) similarly studied how people with diabetes re-articulate and manage information about their risk factors through, among other mechanisms, patient-produced communication and online patient networks. Kelly Pender (2018) extended this focus on patient-generated, materially enacted risk assessment by examining the various embodied and technological practices through which women enact BRCA+ risk, arguing that such risk "should be understood as something that women *do*" (p. 73). Heidi Y. Lawrence (2020) examined the material exigencies of vaccines to locate alternative discourses and deliberative spaces for responding to vaccine skepticism based on more nuanced research about how practitioners, parents, and local communities perceive and experience uncertainties as risks but also as "benefits, questions, or other preoccupations regarding the best way to retain personal health" (p. 103). In his rhetorical-cultural analysis of HIV testing rhetorics and contexts, J. Blake Scott (2003) critiqued identity-based risk communication focused on risky people rather than practices, also advocating for alternative communication that enables people to make nonnormative identifications with risk and vulnerability (p. 116) based on interdependent "needs, concerns, and contexts" (p. 232).

Some TPC scholars have further foregrounded a ***social justice*** approach to ***documentation*** and technology design and use for health-related contexts fraught with risks. In separate studies, Godwin Agboka (2013) and Lucía Durá and colleagues (2019) dimensionalized participatory approaches to creating health-related documentation to more fully account for communities' localized uses and "sociocultural, economic, linguistic, and legal needs" (Agboka, 2013, p. 44); this echoed Sauer's (1996) imploration for technical and professional communicators to more thoroughly investigate stakeholders' local experiences and broader political, scientific, and historical dimensions of their cultural knowledges (p. 326). In other studies, Kristen R. Moore and colleagues (2018) and Maria Novotny and Les Hutchinson (2019) called for TPC specialists to help users repurpose technologies to enable practices of racial justice and women's reproductive empowerment, respectively.

TPC scholars have increasingly called for cultural-political approaches to communication design that respond to environmental risks, too. Donnie Sackey (2020) argued for employing value sensitive design based on environmental justice principles as a means of empowering wearable users. Lynda Olman and Danielle DeVasto (2020) proposed an adaptation of environmental risk visualization to better address hybrid and collective risks for the anthropocene. Aydé Enríquez-Loya and Kendall Léon (2020) offered a "cultural rhetorics approach to environmental justice" through "facilitatory writing" that similarly "engages . . . a constellated terrain of participants and actions" in response to environmental risks associated with "natural" disasters (p. 457).

In another expansion of risk communication's purview, technical and professional communicators also have turned our attention from specific cultural sites and their material and political considerations to transnational and transcultural dimensions and movements. Ding (2014) analyzed what she describes as transcultural, extra-institutional, and unauthorized forms of risk communication (e.g., personal narratives, proclamations) around the emerging SARS epidemic in China and North America; these forms, and the "guerilla" and alternative media in which they circulated, enabled professionals, citizen groups, and other members of transnational publics "to send out risk messages even when professional codes or official orders forbid such communication" (Ding, 2009, p. 344). Erin A. Clark Frost (2013) analyzed the risk communication *after* the Deepwater Horizon disaster, examining the mostly digital work by "complex transcultural networks" of various levels (from local to international) that challenge dominant narratives and understandings. As these studies demonstrate, technical and professional communication scholars have expanded the field's traditional focus of risk communication tensions between risk officials and publics to include ***intercultural communication*** among publics and stakeholders.

The progression of risk communication in technical and professional communication has paralleled broader developments both in the larger interdisciplinary area of risk communication and in technical and professional communication studies. Just as the multidisciplinary field of risk communication has shifted from the transmittal of narrow, technical analyses and assessments of risk to psychological, social, and broader cultural considerations and models, approaches to TPC about risks have expanded to better account for sociocultural (including embodied, material, and political) contexts of risk meaning-making and experience.

Our field has also increasingly developed approaches to risk documentation and design that empower users' consequential participation and redress inequitable harms. Just as TPC has recognized the expanded roles and contributions of technical communicators as authors (Slack et al., 1993), scholars of technical risk communication have expanded our considerations of technical and professional communicators as co-shaping risks and their meanings by learning from, engaging, and facilitating the empowerment of risk stakeholders.

References

Agboka, G. Y. (2013). Participatory localization: A social justice approach to navigating unenfranchised/disenfranchised cultural sites. *Technical Communication Quarterly*, *22*(1), 28–49. https://doi.org/10.1080/10572252.2013.730966.

Anserson (Arduser), L. (2017). *Living chronic: Agency and expertise in the rhetoric of diabetes*. Ohio State University Press. https://doi.org/10.2307/j.ctvw1d7ss.

Carson, R. (1962). *Silent spring*. Houghton Mifflin.

Covello, V., Sandman, P. & Slovic, P. (1998). *Risk communication, risk statistics, and risk comparison: A manual for plant managers*. Chemical Manufacturers Association.

Ding, H. (2009). Rhetorics of alternative media in an emerging epidemic: SARS, censorship, and extra-institutional risk communication. *Technical Communication Quarterly*, *18*(4), 327–350. https://doi.org/10.1080/10572250903149548.

Ding, H. (2013). Transcultural risk communication and viral discourses: Grassroots movements to manage global risks of H1N1 flu pandemic. *Technical Communication Quarterly*, *22*(2), 126–149. https://doi.org/10.1080/10572252.2013.746628.

Ding, H. (2014). *Rhetoric of global epidemics: Transcultural communication about SARS*. Southern Illinois University Press.

Dombrowski, P. M. (1991). The lessons of the Challenger disaster. *IEEE Transactions on Professional Communication*, *43*(4), 211–216. https://doi.org/10.1109/47.108666.

Dura, L., Gonzales, L. & Solis, G. (2019). Creating a bilingual, localized glossary for end-of-life-decision-making in borderland communities. In *Proceedings of the 37th ACM International Conference on the Design of Communication*, Article No. 30, ACM. https://doi.org/10.1145/3328020.3353940.

Enríquez-Loya, A. & Léon, K. (2020). Transdisciplinary rhetorical work in technical writing and composition: Environmental justice issues in California. *College English*, *82*(5), 449–459.

Evia, C. & Patriarca, A. (2012). Beyond compliance: Participatory translation of safety communication for Latino construction workers. *Journal of Business and Technical Communication*, *26*(3), 340–367. https://doi.org/10.1177/1050651912439697.

Field-Springer, K. & Striley, K. (2014). Risk communication: Social construction perspective. In T. L. Thompson (Ed.), *Encyclopedia of health communication* (pp. 1183–1185). SAGE.

Frost, E. A. (2013). Transcultural risk communication on Dauphin Island: An analysis of ironically located responses to the Deepwater Horizon disaster. *Technical Communication Quarterly*, *22*(1), 50–66. https://doi.org/10.1080/10572252.2013.726483.

Giles, T. D. (2010). Communicating the risk of scientific research. *Journal of Technical Writing and Communication*, *40*(3), 265–281. https://doi.org/10.2190/TW.40.3.c.

Goodnight, G. T. (1982). The personal, technical, and public spheres of argument: A speculative inquiry into the art of public deliberation. *Journal of the American Forensics Association, 18*, 214–227. https://doi.org/10.1080/00028533.1982.11951221.

Grabill, J. T. & Simmons, W. M. (1998). Toward a critical rhetoric of risk communication: Producing citizens and the role of technical communicators. *Technical Communication Quarterly, 7*(4), 415–441. https://doi.org/10.1080/10572259809364640.

Herndl, C. G., Fennell, B. A. & Miller, C. R. (1991). Understanding failures in organizations: The accidents at Three Mile Island and the Shuttle Challenger. In C. Bazerman & J. Paradis (Eds.), *Textual dynamics of the professions: Historical and contemporary studies in writing in professional communities* (pp. 279–305). University of Wisconsin Press.

Hilyard, K. M. (2014). Risk communication. In T. L. Thompson (Ed.), *Encyclopedia of health communication* (pp. 1175–1177). SAGE.

Jensen, R. E. (2015). Rhetoric of risk. In H. Cho, T. Reimer & K. A. McComas (Eds.), *The SAGE handbook of risk communication* (pp. 86–97). SAGE. https://doi.org/10.4135/9781483387918.n11.

Katz, S. B. & Miller, C. R. (1996). The low-level radioactive waste siting controversy in North Carolina: Toward a rhetorical model of risk communication. In C. G. Herndl & S. C. Brown (Eds.), *Green culture: Environmental rhetoric in contemporary America* (pp. 111–140). University of Wisconsin Press.

Lawrence, H. Y. (2020). *Vaccine rhetorics*. Ohio State University Press.

McComas, K. A. (2006). Defining moments in risk communication research: 1996–2005. *Journal of Health Communication, 11*(1), 75–91. https://doi.org/10.1080/10810730500461091.

Mirel, B. (1994). Debating nuclear energy: Theories of risk and purposes of communication. *Technical Communication Quarterly, 3*(1), 41–65. https://doi.org/10.1080/10572259409364557.

Moore, K. R., Jones, N., Cundiff, B. S. & Heilig, L. (2018). Contested sites of health risks: Using wearable technologies to intervene in racial oppression. *Communication Design Quarterly, 5*(4), 52–60. https://doi.org/10.1145/3188387.3188392.

Mythen, G. (2004). *Ulrich Beck: A critical introduction to the risk society*. Pluto Press.

Nagelhout, E., Staggers, J. & Tillery, D. (2009). Risk communication, space, and findability in the public sphere: A case study of a physical and online information center. *Journal of Technical Writing and Communication, 39*(3), 227–243. https://doi.org/10.2190/TW.39.3.b.

Novotny, M. & Hutchinson, L. (2019). Data our bodies tell: Towards feminist critical action in fertility and period tracking applications. *Technical Communication Quarterly, 28*(4), 332–360. https://doi.org/10.1080/10572252.2019.1607907.

Olman, L. & DeVasto, D. (2020). Hybrid collectivity: Hacking environmental risk visualization for the anthropocene. *Communication Design Quarterly, 8*(4), 15–28. https://doi.org/10.1145/3431932.3431934.

Pender, K. (2018). *Being at genetic risk: Toward a rhetoric of care*. The Pennsylvania State University Press.

Plough, A. & Krimsky, S. (1987). The emergence of risk communication studies: Social and political context. *Science, Technology & Human Values, 12*(3/4), 4–10.

Powell, D. & Leiss, W. (1997). *Mad cows and mother's milk: The perils of poor risk communication*. McGill-Queen's University Press.

Reamer, D. (2015). "Risk = Probability × Consequences": Probability, uncertainty, and the Nuclear Regulatory Commission's evolving risk communication rhetoric. *Technical Communication Quarterly*, *24*(4), 349–373. https://doi.org/10.1080/10572252.2015.1079334.

Reynolds, B. & Seeger, M. W. (2005). Crisis and emergency risk communication as an integrative model. *Journal of Health Communication*, *10*(1), 43–55. https://doi.org/10.1080/10810730590904571.

Rowan, K. E. (1994). The technical and democratic approaches to risk situations: Their appeal, limitations, and rhetorical alternative. *Argumentation*, *8*, **391–409.** https://doi.org/10.1007/BF00733482.

Sackey, D. J. (2020). One-size-fits-none: A heuristic for proactive value sensitive environmental design. *Technical Communication Quarterly*, *29*(1), 33–48. https://doi.org/10.1080/10572252.2019.1634767.

Sauer, B. (2002). *The rhetoric of risk: Technical documentation in hazardous environments*. Routledge. https://doi.org/10.4324/9781410606815.

Sauer, B. A. (1996). Communicating risk in a cross-cultural context: A cross-cultural comparison of rhetoric and social understandings in U.S. and British mine safety training programs. *Journal of Business and Technical Communication*, *10*(3), 306–329. https://doi.org/10.1177/1050651996010003002.

Scott, J. B. (2003). *Risky rhetoric: AIDS and the cultural practices of HIV testing*. Southern Illinois University Press.

Simmons, W. M. (2007). *Participation and power: Civic discourse in environmental policy decisions*. SUNY Press.

Slack, J. D., Miller, D. J. & Doak, J. (1993). The technical communicator as author: Meaning, power, authority. *Journal of Business and Technical Communication*, *7*(1), 12–36. https://doi.org/10.1177/1050651993007001002.

Stratman, J. F., Boykin, C., Holmes, M. C., Laufer, M. J. & Breen, M. (1995). Communication, metacommunication, and rhetorical stases in the Aspen-EPA Superfund controversy. *Journal of Business and Technical Communication*, *9*(1), 5–41. https://doi.org/10.1177/1050651995009001002.

van Loon, J. (2002). *Risk and technological culture: Towards a sociology of virulence*. Routledge.

Waddell, C. (1995). Defining sustainable development: A case study in environmental communication. *Technical Communication Quarterly*, *4*(2), 201–216. https://doi.org/10.1080/10572259509364597.

Walsh, L. & Walker, K. C. (2016). Perspectives on uncertainty for technical communication scholars. *Technical Communication Quarterly*, *25*(2), 71–86. https://doi.org/10.1080/10572252.2016.1150517.

Welhausen, C. A. (2017). At your own risk: User-contributed flu maps, participatory surveillance, and an emergent DIY risk assessment ethic. *Communication Design Quarterly*, *5*(2). https://dl.acm.org/doi/10.1145/3131201.3131206.

Winsor, D. A. (1988). Communication failures contributing to the Challenger accident: An example for technical communicators. *IEEE Transactions on Professional Communication*, *31*(3), 101–107. https://doi.org/10.1109/47.7814.

Youngblood, S. A. (2012). Balancing the rhetorical tension between right to know and security in risk communication: Ambiguity and avoidance. *Journal of Business and Technical Communication*, *26*(1), 35–64. https://doi.org/10.1177/1050651911421123.

31. Science

Kathryn Northcut
Missouri S&T

Science is a complex term, being defined in the *Oxford English Dictionary* (Oxford University Press, n.d.) in about 6,000 words, and by scholars of technical communication and ***rhetoric*** in even more extensive presentations (e.g., Taylor, 1996; Longo, 2000). Science is expected to observe facts, extrapolate to universal truths, solve problems, and answer our questions about the universe through ***research*** and theories. For technical communicators, *science* can be one of the most important key terms of our careers, entailing a domain of knowledge and activity that supports millions of jobs. In our current landscape featuring the COVID-19 pandemic, catastrophes propagated by climate change, and increased human reliance on ***technology***, science ***literacy*** has become a fundamental need for all citizens.

Science as we understand it today is the distillation of intellectual traditions from multiple civilizations. In the 20th century, science was cemented as a key term of the Anthropocene by scientists themselves, including well-known authors such as Thomas Kuhn, E.O. Wilson, and Stephen Jay Gould. Science inextricably intersects with ***history***, ***knowledge***, research, ***ethics***, ***rhetoric***, and technology. Science is a dominant theme of our age, critical to the understanding of technical communication both as a discipline and a ***profession***, intertwined throughout all the greatest hopes for, and threats to, life on Earth in the 21st century.

The French derivation of the term *science* is glossed as "knowledge, understanding, secular knowledge, knowledge derived from experience, study, or reflection, acquired skill or ability, knowledge as granted by God . . . , the collective body of knowledge in a particular field or sphere . . ." (Oxford University Press, n.d.). These definitions lend an air of authority and immutability to science, an expectation that scientific knowledge is final and absolute. This perception has been challenged extensively in more recent scholarship and literature about science where the nature of the collective or the community is deemed important to viewing the workings of both science and scientific communication as culturally constructed enterprises (Kuhn, 1970). Thomas Kuhn (2000), notably, defined science as follows in *The Road Since Structure*:

> Science is a cognitive empirical investigation of nature that exhibits a unique sort of progress, [which] . . . cannot be further explicated as "approximating closer and closer to reality" . . . rather, progress takes the form of ever-improving technical puzzle-solving ability, operating under strict—though always tradition-bound—standards of success or failure. (p. 2)

DOI: https://doi.org/10.37514/TPC-B.2023.1923.2.31

Kuhn (2000) refers to science as requiring "extraordinarily esoteric" and "often expensive" investigations which make possible "astonishingly precise and detailed knowledge" (p. 3). Kuhn (1970) also addresses the inherent difficulty in defining a concept as robust as science in a single definition such as a *Keywords* entry: "A concept of science drawn from [textbooks] is no more likely to fit the enterprise that produced them than an image of a national culture drawn from a tourist brochure or a language text" (p. 1).

Within the narrower field of technical communication, our research includes excellent scholarship focusing on various aspects of science. New theories of communication are developed based on the ways that science communicates findings and modern thinking. Such new theories include Kenneth Baake's (2003) metaphor harmonics and Maria Gigante's (2018) portal images. Examining contemporary and historical artifacts and ***genres*** in science enables us to better understand the influence of science on technical communication, and the interplay between fields (Brasseur, 2003; Gross et al., 2002). Case studies, pedagogical practices, and communication strategies involved in scientific communication comprise a robust area of scholarship (e.g., Fountain, 2014; Graves, 2013; Walsh, 2013; Yu, 2017; Yu & Northcut, 2018).

Some rhetorical theorists have sought to regularize and norm the ways we describe scientific thinking and logic. For example, Richard Lanham (1991), in the classical rhetorical text *A Handlist of Rhetorical Terms*, refers to "scientific proof" and cites Aristotle's classification of a type of knowledge that develops universally "true conclusions" (p. 122) proven by syllogistic (mathematical or deductive) logic and demonstration. Contemporary and emerging thought, by contrast, focuses on the ephemeral contingency of such "Truth," positing that scientific knowledge is culturally constructed and changes over time, both in response to new data and in response to cultural realities. As Kuhn (1970) theorized, science is paradigmatic, and paradigms are shared bodies of knowledge both reflecting and constituting community members (p. 176). Paradigmatic knowledge changes over time, supplanting the notion of singular, stable scientific Truth; paradigm changes can be abrupt and irregular, not steady and predictable. Such a philosophical bent is reflected in most of our field's rhetorical and critical scholarship about science and science communication. Understanding of paradigmatic changes in sciences is helpful when citizens struggle with what appears to be indecisiveness of scientists facing new phenomena, especially when adherence to ethical research or medical standards is the cause for delay or disagreement.

Canonical 20th century texts expand the argument that science is wholly dependent on and constructed by the human scientists who reify it (e.g., Latour & Woolgar, 1979; Taylor, 1996). Contemporary research builds on those themes. For example, in her articulate analysis of physics laboratory life, Heather Graves (2013) points out how the processes that enable scientific research are products of fallible and vested humans, and the experience of doing or understanding science is inextricably bound to the equipment, processes, and language used (p. 89). In another excellent book about the power of ***visual*** communication, Lee Brasseur (2003) explains both the over-valorization and the dismissal of scientific and technical visual communication through a critical historical lens. Brasseur's book enables students of rhetoric and technical communication to understand how our fields rely on science, while at the same time asking key questions about whether reductive scientific interpretations of the world shortchange humanity.

Further, the reputation of science and scientists has been tainted by a history of crimes against humanity, committed in the name of scientific research, and targeting the most vulnerable. One of the most famous incidents involved Nazis studying legitimate research questions about military operations, but through illegitimate means: painful, humiliating, and often lethal methods of torture carried out on Jewish prisoners at camps including Dachau and Ravensbrueck. The Nuremberg trials of 1946–1947 found 15 defendants guilty and led to the development of the Nuremberg Code, which seeks to proactively protect people from such victimization (Dunn & Chadwick, 2012). In the US, the African American population was exploited for a span of four decades in an extraordinarily long-term medical study of syphilis. Black men with syphilis were tracked by medical professionals, and long after antibiotics were known to cure the disease, were deprived of such treatments (Dunn & Chadwick, 2012). In cases of such abuses of the tools and methods of science, it has sometimes been an instrument of further marginalization of minoritized persons.

The belief that scientists are primarily engaged in "establishing true and absolute descriptions of the nature of things" is losing favor as sociological research reveals that "empirical research rarely makes direct claims about the unmediated nature of the world" (Taber, 2018, p. 6). Today, emphasis is placed on recognizing that the work of science is largely claim, not fact; proposing relationships and hierarchies; identifying laws that may not be final; and, sometimes, promoting and/or protecting the reputation and status of science and scientists collectively and individually.

Scientific communication, similarly, struggles with an identity crisis because it is also expected to be objective, under the faulty assumption that scientists themselves are objective (Yu & Northcut, 2018). Facts (and findings), no matter how important, literally do not speak for themselves. Therefore, scientists face the continuous challenge of first interpreting, then arguing for the importance and morality of their work and the reliability of their findings to each other, to stakeholders, to sponsors, and sometimes even to themselves. Scientists are not equally adept at doing so (Baake, 2003; Woolston, 2020), which is inherently fascinating to fields including technical and scientific communication, linguistics, and journalism. Studying the cultural and communicative processes of science and scientists gave rise to various social science and humanities subdisciplines in the 20th century, including sociology of scientific knowledge, rhetoric of science, and science and technology studies.

Aside from the nature of science, another interesting question with an answer that varies across historical periods is "who is a scientist?" Science was not professionalized until the early 19th century. The gate-keeping functions of professional science (e.g., licensure and formal membership) promote a culture of insiders and outsiders. The culture is reinforced by the requirements of independent federal agencies, such as the National Science Foundation, and the larger federal bureaucracy, such as the Department of Health and Human Services, which oversees the Office of Human Research Protections (OHRP) and the Food and Drug Administration (FDA). Both the OHRP and FDA require that research ethics boards include "scientist" and unambiguously "non-scientist" voting members, although the FDA's own guidance documents are vague about why the distinction is necessary or useful (FDA, 1998).

Other gate-keepers include academic institutions and the cultures of the academic departments within them. Gate-keeping serves to homogenize scientific thinking by requiring common credentials and education of practitioners, but it also tends to reinvent itself in repetitive and potentially damaging ways—for example, through bias and practices that maintain existing power structures (Cole & Hassel, 2017, Northcut, 2017). Scientific communication is an area where the gate-keeping function of jargon has been identified, and many scientific journalists and popularizers (both with and without formal science credentials) endeavor to make scientific knowledge understandable by the interested non-expert ***public*** (Woolston, 2020).

Dividing people through various gate-keeping mechanisms into categories of "scientist" and "non-scientist" feels artificial to social scientists and transdisciplinary workers, and the constructed definition of scientist can serve to alienate non-scientists, presenting science as a clannish, closed culture hostile to outsiders. Placing, and keeping, much of the population on the margins has perpetuated understandings and definitions of science that may haunt us more than they help us.

In our current era of strict credentialing and demarcation of those who are qualified to call themselves scientists, great public tension has emerged between science and politics, starkly apparent during the COVID-19 pandemic. Although scientists (including national academies and the U.S. Centers for Disease Control and Prevention) knew by April 1, 2020 that face coverings (surgical masks, cloth face coverings, and hard plastic shields) were likely to reduce infection rates of the virus, Republican-led state and federal governments were slow to recommend and require them in the US. Initially, alarm about the virus led governments globally to either recommend or force schools, businesses, and transportation to shut down, and travel restrictions were imposed. Reopening began months later, despite little evidence that the virus was less of a global threat, and increased socializing led to outbreaks, particularly in the US. Not until July 2020 did the number of states with a mask mandate exceed the number of states without one, leaving the mask-mandate decision to municipalities and private businesses such as grocery and department stores. Political party identification was shown to be correlated to attitudes about the pandemic (Pew Research Center, 2020). The ongoing impacts of

COVID-19 are attributed by many researchers to result from the failure of elected leaders to encourage scientifically validated precautions such as mask-wearing, at a time when evidence demonstrated efficacy of masks against transmission of a virus that travels and infects primarily as aerosolized particles or airborne droplets (National Academies of Sciences, Engineering, and Medicine, 2020).

The COVID-19 pandemic has clarified the perils of an anti-science population suspicious of, or hostile to, science, and enabled us to imagine benefits that might emerge if science were understood more richly and broadly, and if science were a culture that all citizens, regardless of vocation, were expected to understand, participate in, and critique. The COVID-19 pandemic illustrated the importance of understanding audience when conveying emergent theory (Baake, 2003)—in this case, the theory of transmission of a virus no one had ever studied. We also see the unfortunate consequences of ineffective communications about risk, as COVID cases in 2023 topped 676 million worldwide, and the US, with four percent of the world's population, contains over 15 percent of the cases, and has logged more than its proportion of the deaths (Johns Hopkins, n.d.). Technical communicators possess the academic and professional credentials to be ideally situated to facilitate scientific communication, especially if we are familiar with the history, epistemologies, and cultural studies of science that have shaped the current enterprise.

References

Baake, K. (2003). *Metaphor and knowledge: The challenges of writing science*. SUNY Press.

Brasseur, L. (2003). *Visualizing technical information: A cultural critique*. Baywood.

Cole, K. & Hassel, H. (2017). *Surviving sexism in academia: Strategies for feminist leadership*. Routledge. https://doi.org/10.4324/9781315523217.

Dunn, C. M. & Chadwick, G. L. (2012). *Protecting study volunteers in research: A manual for investigative sites* (4th ed.). CenterWatch.

Fountain, T. K. (2014). *Rhetoric in the flesh: Trained vision, technical expertise, and the gross anatomy lab*. Routledge.

Gigante, M. (2018). *Introducing science through images: Cases of visual popularization*. The University of South Carolina Press. https://doi.org/10.2307/j.ctv6sj8kf.

Gould, S. J. (1989). *Wonderful life: The Burgess Shale and the nature of history*. Norton.

Graves, H. B. (2013). *Rhetoric in(to) science: Style as invention in inquiry*. Hampton.

Gross, A. G., Harmon, J. E. & Reidy, M. (2002). *Communicating science: The scientific artifact from the 17th century to the present*. Parlor Press.

Hauser, G. A. (1999). *Vernacular voices: The rhetoric of publics and public spheres*. University of South Carolina Press.

Johns Hopkins University & Medicine. (n.d.). Retrieved April 11, 2023 from *Coronavirus resource center*. https://coronavirus.jhu.edu.

Kuhn, T. S. (1970). *The structure of scientific revolutions*. University of Chicago Press.

Kuhn, T. S. (2000). *The road since structure: Philosophical essays, 1970–1993, with an autobiographical interview* (J. Conant & J. Haugeland,Eds.). University of Chicago Press.

Lanham, R. A. (1991). *A handlist of rhetorical terms* (2nd ed). University of California Press. https://doi.org/10.1525/9780520912045.

Latour, B. & Woolgar, S. (1979). *Laboratory life: The social construction of scientific facts.* Sage Publications.

Longo, B. (2000). *Spurious coin: A history of science, management, and technical writing.* State University of New York Press.

Morris, S. C. & Caron, J. (2012). *Pikaia gracilens* Walcott, a stem-group chordate from the Middle Cambrian of British Columbia. *Biological Reviews, 87*(2). https://doi.org/10.1111/j.1469-185X.2012.00220.x.

National Academies of Sciences, Engineering, and Medicine. (2020, April 1). *Rapid expert consultation on the possibility of bioaerosol spread of SARS-CoV-2 for the COVID-19 pandemic.* The National Academies Press. https://doi.org/10.17226/25769.

Northcut, K. (2017). Suck it up, Buttercup: Or, why cu*ts leave STEM. In K. Cole & H. Hassel (Eds.), *Surviving sexism in academia: Strategies for feminist leadership* (pp. 98–105). Routledge. https://doi.org/10.4324/9781315523217-10.

Oxford University Press. (n.d.). Science. In *Oxford English Dictionary.* Retrieved March 30, 2021, from https://www.oed.com.

Pew Research Center. (2020, June 25). Republicans, Democrats move even further apart in coronavirus concerns. https://www.pewresearch.org/politics/2020/06/25/republicans-democrats-move-even-further-apart-in-coronavirus-concerns/.

Taber, K. S. (2018). Assigning credit and ensuring accountability: An editor's perspective on authorship. In P. A. Mabrouk & J. N. Currano (Eds.), Credit where credit is due: Respecting authorship and intellectual property (pp. 3–33). American Chemical Society. https://doi.org/10.1021/bk-2018-1291.ch001.

Taylor, C. A. (1996). *Defining science: A rhetoric of demarcation.* University of Wisconsin Press.

United States Department of Health and Human Services. (2019). *IRB registration form.* https://www.hhs.gov/ohrp.

U.S. Food and Drug Administration. (1998). *Institutional Review Boards Frequently Asked Questions.* Guidance for Institutional Review Boards and Clinical Investigators. https://www.fda.gov/regulatory-information/search-fda-guidance-documents/institutional-review-boards-frequently-asked-questionshttps://fda.gov/.

Walsh, L. (2013). *Scientists as prophets: A rhetorical genealogy.* Oxford University Press. https://doi.org/10.1093/acprof:oso/9780199857098.001.0001.

Wilson, E. O. (1998). *Consilience: The unity of knowledge.* Alfred A. Knopf.

Woolston, C. (2020, February 27). Words matter: Jargon alienates readers. *Nature, 579,* 309. https://doi.org/10.1038/d41586-020-00580-w.

Yu, H. (2017). *Communicating genetics: Visualizations and representations.* Palgrave Macmillan. https://doi.org/10.1057/978-1-137-58779-4.

Yu, H. & Northcut, K. (Eds.). (2018). *Scientific communication: Practices, theories, and pedagogies.* Routledge. https://doi.org/10.4324/9781315160191.

32. Social Justice

Natasha N. Jones
Michigan State University

Rebecca Walton
Utah State University

A relatively recent keyword in the field of technical communication (TC), *social justice* extends our field's longer-term focus on critical analysis, which acknowledges the complicity of TC in normalizing and codifying oppression. But social justice has been conflated with "generally good," rather than informing notions of fairness (paradigms of justice) by amplifying the agency of oppressed people (social justice). Some of this conflation may be due to the relative newness of the term within TC. In TC scholarship, the first explicit definition of social justice appeared in 2013 and was borrowed from communication studies (Agboka, 2013). We introduced a field-specific definition two years later (Jones & Walton, 2018) and, with Kristen Moore, further fleshed out the relation of social justice to the field (Walton et al., 2019). Here, we trace that brief ***history*** and tease out nuances in how social justice can inform broader paradigms of justice which underlie our scholarship and activism. Since social justice in TC should engage social justice "in the world," we use contemporary movements to defund/abolish the police as an example of how layering social justice onto broader justice paradigms allows for both flexibility (in selecting the justice paradigms best suited to a particular goal) and precision (in pursuing fairness that accounts for oppression).

Before the keyword *social justice* became widespread in TC, related and overlapping waves of scholarship laid the groundwork for the rise of social justice as a central consideration of the field. For example, the 1990s and early 2000s saw a sociocultural turn in which scholars debunked the myth that TC is neutral (Kynell-Hunt & Savage, 2004; Scott et al., 2007). Much of this early scholarship was pedagogical in focus, calling for TC instructors to equip students to become critical actors within their employing organizations rather than unthinkingly perpetuating harm through their professional practice (e.g., Herndl, 1993).

Another wave of relevant scholarship called for diversifying our academic programs, faculty, and students. These calls for diversity asserted that contributions and expertise of underrepresented groups would improve the field. At the 2004 Council for Programs in Technical and Scientific Communication (CPTSC) national conference, Samantha Blackmon gave a keynote address that explicitly called for increased diversity and inclusion in academic programs, but the call was largely ignored until a wave of similar scholarship less than ten years later provided traction for her arguments. For example, 2011 and 2012 saw several

DOI: https://doi.org/10.37514/TPC-B.2023.1923.2.32

individual journal articles on programmatic diversity (Savage & Mattson, 2011; Savage & Matveeva, 2011) as well as a journal special issue on race and ethnicity in TC (Williams & Pimentel, 2012).

For some time, TC scholarship featured terms such as *social action* (e.g., Savage, 1996), *civic participation* (e.g., Sapp & Crabtree, 2002), *public good* (e.g., Skelton & Andersen, 1993), and *diversity* (e.g., Savage & Matveeva, 2011), until Godwin Agboka's impactful 2013 article. Agboka's article was widely cited, laying the groundwork for conference themes, journal special issues, and award-winning scholarship heralding a "social justice turn" in the field (Haas & Eble, 2018). In 2018, we defined social justice ***research*** in TC as research that "investigates how communication, broadly defined, can amplify the agency of oppressed people—those who are materially, socially, politically, and/or economically under-resourced" (Jones & Walton, 2018, p. 46). We also noted that ***collaboration***, respect, and action are fundamental to social justice work. Therefore, social justice centers the needs of oppressed people by engaging in participatory, strategic action.

Although they are sometimes conflated, justice (a range of paradigms) and social justice (a specific term defined above) differ in important, nuanced ways. A key difference between social justice and broader paradigms of justice is that while social justice actively engages with issues of oppression (recognizing that what constitutes "just" action is inherently affected by social, political, economic, and material affordances and constraints), paradigms of justice are predicated upon "fairness," without necessarily accounting for the effects of oppression on what makes something "fair."[1] Thus, we advocate layering social justice upon paradigms of justice. This layering ensures that marginalized perspectives are centered in the pursuit of fairness.

To engage in socially just action, scholars and practitioners of TC must be explicit and intentional about the paradigm of justice guiding their work. Different paradigms of justice inform and underlie structured societal systems, and each justice paradigm is embedded with specific values that motivate and constrain action. Thus, justice is simultaneously theoretical, applied, and practiced. We review four of the justice paradigms below, illustrating each with examples from efforts in the US to defund and abolish the police.

It is important to note that there are nuances between calls to defund and calls to abolish the police. As Angela Davis (2020) has noted, under the umbrella of the movement to abolish the police, defunding police departments is a step toward abolition. Defunding strategically removes financial support from law enforcement, with full abolition of police departments and the prison-industrial complex being the ultimate goal. However, some activists do ascribe to the belief that defunding, not total abolition, should be the final objective (with funding being reallocated to achieve equity with other publicly funded systems like

1. As Iris Young (1990) notes, oppression can appear in five primary forms: exploitation, marginalization, violence, cultural imperialism, and powerlessness.

education and healthcare). For the purposes of our discussion here, we acknowledge defunding law enforcement as an abolitionist goal.

Distributive justice focuses on the fair allocation of rewards and burdens. Many arguments to defund the police are informed by the distributive justice paradigm: for example, the argument that police budgets are unfairly large and that other public services, such as education, affordable housing, healthcare, and childcare, are underfunded. This argument is also informed by considerations of social justice. After all, those most negatively affected by overfunded police forces and underfunded public services are the marginalized. This example also demonstrates the relevance of distributive justice to TC because public policy, budgets, resource allocation, and civic participation are technical topics, and arguments regarding the just allocation of public funding are often presented in technical ***genres***, such as policy briefs.

Procedural justice requires that the process by which outcomes are determined is fair. A typical context for the procedural justice paradigm is institutional policies and procedures—a context deeply relevant to the field because policies and procedures are documented in TC. One important consideration of procedural justice is transparency: For a process to be fair, it must be known to all relevant stakeholders. Making processes transparent can increase fairness by broadening the range of stakeholders whose interests inform those processes and the policies governing them. For example, when public interests inform procedural documents, such as police use-of-force policies, those policies can be re-envisioned to reflect an ***ethic*** of care focused on protecting vulnerable members of society (Knievel, 2008). This re-envisioning layers social justice (centering marginalized perspectives) onto a procedural justice paradigm (enacting fair processes).

In the context of defunding the police, procedural justice is particularly relevant to budgetary reform. Sources of police funding are myriad and confusing. The opacity regarding how police budgets are planned, approved, funded, and even measured makes it difficult for activists and policymakers to pose reforms (Auxier, 2020) and trace how assets are acquired (Alexander, 2010). This fiscal complexity creates procedural opacity, raising questions about how such procedures can be just when they cannot be widely shared, predicted, or even understood.

Retributive justice paradigms focus on "fair" punishment for crimes and wrongdoing, placing offenders and offense at the center of justice concerns (Walton et al., 2019, p. 42). However, because retributive justice paradigms rely on ideals like "fair" and "equal," these paradigms often fall short—impacting certain groups more negatively than others. The groups that consistently receive harsher punishments are predominantly marginalized populations—often stereotyped as offenders and criminals—who are already at the mercy of biased economic, educational, political, and social systems (Alexander, 2010). TC scholarship can reveal these problems with retributive paradigms: e.g., that "fair punishment" can include death and dehumanization for alleged offenders, especially those who are members of marginalized groups (Moore et al., 2017, p. 43). Offenses such as the murder of Eric

Garner are enabled by a paradigm of justice that focuses on punishment, creating conditions in which agents of the justice system may feel empowered to mete out violent extrajudicial "punishment" by acting as a conglomerated version of judge, jury, and enforcer.[2] It is partially in response to police violence (notably, the murders of George Floyd[3] and Breonna Taylor[4]) that the Abolish the Police movement has reignited. And, given the persistence of police violence in the US, supporters of the movement argue that the current retributive justice system is violent and oppressive by design. Thus, it cannot be reformed and must instead be dismantled.

Restorative justice paradigms ask that offenders, victims, and the impacted community are made "whole" based on ideals of social harmony and peace. Community and collective benefit are at the center of restorative justice paradigms (Walton et al., 2019, p. 44). Because restorative justice requires respectful collaboration that can include redress of wrongs through economic, material, and social means, this particular justice paradigm can closely align with and may be most informed by a social justice orientation. As Angela Davis (2003) argues, reconciliation and restoration can replace retribution (p. 49). However, to move toward restoration and reparation, societal institutions like law enforcement, the legal system, the prison-industrial complex, the healthcare system, and education systems must be wholly reimagined to account for community need, support, and repair. Davis (2003) notes that "the most difficult and urgent challenge today is that of creatively exploring new terrains of justice" (p. 8). For instance, layering social justice upon restorative justice paradigms requires that reparation be initiated at systemic and institutional levels. Social justice "cannot be limited to individual actions or perspectives because the oppressions it targets are structural" (Walton et al., 2019, p. 50).

Embracing Davis' imperative (recently rearticulated in Davis [2020]), we ask, how can technical communicators refrain from requiring oppressed individuals to adapt themselves to society and instead rethink the functioning of society itself as a way of restoring and repairing oppressed communities? This question is timely for the field of TC, as illustrated by an incident from the very week we drafted this keyword entry: A well-respected senior scholar posted a memo by the Department of Homeland Security to the email list for a national TC professional organization, the Association of Teachers of Technical Writing (ATTW).

---

2. Eric Garner was murdered at the hands of officers in the New York City Police Department (NYPD) on July 17, 2014. Garner was placed in an illegal chokehold, and the encounter, during which Garner stated that he could not breathe over 11 times, was recorded and highly publicized. Garner's murderers were not indicted.

3. George Floyd was murdered by Minneapolis police officers in May 2020. A police officer kneeled on Floyd's neck for over nine minutes. Like Eric Garner before him, Floyd pleaded with officers, repeatedly saying, "I can't breathe" for a total of 27 times.

4. Breonna Taylor was murdered by Louisville police officers in March 2020. Officers performed a "no-knock" warrant at the incorrect address (the correct house was over 10 miles away), shooting Taylor eight times in her own home.

The memo announced a new policy that threatened international students studying at U.S. universities with immediate deportation should their classes be moved online in response to the COVID-19 global pandemic. The memo was shared as an example of unethical TC that ATTW members could analyze with students to identify problems with both the policy itself and the memo, highlighting the literal life-and-death stakes of some TC and revealing the complicity of TC in oppression. In revealing oppression directly related to the field, the post demonstrated that social justice is deeply relevant to TC.

But this recognition of relevance is not universal. On the same email thread, a different senior scholar responded with xenophobic comments rejecting the responsibility of educators for their students' wellbeing or educational outcomes. Members of the field immediately spoke out against this oppressive ***rhetoric*** and began to work coalitionally to replace oppressive practices, language, and behaviors. Responses included rejecting the xenophobic comments publicly and in writing by replying all to the listserv, demanding the retraction of an oppressive publication in a TC journal, developing anti-racist resources for editors and reviewers of academic manuscripts, and other efforts.

These efforts offer a complex snapshot of what it can look like for our field to embrace Davis' imperative above. For technical communication scholars this would mean refraining from requiring oppressed individuals to adapt themselves to society. Instead, we should rethink the functioning of society itself to restore and repair community. Specifically in the example used in this chapter, reimaging how our field can be more socially just would look like not expecting international students to accept unjust precarity created by an oppressive policy and rejecting the notion that marginalized TC scholars must simply tolerate racist and otherwise unjust publication practices. We, as a field, would instead publicly call out and refuse to engage in or entertain xenophobic comments and move to rethink academic publication practices to intentionally cultivate more just and inclusive norms. This example also illustrates some broader implications for the field now that the keyword social justice has entered our disciplinary lexicon. Firstly, recognizing injustice and TC's complicity in it is a starting place for action, not an end goal. Secondly, the actions necessary to "amplify the agency of oppressed people" (Jones & Walton, 2018, p. 46) are contextual, complex, and varied, and therefore require the work of coalitions. And thirdly, layering social justice onto explicitly identified paradigms of justice offers a simultaneously theoretical and applied strategy for centering the marginalized.

References

Agboka, G. Y. (2013). Participatory localization: A social justice approach to navigating unenfranchised/disenfranchised cultural sites. *Technical Communication Quarterly*, *22*(1), 28–49. https://doi.org/10.1080/10572252.2013.730966.

Alexander, M. (2010). *The new Jim Crow: Mass incarceration in the age of colorblindness.* The New Press.
Auxier, R. C. (2020, June 9). *What police spending data can (and cannot) explain amid calls to defund the police.* Urban Institute. https://www.urban.org/urban-wire/what-police-spending-data-can-and-cannot-explain-amid-calls-defund-police.
Blackmon, S. (2004). Which came first? On minority recruitment and retention in the academy. In *CPTSC Proceedings* (pp. 1–3).
Davis, A. (2003). *Are prisons obsolete?* Seven Stories Press.
Davis, A. (2020, June 12). *Angela Davis on abolition, calls to defund police, toppled racist statues & voting in 2020 election.* DemocracyNow! https://www.democracynow.org/2020/6/12/angela_davis_on_abolition_calls_to.
Haas, A. M. & Eble, M. F. (Eds.). (2018). *Key theoretical frameworks: Teaching technical communication in the twenty-first century.* Utah State University Press.
Herndl, C. G. (1993). Teaching discourse and reproducing culture: A critique of research and pedagogy in professional and non-academic writing. *College Composition and Communication, 44*(3), 349–363. https://doi.org/10.2307/358988.
Jones, N. & Walton, R. (2018). Using narratives to foster critical thinking about diversity and social justice. In M. Eble & A. Haas (Eds.), *Key theoretical frameworks for teaching technical communication in the 21st century* (pp. 241–267). Utah State University Press.
Knievel, M. S. (2008). Rupturing context, resituating genre: A study of use-of-force policy in the wake of a controversial shooting. *Journal of Business and Technical Communication, 22*(3), 330–363. https://doi.org/10.1177/1050651908315
Kynell-Hunt, T. & Savage, G. (Eds.). (2004). *Power and legitimacy in technical communication: Strategies for professional status* (Vol. 2). Baywood.
Moore, K., Jones, N., Cundiff, B. & Heilig, L. (2017). Contested sites of health risks: Using wearable technologies to intervene in racial oppression. *Communication Design Quarterly, 5*(4), 52–60. https://doi.org/10.1145/3188387.3188392.
Sapp, D. A. & Crabtree, R. D. (2002). A laboratory in citizenship: Service learning in the technical communication classroom. *Technical Communication Quarterly, 11*(4), 411–432. https://doi.org/10.1207/s15427625tcq1104_3.
Savage, G. J. (1996). Redefining the responsibilities of teachers and the social position of the technical communicator. *Technical Communication Quarterly, 5*(3), 309–327. https://doi.org/10.1207/s15427625tcq0503_4.
Savage, G. & Mattson, K. (2011). Perceptions of racial and ethnic diversity in technical communication programs. *Programmatic Perspectives, 3*(1), 5–57.
Savage, G. & Matveeva, N. (2011). Toward racial and ethnic diversity in technical communication programs. *Programmatic Perspectives, 3*(1), 58–85.
Scott, J. B., Longo, B. & Wills, K. V. (Eds.). (2007). *Critical power tools: Technical communication and cultural studies.* SUNY Press. https://doi.org/10.1177/0021943606298.
Skelton, T. & Andersen, S. (1993). Guest editorial: Professionalism in technical communication. *Technical Communication, 40*(2), 202–207.
Walton, R., Moore, K. & Jones, N. (2019). *After the social justice turn: Building coalitions for action.* Routledge. https://doi.org/10.4324/9780429198748-3.
Williams, M. F. & Pimentel, O. (Eds.). (2012). Race, ethnicity, and technical communication [Special issue]. *Journal of Business and Technical Communication, 26*(3).
Young, I. M. (1990). *Justice and the politics of difference.* Princeton University Press.

33. Social Media

Liza Potts
Michigan State University

Michael Trice
Massachusetts Institute of Technology

Social media describes a diverse, and not always cohesive, array of platforms wherein participants can interact with each other in digital spaces meant to communicate across space and time. Definitions of *social media* vary, but commonly categorize the ***technology*** based on technical features in combination with social purpose and multitextual possibilities (boyd & Ellison, 2007; Kimme Hea, 2014; Vie, 2008; Zittrain, 2008). One of the most often cited definitions arises from dannah boyd and Nicole B. Ellison (2007), who define social networking sites as spaces that

> (1) construct a public or semi-public profile within a bounded system, (2) articulate a list of other users with whom they share a connection, and (3) view and traverse their list of connections and those made by others within the system. (p. 211)

While this functional description captures apps like Facebook, Twitter, and Instagram that focus on individual participant feeds, it perhaps is overly exclusionary in regards to topic-centered social sites, like Reddit, Wikipedia, and multitudes of fora across the internet. For this entry, we consider the ***history*** and present of both topic- and participant-centered social media. We also take cues from Amy C. Kimme Hea's focus on social media as cultural practices.

The 20th century predecessors of social media included systems used primarily by folks such as academics, technologists, the military, hobbyists, and media fans. While this group is significantly smaller than today's social media userbase, these early systems created space for the exchange of ***information***, ideas, and materials that hint towards ways in which social media would eventually be deployed. Using technologies such as telenet, dial-up bulletin board systems, and USENET discussion groups, these users were able to communicate with others who shared their interests. Our field explored these earlier incarnations of social media through work on technology and writing (Bolter, 1991), technical communication (Gurak, 2001), and technology and society (Warnick, 2001). Many studies written during this era focused on the ways in which these technologies altered our writing processes. Opening up these discussions would later lead to ***research*** on other tools such as wikis, video, and social networks.

Before they were called social media, these technologies were referred to as social software in the late 1990s and early 2000s. This emphasis on the software

DOI: https://doi.org/10.37514/TPC-B.2023.1923.2.33

itself is illustrative of the digital skills needed to implement turn-of-the-century social tools such as blogs, wikis, and forums. Technical communicators took note, both in research (Gurak et al., 2004) and practice (Barton & Cummings, 2008; Jones, 2009; Mader, 2009). The connections between places, cultures, spaces, and peoples were illustrated through multiple histories of multimedia, hypertext, and the many digital antecedents of the social web (Ball, 2012; Haas, 2007; Manion & Self, 2012) that helped us to understand how hypertext holds meaning.

At that time, the term "platform" often referred to operating systems, such as Windows, MacOS, and Linux. Over the past several years, the term "platform" began to refer to social media spaces such as Twitter, Instagram, Facebook, Twitch, Snapchat, and WhatsApp. In 2010, Tarleton Gillespie would challenge the ***rhetorical*** use of "platform," pulling at its computational, figurative, political, and architectural meanings as a way to illustrate the tensions of corporate social media platforms striving to be also seen as civic and user-generated platforms. Gillespie highlighted the need to consider how the metaphor of "platform" served numerous rhetorical purposes in the way participants, companies, and even the government shifted meanings depending upon purpose and context. Over time, this would lead technical communication to consider the rhetoric of platforms (Edwards & Gelms, 2018; Jones, 2014) as well as the ***ethics*** of platforms (Cagle, 2019; Sano-Franchini, 2018).

Around the same time, the term "Web 2.0" delineated the change from static webpages to interactive websites, deploying techniques like AJAX that allowed for more advanced tool building and the beginnings of today's social media platforms. This shift at the turn of the century opened the possibilities of participatory cultures (Jenkins, 2008), propelling us forward into online spaces where user-generated content became an area of study, application, and ***pedagogy*** (Balzhiser et al., 2011; Barton & Cumming, 2008), bringing about a shift in the distribution of agency, control, and content.

Early discussions around these concerns appeared in books (Spilka, 2009), articles, and blogs written by researchers and practitioners (Hart-Davidson et al., 2007; Sidler & Jones, 2008). This focus on agency and content could also be seen in concepts of delivery particularly suited to considering how messages adapt across social networks and platforms, such as Jim Ridolfo and Danielle Nicole DeVoss' (2009) rhetorical velocity. Rhetorical velocity held that technical communicators and creatives were accountable for anticipating and theorizing the manner in which third parties might utilize content. In many ways, it applied the principle of single sourced adaptation within organization to a broader cultural and social landscape that would anticipate the rise of both memetic content and cross-platform branding and activism.

Notably, by the turn of the century, many technical communicators were already connecting the dots between technical writing and usability (Redish, 2010), encouraging us to use our skills to improve interfaces and policies beyond traditional outputs such as ***documentation***. From there, works focused more on technical communication and social media, pointing to its use in workplace settings (Katajisto, 2010), mixed spaces (Johnson-Eilola, 2004), and across cultures (Sun, 2012). Entire special issues were dedicated to understanding social media and technical communication (Dyrud, 2012; Geisler, 2011; Kimme Hea, 2014). These special issues would bring social media into long-standing discussion in technical communication about pedagogy, ***knowledge*** work, diversity, and rhetorical reach—while engaging with a wide variety of platforms, including Reddit, Facebook, Twitter, and Wikipedia. They might also highlight a deficit in how we were slower to sites like YouTube, image boards, and GitHub, though some progress would occur over time (Winter & Salter, 2020).

Synchronously, more scholarship addressing issues of accountability regarding racism and technology entered our conversations more visibly (Haas, 2012; Williams & Pimentel, 2014) and connected scholarship across fields (Nakamura, 2007). These trends would foreshadow the move into application accountability in technical communication that would arise in the most recent decade of work. Indeed, the ethics of social media has also become a central focus of technical communication. These examinations include how the ***design*** of social media platform interfaces generates political and individual discord (Muhlhauser & Schafer, 2020; Sano-Franchni, 2018), the way surveillance is incorporated into social media (Cagle, 2019), and the impact of social media as activism and aggression (Chen & Wang, 2020; Potts et al., 2019; Reyman & Sparby, 2019). One of the central recent ethical movements has been technical communication's ***social justice*** turn, which has impacted social activism online via concepts of rhetorical agency (Jones & Walton, 2017) and ethics of care (Colton et al., 2017).

Increasingly, technical communication in social media has spanned across topics relating to knowledge work and content strategy, including ***health communication***, disaster communication, environmental activism, and social justice. Within various strands of interests within technical communication today—***user experience***, medical, disaster, and environmental activism framed within the need for social justice and advocacy (Jones, 2016, Edenfield et al., 2019)—social media plays a role as both a conduit for communication among researchers and practitioners and a site of study for our field. It has also included an emphasis on ***genre*** use and context to help us better understand how digitals can empower activity (Ferro & Zachry, 2013, Trice, 2015), support emergency management (Angeli, 2018), and design for global use (Sun, 2020) and platform ideologies (Wang & Gu, 2016).

Technical communication scholars are currently exploring issues concerning the owners, moderators, designers, users, and policies that constitute social media platforms. These perspectives allow us to research the user experience architecture of these platform ***structures***, the ways in which platform leaders position their organizations through their policies, and how participants on these systems are able to communicate across these networks. Our work seeks to understand how social media intersects and affects the outcomes of social movements, elections, disasters, environmental policies, social justice, and the everyday lives of individuals across the globe.

Our classrooms demonstrate how technical communication puts these foci into praxis with emphases on using the tools available to both build community within our classrooms (Kaufer et.al., 2011) and prepare students for their futures (Maggiani, 2011) as professionals and citizens. As pedagogy has long been essential in the field of technical communication, social media has smoothly entered that conversation as a means for students to demonstrate professionalism, rhetorical agency, activism, and civic leadership. When it comes to teaching, technical communication focuses upon social media as praxis and skill development (Daer & Potts, 2014) and "as cultural practices that shape and are shaped by political, social, and cultural conditions" (Kimme Hea, 2014, p. 2). Scholars like Melody Bowdon (2014) have focused on the importance of teaching ethos as a factor in online communication, while others have focused on practitioner praxis (Pigg, 2014) and service learning (Melton & Hicks, 2011).

Looking forward, perhaps one of the most important contributions technical communication researchers can make to social media is in terms of policy. Our rich backgrounds and understanding of rhetoric, design, activism, and social justice uniquely situate our work to make an impact on the ways in which platforms, governments, and organizations deploy these systems, employ design patterns, surveil users, collect personal and ***public*** data, and distribute or sell such data. These interfaces and the data these organizations collect are used to enforce social, political, and economic policies across the globe. The role of moderation, parameters of ***accessibility,*** and the rhetorical impact of knowledge-making systems upon society and specific communities are areas that technical communication has deep experience addressing. The future of social media will depend upon addressing these areas, and technical communicators must be present in those decisions.

References

Angeli, E. L. (2018). *Rhetorical work in emergency medical services: Communicating in the unpredictable workplace*. Routledge. https://doi.org/10.4324/9781315104881.

Ball, C. E. (2012). Assessing scholarly multimedia: A rhetorical genre studies approach. *Technical Communication Quarterly*, *21*(1), 61–77. https://doi.org/10.1080/10572252.2012.626390.

Balzhiser, D., Polk, J. D., Grover, M., Lauer, E., McNeely, S. & Zmikly, J. (2011). The Facebook papers. *Kairos: A Journal of Rhetoric, Technology, and Pedagogy, 16(1)*. http://www.technorhetoric.net/16.1/praxis/balzhiser-et-al.

Barton, M. & Cummings, R. E. (2008). *Wiki writing: Collaborative learning in the college classroom*. University of Michigan Press.

Bolter, J. D. (1991). *Writing space*. Erlbaum.

Bowdon, M. A. (2014). Tweeting an ethos: Emergency messaging, social media, and teaching technical communication. *Technical Communication Quarterly*, *23*(1), 35–54. https://doi.org/10.1080/10572252.2014.850853.

boyd, d. m. & Ellison, N. B. (2007). Social network sites: Definition, history, and scholarship. *Journal of Computer-Mediated Communication, 13*(1), 210–230. https://doi.org/10.1111/j.1083-6101.2007.00393.x.

Cagle, L. E. (2019). Surveilling strangers: The disciplinary biopower of digital genre assemblages. *Computers and Composition, 52*, 67–78. https://doi.org/10.1016/j.compcom.2019.01.006.

Chen, C. & Wang, X. (2020). # Metoo in China: Affordances and constraints of social media platforms. In J. Jones & M. Trice (Eds.), *Platforms, protests, and the challenge of networked democracy* (pp. 253–269). Palgrave Macmillan. https://doi.org/10.1007/978-3-030-36525-7_14.

Colton, J. S., Holmes, S. & Walwema, J. (2017). From NoobGuides to #OpKKK: Ethics of Anonymous' tactical technical communication. *Technical Communication Quarterly, 26*(1), 59–75. https://doi.org/10.1080/10572252.2016.1257743.

Daer, A. R. & Potts, L. (2014). Teaching and learning with social media. *Programmatic Perspectives, 6*(2), 21–40.

Dyrud, M. A. (2012). Posting, tweeting, and rejuvenating the classroom. *Business Communication Quarterly, 75*(1), 61–63. https://doi.org/10.1177/1080569911432738.

Edenfield, A. C., Holmes, S. & Colton, J. S. (2019). Queering tactical technical communication: DIY HRT. *Technical Communication Quarterly, 28*(3), 177–191. https://doi.org/10.1080/10572252.2019.1607906.

Edwards, D. W. & Gelms, B. (2018). Special issue on the rhetoric of platforms. *Present Tense, 6*(3). Retrieved from http://www.presenttensejournal.org/editorial/vol-6-3-special-issue-on-the-rhetoric-of-platforms/.

Farkas, D. & Farkas, J. B. (2002). *Principles of web design*. Longman Publishing Group.

Ferro, T. & Zachry, M. (2013). Technical communication unbound: Knowledge work, social media, and emergent communicative practices. *Technical Communication Quarterly, 23*(1), 6–21. https://doi.org/10.1080/10572252.2014.850843.

Geisler, C. (2011). IText revisited. *Journal of Business and Technical Communication, 25*(3), 251–255. https://doi.org/10.1177/1050651911400701.

Gillespie, T. (2010). The politics of 'platforms.' *New Media & Society, 12*(3), 347–364. https://doi.org/10.1177/1461444809342738.

Gurak, L. J. (2001). *Cyberliteracy: Navigating the internet with awareness*. Yale University Press.

Gurak, L., Antonijevic, S., Johnson, L., Ratliff, C. & Reyman, J. (Eds.). (2004). *Into the blogosphere*. University of Minnesota. https://conservancy.umn.edu/handle/11299/172275.

Haas, A. M. (2007). Wampum as hypertext: An American Indian intellectual tradition of multimedia theory and practice. *Studies in American Indian Literatures, 19*, 77–100. https://doi.org/10.1353/ail.2008.0005.

Haas, A. (2012). Race, rhetoric, and technology: A case study of decolonial technical communication theory, methodology, and pedagogy. *Journal of Business and Technical Communication, 26*(3), 277–310. https://doi.org/10.1177/1050651912439539.

Hart-Davidson, W., Bernhardt, G., McLeod, M., Rife, M. & Grabill. J. T. (2007). Coming to content management: Inventing infrastructure for organizational knowledge work. *Technical Communication Quarterly, 17*(1), 10–34. https://doi.org/10.1080/10572250701588608.

Jenkins, H. (2008). *Convergence culture: Where old and new media collide*. New York University Press.

Johnson-Eilola, J. (2004). *Datacloud: Toward a new theory of online work*. Hampton Press.

Jones, J. (2009). Patterns of revision in online writing: A study of Wikipedia's featured articles. *Written Communication*, *25*(2), 262–289. https://doi.org/10.1177/0741088307312940.

Jones, J. (2014). Switching in Twitter's hashtagged exchanges. *Journal of Business and Technical Communication*, *28*(1), 83–108.

Jones, N. (2016). The technical communicator as advocate: Integrating a social justice approach in technical communication. *Journal of Technical Writing and Communication*, *46*(3), 342–361. https://doi.org/10.1177/0047281616639472.

Jones, N. N. & Walton, R. (2017). Using narratives to foster critical thinking about diversity and social justice. In A.M. Haas & M.F. Eble (Eds.), *Key theoretical frameworks: Teaching technical communication in the twenty-first century* (pp. 241–267). University of Colorado Press.

Katajisto, L. (2010). Implementing social media in technical communication. In *2010 IEEE International Professional Communication Conference* (pp. 236–242). IEEE. https://doi.org/10.1109/IPCC.2010.5530019.

Kaufer, D., Gunawardena, A., Tan, A. & Cheek, A. (2011). Bringing social media to the writing classroom: Classroom salon. *Journal of Business and Technical Communication*, *25*(3), 299–321. https://doi.org/10.1177/1050651911400703.

Kimme Hea, A. C. (2014) Social media in technical communication. *Technical Communication Quarterly*, *23*(1), 1–5. https://doi.org/10.1080/10572252.2014.850841.

Mader, S. (2009, January). Your wiki isn't Wikipedia: How to use it for technical communication. *Intercom*, *56*, 14–15.

Maggiani, R. (2011). Making the time for social media. *Intercom*, *58*(10), 33–34.

Manion, C. E. & Selfe, R. D. (2012). Sharing an assessment ecology: Digital media, wikis, and the social work of knowledge. *Technical Communication Quarterly*, *21*(1), 25–45. https://doi.org/10.1080/10572252.2012.626756.

Melton, J. & Hicks, N. (2011). Integrating social and traditional media in the client project. *Business Communication Quarterly*, *74*(4), 494–504. https://doi.org/10.1177/1080569911423959.

Muhlhauser, P. & Schafer, D. (2020). Tweeting inequity: @ realDonaldTrump and the world leader exception. In J. Jones & M. Trice (Eds.), *Platforms, protests, and the challenge of networked democracy* (pp. 199–212). Palgrave Macmillan. https://doi.org/10.1007/978-3-030-36525-7_11.

Nakamura, L. (2007). Digitizing race: Visual cultures of the internet. Minneapolis University of Minnesota Press.

Pigg, S. (2014). Coordinating constant invention: Social media's role in distributed work. *Technical Communication Quarterly*, *23*(2), 69–87. https://doi.org/10.1080/10572252.2013.796545.

Potts, L. (2013). *Social media in disaster response: How experience architects can build for participation*. Routledge. https://doi.org/10.4324/9780203366905.

Potts, L., Small, R. & Trice, M. (2019). Boycotting the knowledge makers: How Reddit demonstrates the rise of media blacklists and source rejection in online communities. *IEEE Transactions on Professional Communication*, *62*(4), 351–363. https://doi.org/10.1109/TPC.2019.2946942.

Redish, J. (2010). Technical communication and usability: Intertwined strands and mutual influences commentary. *IEEE Transactions on Professional Communication*, *53*(3), 191–201. https://doi.org/10.1109/TPC.2010.2052861.

Reyman, J. & Sparby, E. M. (Eds.). (2019). *Digital ethics: Rhetoric and responsibility in online aggression*. Routledge. https://doi.org/10.4324/9780429266140.

Ridolfo, J. & DeVoss, D. N. (2009). Composing for recomposition: Rhetorical velocity and delivery. *Kairos: A Journal of Rhetoric, Technology, and Pedagogy*, *13*(2), n2.

Sano-Franchini, J. (2018). Designing outrage, programming discord: A critical interface analysis of Facebook as a campaign technology. *Technical Communication*, *65*(4), 387–410.

Sidler, M. & Jones, N. (2008). Genetics interfaces: Representing science and enacting public discourse in online spaces. *Technical Communication Quarterly*, *18*(1), 28–48. https://doi.org/10.1080/10572250802437317.

Spilka, R. (Ed.). (2009). *Digital literacy for technical communication: 21st century theory and practice.* Routledge. https://doi.org/10.4324/9780203866115.

Sun, H. (2012). *Cross-cultural technology design: Creating culture-sensitive technology for local users*. Oxford University Press. https://doi.org/10.1093/acprof:oso/9780199744763.001.0001.

Sun, H. (2020). *Global social media design: Bridging differences across cultures*. Oxford University Press. https://doi.org/10.1093/oso/9780190845582.001.0001.

Trice, M. (2015). Putting GamerGate in context: How group documentation informs social media activity. In *SIGDOC '15: Proceedings of the 33rd Annual International Conference on the Design of Communication*. ACM. https://doi.org/10.1145/2775441.2775471.

Vie, S. (2008). Digital divide 2.0: "Generation M" and online social networking sites in the composition classroom. *Computers and Composition*, *25*(1), 9–23. https://doi.org/10.1016/j.compcom.2007.09.004.

Wang, X. & Gu, B. (2016). The communication design of WeChat: Ideological as well as technical aspects of social media. *Communication Design Quarterly Review*, *4*(1), 23–35. https://doi.org/10.1145/2875501.2875503.

Warnick, B. (2001). *Critical literacy in a digital era: Technology, rhetoric, and the public interest.* Routledge. https://doi.org/10.4324/9781410603838.

Williams, M. F. & Pimentel, O. (Eds.). (2014). *Communicating race, ethnicity, and identity in technical communication*. Baywood.

Winter, R. & Salter, A. (2020). DeepFakes: Uncovering hardcore open source on GitHub. *Porn Studies*, *7*(4), 382–397. https://doi.org/10.1080/23268743.2019.1642794.

Zittrain, J. (2008). *The future of the internet and how to stop it.* Yale University Press.

34. Structure

Carlos Evia
Virginia Tech

The implementation of structural moves in ***public*** and professional discourse has been practiced and studied in disciplines related to ***rhetoric*** for centuries. In traditional rhetorical treatises, for example, persuasive speeches were not presented as amorphous sequences of words. Instead, they were assembled according to specific structures that could include sections announcing, explaining, outlining, supporting, and summarizing the parts of a speech (Cicero, 2014). Each component of those structures served a purpose that eventually enabled the production of specific modes of discourse beyond persuasive oratory. This process of identifying, documenting, and implementing common structures to preserve order and rules in discourse established a longstanding tradition of applied rhetoric in writing studies. Technical and professional communication continues this tradition, as some of its scholars acknowledge that "many discourse conventions are, in fact, formalizations of rhetorical moves" (Flower, 1989, p. 34).

For technical communicators, the formalization of rhetorical moves into common structures enabled the production of ***genre***-based documents. Instead of a disjointed collection of paragraphs, a written ***proposal*** can have structural components that *propose* something to a specific audience, a report can have structural sections that *report* on a situation for interested readers, and a set of instructions can *instruct* users on how to accomplish a series of tasks. The production of technical communication genres with established conventions and expected components established the field's importance in the computing industry, as corporate and academic authors published guidelines for structuring technical ***documentation*** and manuals for software and hardware in the 1970s and 1980s (Cohen & Cunningham, 1984; Price, 1984; Rigo, 1976).

Applied at a presentational level, structures in technical communication can create content templates, which have been described as "a kind of wizard for content development" (Kissane, 2009). Content templates can establish that, for example, every section in a quick start guide for a new computer *should* have a title, a paragraph, and a numbered list. Content templates can be implemented as formatting structures in most desktop publishing software applications or, for online publication, with presentational tags from Hypertext Markup Language (HTML), which is foundational to most web-aimed ***content management*** solutions.

For technical communicators, the main benefit of using templates to structure content is the availability of pre-determined styles. Writers "don't spend time

DOI: https://doi.org/10.37514/TPC-B.2023.1923.2.34

figuring out how to create particular formatting—they apply ***styles*** to add formatting" (Pringle & O'Keefe, 2009, p. 41). For consumers of technical content created with a template, the main benefits are defined structural patterns that keep content consistent and make it easier to skim or browse.

At a semantic level, structure supports the practice known as structured authoring, which is "a publishing workflow that lets you define and enforce consistent organization of ***information*** in documents, whether printed or online" (O'Keefe & Pringle, 2017, p. 2). Beyond what a template can establish, structured authoring dictates that content *must* adhere to a specific structure. Structured content "clearly indicates not only the parts of the discourse (the titles, sections, lists, tables, and phrases that represent organization) but also the semantic intent of those containers" (Day, 2016, p. 51). Therefore, if a template allows formatting of a quick start guide, structured authoring can specify that the title is a section heading, the paragraph is an introduction, and the numbered list is a series of steps (see Figure 34.1).

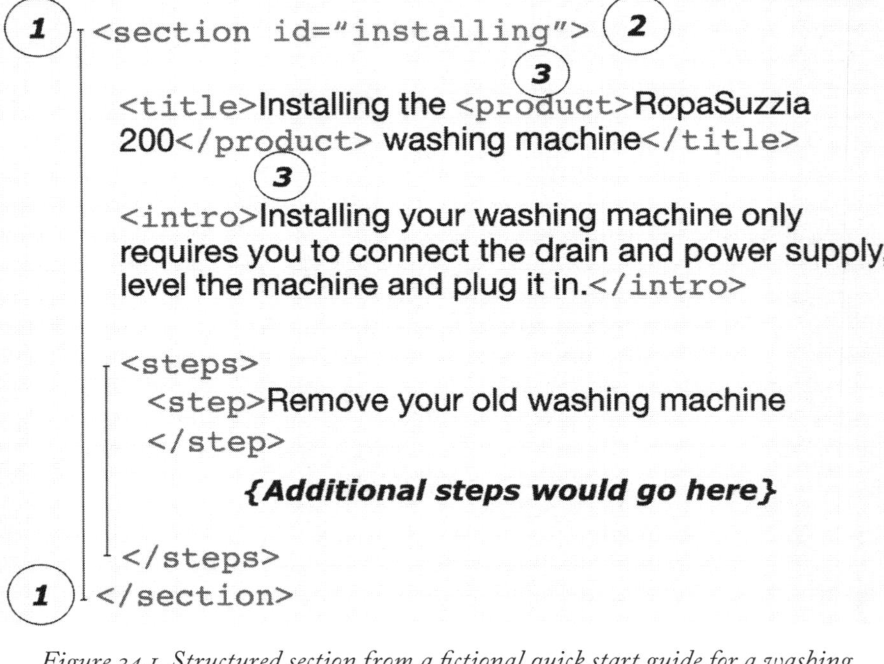

Figure 34.1. Structured section from a fictional quick start guide for a washing machine. No. 1 shows markup tags (using Extensible Markup Language) that describe the code snippet as a section in what could be a larger guide. The opening bracket for the section opens at the beginning of the snippet and closes at the end. No. 2 shows an attribute that gives a unique identifier to the section. The section element contains block sub-elements for title, introduction, and steps. No. 3 shows an inline sub-element referencing the product name.

For the past two decades, structured authoring in technical communication has frequently involved implementations of the Extensible Markup Language (XML). Particularly, the Darwin Information Typing Architecture (DITA) has become one of the main XML grammars used for technical communication purposes. DITA started as "a technical documentation authoring and publishing architecture that is based on principles of modular reuse and extensibility" (Priestley et al., 2001, p. 352) at IBM. Since 2004, DITA has been an open standard maintained by the nonprofit Organization for the Advancement of Structured Information Standards (OASIS).

The modular structure of DITA is based on a generic topic type that can describe almost any content. In a DITA authoring environment, writers create "technical content by assembling topic-oriented information types or blocks of information that serve particular functions in a document" (Swarts, 2010, p. 133). Over the years, the DITA standard has specialized the generic topic into information types that "represent the vast majority of content produced to support users of technical information" (Hackos, 2011, p. 7). These topic types for structuring technical content include concept, task, reference, glossary, and troubleshooting. The DITA standard also includes topic types designed for structuring learning and training projects: learning plan, learning overview, learning content, learning summary, and learning assessment.

Besides the preestablished topic types for technical content and learning and training projects, DITA topics can be customized (in a process known as specialization) to create information types unique to any domain. This exercise in markup flexibility is a direct application of both the *extensible* part of XML and the *Darwin* element in DITA: XML elements can be *extended*, and DITA information types can *evolve* to structure diverse content needs. For example, a DITA specialization for music composition could have a topic type for song with predetermined elements for intro, chorus, and bridge.

For technical authors, potential benefits of structured authoring in a workflow using DITA (or a similar standard) include streamlining the content creation process, increasing the quality of content by standardizing it, and allowing authors to leverage content in many different ways, which include reusing it, publishing it in different formats, and translating it (Samuels, 2014). From a business perspective, DITA can lead to promoting the reuse of information quickly and easily across multiple deliverables, which leads to reducing the cost of maintaining, updating, and localizing information (Hackos, 2011).

The reuse capabilities of structured content are the strongest selling points of a standard like DITA. Kristen Eberlein (2016), chair of the DITA Technical Committee with OASIS, defines reuse as the "practice of using content components in multiple information products" (p. 54). She adds that in many technical communication workflows, "efficient content reuse does not involve copy-and-pasting; instead it uses transclusion, whereby content is authored in one location and used by reference in other locations" (Eberlein, 2016, p. 55)

Content structured in DITA or DITA-like methodologies also opens the possibility of single sourcing, which can be defined as the practice of "creating content once, planning for its reuse in multiple places, contexts, and output channels" (White, 2016 p. 56). The tags of an XML-based grammar like DITA also can make content behave like data; as a result, structured content is computable and allows machine processing (Day, 2016 p. 50). Structured content can include metadata, which is defined as "'data about data,' which means data that isn't the primary purpose of the content object, but serves some secondary purpose" (Barker, 2016, p. 92). With an appropriate combination of structure and metadata, for example, the same task on how to configure a new computer can include introductory steps for inexperienced users and advanced steps for expert users (e.g., *<step audience="introductory"> Turn on the computer</step>* and *<step audience="advanced"> Replace the motherboard</step>* can be included on the same task). Publishing instructions would then include filters and routines to produce deliverables aimed at either inexperienced or advanced users that occlude (but do not delete from the structured source) content that would be irrelevant for the intended audience group.

Despite its actual and potential benefits for content creators and their business supervisors, the implementation and enforcement of structure in technical communication authoring workflows is not without its challenges. A major challenge to widespread adoption is the separation of content and presentation required by workflows based on DITA or a similar standard. This separation "can create philosophical and cognitive dissonance for technical communicators trained to think of information as content that is inherently linked to presentation" (Clark, 2007, p. 36). According to some, writers separating content from presentation "will have no control over the context in which their information appears or the uses to which it may be put" (Gu & Pullman, 2009, p. 6). Adopting templates in desktop publishing applications could, therefore, be an effective introduction to structure for novice technical communication practitioners, and it might be enough for situations in which content reuse and single sourcing are not required. If the reuse needs of a project change or evolve, commercial and open-source tools can relatively easily convert template-based documents to structured content using DITA or a similar standard.

Another challenge is the perceived loss of creativity for authors using a structured content type as opposed to a writing environment without restrictions, or "the perception that XML forces writers into creating cookie-cutter topics rather than useful technical information" (O'Keefe, 2010, p. 37). Taken to its most dangerous extreme, the implementation of structure in technical communication could lead to the standardization of cultural products that Theodor Adorno (1991) presaged. However, taken to its most beneficial extreme, structured content workflows could produce information schemas like those proposed by J.C.R. Licklider (1965) for cataloging cultural artifacts, which revolutionized the ways in which ***technology*** helps librarians gather, index, organize, store, and distribute print and digital content. Some scholars tackle this challenge as an opportunity to acknowledge that "while the technology can hamper some elements of creativity, it can also open up new possibilities for rhetorical expression, for writing content that can be assembled into new meaningful forms" (Swarts, 2020, p. 171).

The evolution of structure in technical communication is leading to the development of more flexible methodologies and standards (e.g., Markdown, JSON, and proprietary solutions for separating content from presentation). Although they do not provide all the capabilities of XML, they can replicate most of the transclusion and single sourcing features of DITA (Evia et al., 2018). Evolutionary trends also include the practice of content-as-a-service (CaaS), which "focuses on managing structured content into feeds that other applications and properties can consume" ([A], 2017). In a CaaS-based workflow, structured content does not necessarily inherit formatting and processing rules from the same organization where it is developed, but it is available for use in different contexts and environments via online information requests.

As an explicit change of tone in speech or a new section in a piece of user documentation, or behind the scenes as a command for a computer request sending content to a voice application, structure is essential to technical communication. Audiences and authors will continue evolving, and their use of technology will no doubt become more sophisticated and complex over time. Regardless of medium and technology, a well-structured document will always be a more effective piece of communication than a disorganized blob of words.

References

[A]. (2017, March 2). *CaaS: What is content-as-a-service?* https://simplea.com/Articles/what-is-content-as-a-service.

Adorno, T. W. (1991). *The culture industry: Selected essays on mass culture.* Routledge.

Barker, D. (2016). *Web content management: Systems, features, and best practices.* O'Reilly.

Cicero, M. T. (2014). *De inventione; De optimo genere oratorum; Topica.* Harvard University Press.

Clark, D. (2007). Content management and the separation of presentation and content. *Technical Communication Quarterly, 17*(1), 35–60. https://doi.org/10.1080/10572250701588624.

Cohen, G. & Cunningham, D. H. (1984). *Creating technical manuals: A step-by-step approach to writing user-friendly instructions.* McGraw-Hill.

Day, D. (2016). Structured content. In R. Gallon (Ed.), *The language of technical communication* (pp. 50–51). XML Press.

Eberlein, K. J. (2016). Content reuse. In R. Gallon (Ed.), *The language of technical communication* (pp. 54–55). Laguna Hills, CA: XML Press.

Evia, C., Eberlein, K. & Houser, A. (2018). *Lightweight DITA: An introduction. Version 1.0.* OASIS.

Flower, L. (1989). Rhetorical problem solving: Cognition and professional writing. In M. Kogen (Ed.), *Writing in the business professions* (pp. 3–36). National Council of Teachers of English.

Gu, B. & Pullman, G. (2009). Introduction: Mapping out the key parameters of content management. In G. Pullman & B. Gu (Eds.), *Content management: Bridging the gap between theory and practice* (pp. 1–12). Baywood.

Hackos, J. T. (2011). *Introduction to DITA: A user guide to the Darwin Information Typing Architecture including DITA 1.2* (2nd ed.). Comtech Services, Inc.

Kissane, E. (2009, July 7). Content templates to the rescue. *A List Apart*. https://alistapart.com/article/content-templates-to-the-rescue/.

Licklider, J.C.R. (1965). *Libraries of the future*. MIT Press.

O'Keefe, S. (2010). XML: The death of creativity in technical writing? *Intercom*, *57*(2), 36–37.

O'Keefe, S. & Pringle, A. (2017). *Structured authoring and XML*. Scriptorium.

Price, J. (1984). *How to write a computer manual: A handbook of software documentation*. Benjamin Cummings.

Priestley, M., Hargis, G. & Carpenter, S. (2001). DITA: An XML-based technical documentation authoring and publishing architecture. *Technical Communication*, *48*(3), 352–367.

Pringle, A. & O'Keefe, S. (2009). *Technical writing 101: A real-world guide to planning and writing technical content*. Scriptorium Press.

Rigo, J. (1976). User manual outline. *Asterisk*, *2*(8), 7–10.

Samuels, J. (2014, February 3). What is DITA? *TechWhirl*. http://techwhirl.com/what-is-dita/.

Swarts, J. (2010). Recycled writing: Assembling actor networks from reusable content. *Journal of Business and Technical Communication*, *24*(2), 127–163. https://doi.org/10.1177/1050651909353307.

Swarts, J. (2020). Writing about structure in DITA. In T. Bridgeford (Ed.), *Teaching content management in technical and professional communication* (pp. 155–173). Routledge. https://doi.org/10.4324/9780429059612-9.

White, L. W. (2016). Single sourcing. In R. Gallon (Ed.), *The language of technical communication* (pp. 54–55). XML Press.

35. Style

Jonathan Buehl
The Ohio State University

Technical communicators often focus on style—the word- and sentence-level choices directing how readers will receive and understand a text. For example, revising to remove distracting "wordiness," focusing on action verbs in instructions, conforming citations to a style guide, placing information at the ends of sentences to create cohesion, or writing in registers that either signal expertise (e.g., scientific writing) or communicate expertise to non-experts (e.g., "***plain language***"). However, style is more than close attention to grammar, syntax, and vocabulary, as Dan Jones (1998) noted in a technical communication textbook dedicated entirely to style:

> Style affects almost all other elements of writing. Style is your choices of words, phrases, clauses, and sentences and how you connect these sentences. Style is the unity and coherence of your paragraphs and larger segments. Style is your tone—your attitude toward your subject, your audience, and yourself—in what you write. (p. 3)

This chapter considers the complexities of style in technical and professional communication (TPC) by examining multiple ways scholars have defined style, by identifying stylistic traditions in TPC, and by considering how style connects with TPC issues related to ***knowledge***, ***ethics***, justice, and inclusion.

Categorical definitions have considered both the categories to which style belongs and how style itself can be categorized. Style is one of the so-called canons of ***rhetoric***—traditionally, the five activities constituting rhetorical performance. The others are invention (identifying arguments), arrangement (organizing arguments), memory (remembering a text and making it memorable), and delivery (the material performance of a text). Rhetorical theorists continue to discuss and question the boundaries between canonical categories, approaching them not as steps in a rigid process (first we invent, then we arrange, then we choose a style, etc.) but as interrelated, co-constitutive activities. For example, Jeanne Fahnestock (2002, 2004) has demonstrated how rhetorical figures in scientific communication serve as more than mere ornamental flourishes—they are structures used to develop, epitomize, and reinforce lines of reasoning (see also Graves, 2005). Similarly, Paul Butler's (2008) term "inventional style" acknowledges the fuzzy boundaries and connections between generating ideas (invention) and choosing the words to express them (style). The TPC takeaway is that style is not just a late-stage activity (e.g., part of copyediting or proofreading);

DOI: https://doi.org/10.37514/TPC-B.2023.1923.2.35

rather, it is an integral aspect of communication requiring attention at different stages of a project.

Although canon-based approaches focus on the taxonomies to which style belongs, other definitions classify styles into operational types. For example, Nora Bacon (2015) identifies five ways people invoke "style" to describe particular aspects of language use:

- Style 1—Individual Style: "the sound of [an author's] voice on the page" (p. 292)
- Style 2—House Style: the conventions articulated and enforced by a community of editors to achieve consistency; e.g., MLA style, APA style, or a style codified in a company's style guide or a project's style sheet
- Style 3—Usage: a stylistic focus on linguistic etiquette; e.g., injunctions to be precise with such distinctions as "effect" vs. "affect" or to avoid passive voice
- Style 4—Plain Style: an approach privileging clarity and conciseness; e.g., the advice of William Strunk, Jr. and E.B. White's *The Elements of Style*
- Style 5—Elaborated Style: an approach focused on "sentence variety, syntactic dexterity, and artfulness," such as the creative use of rhetorical figures (p. 292)

To Bacon's list, I add a sixth variant of style invoked in TPC contexts: Style 6—Structural Style or Technologized Style: the digital features facilitating how computers present ***content***, such as Microsoft Word styles or the style sheets that transform XML ***structured*** content into deliverables. For example, an XML transformation might specify that top-level headings should appear in Times font for a PDF but Arial font for a web page presenting the same content. Style 6 also highlights a point relevant to Bacon's other types of style: Style includes choices about words themselves as well as formatting, ***design***, and other nonverbal elements that nonetheless shape how words are perceived.

Other means of categorizing styles focus on the occasions of their use. For example, classical rhetorical theorists identified three types or "levels" of style, each associated with a specific purpose:

- the low or plain style, to be used to instruct an ***audience***
- the middle style, to be used to move or persuade an audience
- the high or grand style, to be used to please an audience

These levels persist in such contemporary stylistic distinctions as "colloquial," "standard," and "formal" (Fahnestock, 2011, p. 81). As Russell Willerton (2015) explains, plain style has long been associated with technical communication, with calls to use "plain English" for expert and non-expert audiences dating back to the 14th century (p. 3). It is important to note that the boundaries between these levels are not hard lines—even for Cicero, the Roman orator often credited with the leveling concept (von Albrecht, 2003, pp. 20–25). Moreover, these levels are not hierarchical—i.e., a grand style is not qualitatively better than a plain style. As Michael von Albrecht (2003) observes, the real innovation of the levels approach is its recognition of "a close interrelation between subject and style" (p. 22).

The idea of stylistic "levels" took a quantitative turn in the 20th century, when researchers developed so-called readability formulas to rate texts for reader comprehension. For example, the Gunning Fog Index and the Flesch-Kincaid Grade Level test calculate the "grade level" of a passage (e.g., a score of 9 indicates a ninth-grade reading level). The Flesch Reading Ease test assigns a score ranging from 0 to 100, with higher scores associated with greater readability. Although they all use the same metrics (sentences, words, and syllables), each varies in how those features factor into the readability calculation. Table 35.1 demonstrates how these formulas evaluate passages. The first example, an abstract from a scientific article, has grade-level scores of 18.2 and 17.4 (i.e., graduate school) and a low reading ease score of 13.4 ("very difficult"). The other examples are from websites written for the general ***public***. They present similar content on how COVID-19 spreads, but they demonstrate lower grade levels and higher reading ease scores.

Readability formulas can help less-experienced writers focus on word- and sentence-level revisions; however, relying on readability scores as indications of "good" writing is potentially problematic (Selzer, 1981, 1983; Redish, 2000; Redish & Selzer, 1985). Indeed, using shorter words and more-but-shorter sentences will not necessarily result in a better text. For example, if only the italicized parenthetical statements were deleted from the second example in Table 35.1, the Flesch-Kincaid grade level would drop from 15.2 to 13.2; however, important clarifying information would be lost. Janice (Ginny) Redish (2000) has proposed that usability testing (also known as ***user experience*** testing) is a better approach for assessing reader comprehension.

Another approach for defining style focuses on valued attributes of discourse. Classical Greek theorists identified five "virtues" of style: clarity, correctness, vividness (*enargeia*), appropriateness, and ornateness (Burton, 2007b). Other theorists valued other virtues; for example, the Byzantine theorist Hermogenes included grandeur, beauty, rapidity, character, sincerity, and force along with clarity in his list of stylistic virtues (Burton, 2007a). Similar values-based typologies of style have long been commonplace in professional communication textbooks (Carbone, 1994). For example, Sada A. Harbarger's (1923) *English for Engineers*—which Robert J. Connors (1982) identifies as the first modern technical communication textbook—promoted three virtues for engineering writing: clearness, conciseness, and emphasis (Harbarger, 1923, p. 23). Similar lists persist today and are often expressed through the common mnemonic device of "the [insert number] Cs" of effective writing: for example, clarity, coherence, conciseness (Wasko, 2011) or consideration, clarity, conciseness, coherence, correctness, confidence

(Howe Writing Initiative, n.d.). (See Carbone [1994] for the long history of "the Cs" mnemonic in business writing texts.) Stylistic "virtues" are often presented as universal traits; however, they are scalar and contingent values. For example, a passage offering an appropriate level of detail for one context might be too wordy for others. Similarly, a maximally concise passage might be considered curt or even rude by some readers.

Another traditional approach is to name styles based on sets of features. For example, the "plain language" style is a specific variation of plain style that emerged from the plain language movement (Mazur, 2000; Willerton, 2015, this volume). It is often contrasted with bureaucratic style (Shuy, 1998), which needlessly obfuscates ***information*** through unnecessarily complex phrasing, insider vocabulary, and unclear agency. Conversely, plain language principles regarding organization (e.g., "address separate audiences separately"), verbs (e.g., "use the active voice"), nouns (e.g., "don't turn verbs into nouns"), sentences (e.g., "keep subject, verb, and object close together"), and paragraphs (e.g., "cover only one point in each paragraph") are meant to increase the chances that readers can find, understand, and use the information in a document (Plain Language Action and Information Network, 2011). Like plain language, writing with "you attitude" attends to the needs of the reader through such strategies as preferring "you" as a sentence subject when addressing what readers can gain or must do; however, it also protects the reader's ego through careful attention to avoiding negative language (Hotchkiss and Drew, 1916; Locker, 1995).

Although scientific style's purposeful use of passive voice and nominalizations might seem like the antithesis of plain language, the two styles are otherwise compatible (see Gopen & Swan, 1990; Green, 2013). Moreover, the "grammatical problems" that make scientific language challenging for non-experts—such as lexical density, complex noun phrases (e.g., "severe acute respiratory syndrome coronavirus"), interlocking definitions, and implicit taxonomies—are actually discursive features that have evolved to facilitate communication between experts who share a common base of knowledge (Halliday, 1993a, 1993b). This "scientific writing" for expert readers is often contrasted with "science writing," which can refer to a range of styles used to accommodate ***science*** for non-experts (see Buehl, 2013; Fahnestock, 1998).

Although categorical and descriptive approaches can help technical communicators understand style, they do not address the range of epistemological and ethical entailments related to both definitions of style and stylistic choices. Approaches to style vary in their epistemological assumptions about the relationships between language, knowledge, and reality. Linguists Geoffrey Leech and Michael Short (2007) identify three main philosophies:

- "Dualism": style is merely the manner in which content is expressed
- "Monism": style and content are inseparable
- "Pluralism": language simultaneously performs different functions

Table 35.1. Comparing Popular Readability Formulas*

Passage 1 (Expert Audience): Severe acute respiratory syndrome coronavirus 2 (SARS-CoV-2) has spread rapidly throughout the world since the first cases of coronavirus disease 2019 (COVID-19) were observed in December 2019 in Wuhan, China. It has been suspected that infected persons who remain asymptomatic play a significant role in the ongoing pandemic, but their relative number and effect have been uncertain. The authors sought to review and synthesize the available evidence on asymptomatic SARS-CoV-2 infection. Asymptomatic persons seem to account for approximately 40% to 45% of SARS-CoV-2 infections, and they can transmit the virus to others for an extended period, perhaps longer than 14 days. Asymptomatic infection may be associated with subclinical lung abnormalities, as detected by computed tomography. Because of the high risk for silent spread by asymptomatic persons, it is imperative that testing programs include those without symptoms. To supplement conventional diagnostic testing, which is constrained by capacity, cost, and its one-off nature, innovative tactics for public health surveillance, such as crowdsourcing digital wearable data and monitoring sewage sludge, might be helpful. (**Source:** Oran & Topal, 2020. "Prevalence of Asymptomatic SARS-CoV-2 Infection: A Narrative Review.") **Gunning Fog Index (Grade Level):** 18.4 | **Flesch-Kincaid Grade Level:** 17.4 | **Flesch Reading Ease:** 13.3
Passage 2 (Public Audience - General): COVID-19 spreads mainly from person to person through respiratory droplets produced when an infected person coughs, sneezes, talks, or raises their voice *(e.g., while shouting, chanting, or singing)*. These droplets can land in the mouths or noses of people who are nearby or possibly be inhaled into the lungs. Recent studies show that a significant portion of individuals with COVID-19 lack symptoms *(are "asymptomatic")* and that even those who eventually develop symptoms *(are "pre-symptomatic")* can transmit the virus to others before showing symptoms. (**Source:** Centers for Disease Control and Prevention, 2020. "About Cloth Face Coverings." Emphasis added.) **Gunning Fog Index (Grade Level):** 10.5 | **Flesch-Kincaid Grade Level:** 15.2 | **Flesch Reading Ease:** 35
Passage 3 (Public Audience - Parents): Most commonly, the virus that causes COVID-19 enters people's bodies when it's on their hands and they touch their mouths, noses or eyes. A virus is so tiny that you can't see it. This is why it's important to wash your hands often and try not to touch your mouth, nose or eyes. If someone who has the infection coughs or sneezes on you from a close distance — closer than six feet — then that also can spread the virus. (**Source:** Mayo Clinic, 2020. "How to Talk to Your Kids about COVID-19") **Gunning Fog Index (Grade Level):** 10.1 | **Flesch-Kincaid Grade Level:** 7.6 | **Flesch Reading Ease:** 75.4

* *Each of the passages describes similar content on the spread of COVID-19 but for very different audiences—scientific experts, the general public, and parents of small children. Each passage has been scored according to three popular readability formulas:*

- *Gunning Fog Index = 0.4 [(total words / total sentences) + 100 (complex words / total words)]*
    - *"Complex words": Words with more than three syllables (excluding proper nouns, "familiar jargon," and compound words)*
- *Flesch-Kincaid Grade Level = 0.39 (total words / total sentences) + 11.8 (total syllables / total words – 15.59)*
- *Flesch Reading Ease Score = 206.835 – 1.015 (total words / total sentences) – 84.6 (total syllables / total words)*

Understanding these distinctions is important because technical communicators might encounter people with particularly rigid views of language; for example, an "objective" style represents objective thinking.

Style is often discussed in relation to ethics, the politics of language, and relationships between language, power, and identity. Although TPC discourses are often regarded as objective or neutral, a seemingly neutral style does not necessarily mean a text is ideologically neutral or ethical. As Steven B. Katz (1992) demonstrated, the Nazis wrote clear and precise ***documentation*** of their technologies of genocide. Similarly, Nigerian military officers wrote in precise, audience-appropriate vocabulary about murdering innocent civilians to benefit an oil company (Agboka, 2018). As Michael J. Zerbe (2007) has observed, scientific discourse is the dominant "power" discourse of our time, and thus, it is crucial for students to be able to read, write, and critique it. However, we also have an obligation to help students recognize and navigate stylistic diversity without marginalizing specific dialects (Conference on College Composition and Communication, 1974; Wilson & Crow, 2017). In TPC classes, we often task students with performing styles typical for contemporary workplaces; however, "standard" styles should not be held out as objectively standard or ideal. Rather, they are sets of discursive moves that have become conventionalized as appropriate and expected for particular contexts. And "standard" conventions evolve as contexts evolve.

Consider, for example, shifts in conventions regarding gender and language. It was once acceptable to use masculine pronouns and male terms generically (e.g., "Each applicant must sign his name."). Most style guides now promote the use of sex-inclusive language ("Each applicant should sign his or her name.") or gender-neutral language ("Applicants must sign their names."). However, specific guidance on removing gender bias varies widely. For example, *The IBM Style Guide* (2012) discourages using plural pronouns as gender-neutral replacements for singular nouns ("Each applicant must sign their name."). The *Microsoft Writing Style Guide* (2020) states, "it's OK to use a plural pronoun (*they, their,* or *them*) in generic references to a single person" if there's no other option, while the *Mailchimp Style Guide* (2020) explicitly permits the singular "they." As Allen Smith (2020) observes, more and more companies are updating employee handbooks with gender-neutral pronouns to make these documents more inclusive of nonbinary individuals. Although approving of the singular "they" is the more common stylistic change, some companies (including the financial firm Goldman Sachs [2019]) openly support other singular nonbinary pronouns (ze / zer / zirs or ze / zem / zes). Such changes in stylistic conventions have ***social justice*** implications for professional communication and can support commitments to inclusion.

Calls for language diversity are other sites where style intersects with inclusive communication practices. As the field expands its understanding of the sites of TPC activity, the range of styles that "count" as technical and professional communication are also expanding. For example, in describing the possibilities of hip-hop pedagogies for TPC, Marcos del Hierro (2018) observes how rap songs can communicate technical information through hip-hop styles. Krystle Danuz (2014) noted how Spanglish—the often-disparaged dialectal blend of Spanish and English—can actually be more effective than writing in a "standard" professional style when communicating technical information to some multilingual readers. As Temptaous T. Mckoy (2019) has demonstrated, even TPC scholarship can be performed effectively and insightfully through a diverse range of styles, which for Mckoy include "traditional" academic prose as well as African American Vernacular English (AAVE) and ***multimodal*** trap-music videos. In short, recognizing linguistic and stylistic diversity is entirely compatible with the core goal of TPC (as a field and as a ***profession***)—to share expertise effectively with diverse audiences.

To conclude with a stylistic flourish, just as style affects all aspects of writing, all aspects of writing affect style. Categorical, descriptive, operational, epistemological, ethical, and inclusive perspectives on style can help TPC scholars, students, and practitioners make meaningful choices to craft effective and ethical texts.

References

Agboka, G. Y. (2018). Indigenous contexts, new questions: Integrating human rights perspectives in technical communication. In A. M. Haas & M. F. Eble (Eds.), *Key theoretical frameworks: Teaching technical communication in the twenty-first century* (pp. 114–137). Utah State University Press.

Bacon, N. (2015). Cross-disciplinary approaches to style. *College Composition and Communication, 67*(2), 290–303.

Buehl, J. (2013). Style and the professional writing curriculum: Teaching stylistic fluency through science writing. In M. Duncan and S. M. Vanguri (Eds.), *The centrality of style* (pp. 279–308). The WAC Clearinghouse; Parlor Press. https://doi.org/10.37514/PER-B.2013.0476.2.17.

Burton, G. O. (2007a). *Hermogenes'* On Style. Silva rhetoricae: The forest of rhetoric. http://rhetoric.byu.edu/.

Burton, G. O. (2007b). *Virtues of style.* Silva rhetoricae: The forest of rhetoric. http://rhetoric.byu.edu/.

Butler, P. (2008). *Out of style: Reanimating stylistic study in composition and rhetoric.* Utah State University Press. https://doi.org/10.2307/j.ctt4cgmzv.

Carbone, M. T. (1994). The history and development of business communication principles: 1776–1916. *The Journal of Business Communication, 31*(3), 173–193. https://doi.org/10.1177/002194369403100302.

Centers for Disease Control and Prevention. (2020). *About cloth face coverings.* https://stacks.cdc.gov/view/cdc/89934.

Conference on College Composition and Communication. (1974). Students' right to their own language. *College Composition and Communication, 25*(3), 1–32. https://doi.org/10.2307/356219.

Connors, R. J. (1982). The rise of technical writing instruction in America. *Journal of Technical Writing and Communication, 12*(4), 329–352. https://doi.org/10.1177/004728168201200406.

Danuz, K. (2014). *Spanglish: A new communication tool.* In M. F. Williams & O. Pimentel (Eds.), *Communicating race, ethnicity, and identity in technical communication* (pp. 121–132). Baywood Publishing Co.

Del Hierro, M. (2018). Stayin' on our grind : What hiphop pedagogies offer to technical writing. In A. M. Haas & M. F. Eble (Eds.), *Key theoretical frameworks: Teaching technical communication in the twenty-first century* (pp. 163–184). Utah State University Press. https://doi.org/10.7330/9781607327585.c007.

DeRespinis, F., Hayward, P., Jenkins, J., Laird, A., McDonald, L. & Radzinski, E. (2012). *The IBM style guide: conventions for writers and editors.* IBM Press.

Fahnestock, J. (1998). Accommodating science: The rhetorical life of scientific facts. *Written Communication*, *15*(3), 330–350. https://doi.org/10.1177/0741088398015003006.

Fahnestock, J. (2002). *Rhetorical figures in science*. Oxford University Press.

Fahnestock, J. (2004). Preserving the figure: Consistency in the presentation of scientific arguments. *Written Communication*, *21*(1), 6–31. https://doi.org/10.1177/07410883 03261034.

Fahnestock, J. (2011). *Rhetorical style: The uses of language in persuasion*. Oxford University Press. https://doi.org/10.1093/acprof:oso/9780199764129.001.0001.

Goldman Sachs. (2019). *Bringing your authentic self to work: Pronouns*. https://www.gold mansachs.com/careers/blog/posts/bring-your-authentic-self-to-work-pronouns.html.

Gopen, G. D. & Swan, J. A. (1990). The science of scientific writing. *American Scientist*, *78*(6), 550–558.

Graves, H. B. (2005). *Rhetoric in (to) science: Style as invention in inquiry*. Hampton Press.

Greene, A. E. (2013). *Writing science in plain English*. University of Chicago Press. https://doi.org/10.7208/chicago/9780226026404.001.0001.

Halliday, M.A.K. (1993a). On the language of physical science. In M.A.K Halliday & J. R. Martin, *Writing science: Literacy and discursive power* (pp. 54–68). University of Pittsburgh Press. (Original work published 1988)

Halliday, M.A.K. (1993b). Some grammatical problems in scientific English. In M.A.K Halliday & J. R. Martin, *Writing science: Literacy and discursive power* (pp. 69–85). University of Pittsburgh Press. (Original work published 1989)

Harbarger, S. A. (1923). *English for engineers.* McGraw-Hill.

Hotchkiss, G. B. & Drew, C. A. (1916). *Business English: Its principles and practice*. American Book Company.

Howe Writing Initiative. (n.d). *The six Cs of business communication*. University of Miami – Farmer School of Business. http://www.fsb.miamioh.edu/fsb/content/programs /howe-writing-initiative/HWI-handout-CsofBusComm.html.

Jones, D. (1998). *Technical writing style.* Pearson.

Katz, S. B. (1992). The ethic of expediency: Classical rhetoric, technology, and the Holocaust. *College English*, *54*(3), 255–275. https://doi.org/10.2307/378062.

Leech, G. N. & Short, M. (2007). *Style in fiction: A linguistic introduction to English fictional prose*. Routledge.

Locker, K. O. (1995). *Business and administrative communication*. McGraw-Hill.

Mailchimp content style guide. (2020). https://styleguide.mailchimp.com/.

Mayo Clinic. (2020). *How to talk to your kids about COVID-19*. https://www.mayoclinic .org/diseases-conditions/coronavirus/in-depth/kids-covid-19/art-20482508.

Mazur, B. (2000). Revisiting plain language. *Technical Communication*, *47*(2), 205–211.

Mckoy, T. T. (2019). *Y'all call it technical and professional communication, we call it #ForTheCulture: The use of amplification rhetorics in Black communities and their implications for technical and professional communication studies* [Doctoral dissertation, East Carolina University]. http://hdl.handle.net/10342/7421.

Microsoft Corporation. (2020). *Microsoft writing style guide*. https://docs.microsoft.com/en-us/style-guide/welcome/.

Plain Language Action and Information Network. (2011). Federal plain language guidelines. *plainlanguage.gov*. https://www.plainlanguage.gov/media/FederalPLGuidelines.pdf.

Oran, D. P. & Topol, E. J. (2020). Prevalence of asymptomatic SARS-CoV-2 infection: A narrative review. *Annals of Internal Medicine*, *173*(5), 362–367. https://doi.org/10.7326/M20-3012.

Redish, J. (2000). Readability formulas have even more limitations than Klare discusses. *ACM Journal of Computer Documentation (JCD)*, *24*(3), 132–137. https://doi.org/10.1145/344599.344637.

Redish, J. C. & Selzer, J. (1985). The place of readability formulas in technical communication. *Technical Communication*, *32*(4), 46–52.

Selzer, J. (1981). Readability is a four-letter word. *The Journal of Business Communication*, *18*(4), 23–34. https://doi.org/10.1177/002194368101800403.

Selzer, J. (1983). What constitutes a "readable" technical style? In P. V. Anderson, R. J. Brockman, C. R. Miller (Eds.), *New essays in technical and scientific communication* (pp. 71–89). Routledge. https://doi.org/10.4324/9781315224060-7.

Shuy, R. W. (1998). *Bureaucratic language in government and business*. Georgetown University Press.

Smith, A. (2020, February 9). *More employee handbooks replace 'he' and 'she' with 'they'*. SHRM. https://www.shrm.org/resourcesandtools/legal-and-compliance/employment-law/pages/handbooks-gender-neutral-pronouns.aspx.

von Albrecht, M. (2017). *Cicero's Style: A Synopsis*. Brill.

Wasko, B. (2011, November 2). The three C's of solid writing. *WriteAtHome.com*. http://blog.writeathome.com/index.php/2011/11/the-three-cs-of-solid-writing/.

Willerton, R. (2015). *Plain language and ethical action: A dialogic approach to technical content in the twenty-first century*. Routledge. https://doi.org/10.4324/9781315796956.

Wilson, N. and Crow, A. (2014). A response to "Students' Right to their Own Language" (Eds.), *Communicating race, ethnicity, and identity in technical communication* (pp.113–119). Baywood Publishing Co. https://doi.org/10.4324/9781315232584.

Zerbe, M. J. (2007). *Composition and the rhetoric of science: Engaging the dominant discourse*. Southern Illinois University Press.

36. Technology

Bernadette Longo
New Jersey Institute of Technology

The root of the word *technology* is the Greek term *tekhne*, which Aristotle defined as an art or "reasoned habit of mind in making something" (1991, p. 320). In 17th-century post-classical Latin, the term *technologia* (Greek *tekhne* + Latin *logia*, the study of) was used to describe the systematic study of an art or practical craft (R. Williams, 1985, p. 315). By the 18th century, *technology* was not only a study of practical arts, but particularly of the mechanical arts and applied ***sciences*** (Oxford University Press, n.d.). By the mid-19th century, this term implied not only the study, but also the active application, of mechanical arts, especially in manufacturing and industry (Technology, 2020). By the 20th century, the use of the term had expanded to include the products of people applying mechanical arts in manufacturing and industry (Oxford University Press, n.d.). In this sense of the term, technology can mean both the ***knowledge*** to make a mechanical object, as well as the object itself, as in this sentence: "Technology is starting to behave in intelligent and unpredictable ways that even its creators don't understand" (Bridle, 2018, p. 1). This contemporary sense of the word *technology* blurs the boundary between the person who has the knowledge and ability to make an object and that human-made object itself. As Steven B. Katz (1992) argued, "Technology becomes both a means and an end in itself" (p. 266), thus creating ethical implications that technical communicators should consider as they work with and write about technologies.

In *Nicomachean Ethics,* Aristotle argued that a person could lead a good life by pursuing virtuous knowledge and carrying out virtuous acts. He discussed *tekhne* as an intellectual virtue comprising one element of a good life. If the person who had technical knowledge was virtuous, the product of that *tekhne* would result in civic good: "The first principle is in the maker but not in what is made" (as cited in Kennedy, 1991, p. 289). In this early sense, the product of *tekhne* was the result of human agency, and the product could be evaluated according to the nature of its human creator. Thus, the product of *tekhne*—or what we might today call "a technology"—reflected its human maker and was under human control. Written communication can be considered to be an early technology in this sense (e.g., Havelock, 1986; Ong, 1992; Postman, 1993).

As relationships between humans and technology have evolved, the question of who is in control of technology has become contested. For example, this definition of how a thermostat works gives agency to the device: "While a thermometer is a tool to read a room's temperature, a thermostat is able to control it" (Hometree, n.d.). This attribution of agency to a technology in technical communication

DOI: https://doi.org/10.37514/TPC-B.2023.1923.2.36

discourse is so naturalized as to appear as common sense. Yet the implications of placing this device in the subject position of a sentence open the door to metaphorically considering a technology as having independent agency to carry out actions in the world. This metaphorical need to place a device in the subject position of an active verb points to a limitation in the English language inherent in defining a *technology* as both the human know-how and the object created by that know-how: It confuses subject and object in the text. Take, for example, this account of what happened when software engineers added a "Like" button to the Facebook interface: "The 'like' button, it turns out, transformed the social media experience" (Newport, 2020, p. 51). Cal Newport's attribution of agency to a ***social media*** feature aptly illustrates his exploration of technological determinism. This sentence attributes the transformation of users' experiences to a software feature, not to the people who programmed the feature. The "Like" button is the hero of this small story about technology and society. When people read text, they look for stories. Technical and professional communicators provide these stories about people and technologies, as well as determining the subjects taking actions in these stories.

A new technology can change the way that people view (im)possible relations between humans and machines. What seemed impossible in the past —that machines can learn and make independent decisions impacting people's lives—is now a relationship that seems natural. When intelligent machines can have linguistic agency in sentences, people are taught to consider machines as actors in the physical world. When intelligent machines then have actual agency in that physical world, distinctions between *technology* and *human* become blurred. As Langdon Winner (1992) observed, "the nature of man's own creations has now emerged as a source of genuine perplexity" (p. 5). He continued, technology is "the totality of rational methods . . . that stands at the center of modern culture. . . . Some of the most intriguing new technologies have to do with the alteration of psychological or spiritual states" (Winner, 1992, p. 9), especially when we consider intelligent systems that can learn and act autonomously. Machine learning has already been implemented to take on some commercial operations as described by technical writer Jennifer Kite-Powell (2017):

> Bots can already be trained to answer and respond to simple queries. Over time, Bots will be able to respond to more complex queries and their ability to solve complex problems will continue to increase, allowing them to interact in more meaningful ways with customers. (n.p.)

In this example, an intelligent technology is acting in the physical world, as well as being represented linguistically as an agent acting in a sentence. Once a technology can take actions that impact people in a physical world, ethical questions arise, especially regarding technological systems that have the potential for lethal outcomes. When a technology can act independently and potentially take an action that can kill a human, who is responsible for that action? Winner (1992) argued that "Autonomous technology is ultimately nothing more or less than the question of human autonomy held up to a different light" (p. 43). Technical and professional communicators are necessarily implicated in these ethical relationships when we write about technologies.

If an intelligent technology can take independent action similar to a person, can the consequences of that action be judged by the same ethical principles whether it is taken by a machine or a person? In considering human actions, Keith Abney (2012) distinguished between actions taken through instinct and those taken after deliberation. He concluded that machines are not subject to ethical judgement because their actions are programmed and therefore instinctual, not deliberative (Abney, 2012, p. 46). The question remains unanswered, though: "Who is responsible for an action taken by an intelligent technology?" The ***designer***? The programmer? The operator? The technical communicator who enables the operator to use the machine? This question comes into sharp focus when we consider intelligent military systems known as "lethal autonomous weapons" that are designed to fight, defend, and kill. This lethal defense technology is undoubtedly embedded in a complex network of people who design, produce, and implement the system, as well as people who are targets of the system. As Winner (1992) argued, such autonomous technological systems seem out of the control of any one person or group of people. More than a question of direct implementation, the question of responsibility becomes more about underlying values than direct action. When the technological system is so complex as to be beyond the control of any one human organization, the implication is that the values embodied in the technology are social values.

Technical and professional communicators participate in systems of social values when we give voice to technological knowledge. What is our responsibility in this knowledge/power system? Although technical communication has historically been viewed as functional and instrumental, more recent cultural studies conclude that technique and correctness in themselves do not represent the influence that technical communicators exert on people's understanding of their (im)possible relations with technologies (e.g.,; Jones, 2016; Jones et al., 2016; Longo, 1998, 2000; Slack et al., 1993). Because technical and professional writers work within institutions, such as businesses, governmental agencies, and academia, our practices "serve to (de)stabilize important rational and scientific knowledge/power structures in our culture" (Longo, 2006, p. 22). We work at the intersection of institutions and publics; whose interests do we serve? "Only when technical communicators accept responsibility as authors within our cultural context can we begin to understand and control our practices and the technologies in which we are complicit" (Longo, 2006, p. 22). Only when we look for the interests of people whose experiences have traditionally been marginalized because they threatened to destabilize the dominant knowledge/power system—such as the half of the world's population who are currently not connected to the internet or people who live with very low incomes that do not allow them full online access to opportunities and services—can technical communicators add ***social justice*** concerns to our professional values and our "reasoned habit of mind in making something" (Aristotle, 1991, p. 320).

Technology as a means and an end becomes in itself a rationale for action, since it shapes a society's values while it is, in turn, shaped by those values. Neil Postman (1993) argued that "every culture must negotiate with technology" (p. 5) because "radical technologies create new definitions of old terms . . . that have deep-rooted meanings" (p. 8), such as *human* and *technology*. Postman further argued that a technology "creates the 'conditions of intercourse' by which we relate to each" (p. 14). In examining one documentary example of how society shapes and is shaped by technological values, Katz (1992) asked how some people in the Third Reich could come to view other people as subhuman objects for extermination. He determined that their rationale was "grounded not in the arrogance of a personal belief in one's superiority, but rather in a cultural and ethical norm of technology . . . the ethic of technological expediency" (Katz, 1992, p. 265). On a textual scale, this case illustrated the importance of word choice and syntax in reflecting cultural values. On a societal scale, it illustrated an ethical system in which humans and technologies were intertwined in institutional systems with far-reaching consequences for people's lives.

As long as the word *technology* obscures human and machine agency, the use of this term contains the possibility of ethical ambiguity. This term can also reveal societal values that place convenience and practicality over the messiness of human nature (e.g., Dilger, 2006). As technical and professional communicators are increasingly called upon to consider questions of social justice as well as institutional stability (e.g., Haas & Elbe, 2018; Walton et al., 2019; Williams & Pimentel, 2012), we should use the word *technology* with caution because adopting a machine-based ethic has important, life-and-death implications for other people and the world we perpetuate. We should use what Natasha N. Jones and Miriam F. Williams (2020) call the "just use of imagination" to safeguard the humanity of all people and counteract oppressive practices that could be contained in relationships between humans and machines.

References

Abney, K. (2012). Robotics, ethical theory, and metaethics: A guide for the perplexed. In P. Lin, K. Abney & G. A. Bekey (Eds.), *Robot ethics: The ethical and social implications of robotics* (pp. 35–52). The MIT Press.

Aristotle. (1991). *On rhetoric: A theory of civic discourse* (G. A. Kennedy, Trans.). Oxford University Press.

Bridle, J. (2018, June 15). Rise of the machines: Has technology evolved beyond our control? *The Guardian*. https://www.theguardian.com/books/2018/jun/15/rise-of-the-machines-has-technology-evolved-beyond-our-control-

Dilger, B. (2006). Extreme usability and technical communication. In J. B. Scott, B. Longo & K. V. Wills. (Eds.), *Critical power tools: Technical communication and cultural studies* (pp. 47–70). State University of New York Press.

Haas, A. M. & Eble, M. F. (Eds.). (2018). *Key theoretical frameworks: Teaching technical communication in the twenty-first century*. Utah State University Press.

Havelock, E. A. (1986). *The muse learns to write: Reflections on orality and literacy from antiquity to the present*. Yale University Press.

Hometree. (n.d.). *How does a thermostat work?* https://www.hometree.co.uk/energy-advice/central-heating/how-does-a-thermostat-work.html.

Jones, N. N. (2016). The technical communicator as advocate: Integrating a social justice approach in technical communication. *Journal of Technical Writing and Communication, 46*(3), 342–361. https://doi.org/10.1177/0047281616639472.

Jones, N. N., Moore, K. R. & Walton, R. (2016). Disrupting the past to disrupt the future: An antenarrative of technical communication. *Technical Communication Quarterly, 25*(4), 211–229. https://doi.org/10.1080/10572252.2016.1224655.

Jones, N. N. & Williams, M. F. (2020). *The just use of imagination: A call to action*. ATTW. https://attw.org/blog/the-just-use-of-imagination-a-call-to-action/.

Katz, S. B. (1992). The ethic of expediency: Classical rhetoric, technology and the Holocaust. *College English, 54*(3), 255–275. https://doi.org/10.2307/378062.

Kite-Powell, J. (2017, December 29). The next technology shift: The internet of actions. *Forbes*. https://www.forbes.com/sites/jenniferhicks/2017/12/29/the-next-technology-shift-the-internet-of-actions/?sh=27adcde32270.

Longo, B. (1998). An approach for applying cultural study theory to technical writing research. *Technical Communication Quarterly, 7*(1), 53–73. https://doi.org/10.1080/10572259809364617.

Longo, B. (2000). *Spurious coin: A history of science, management, and technical writing*. State University of New York Press.

Longo, B. (2006). Theory. In J. B. Scott, B. Longo & K. V. Wills. (Eds.), *Critical power tools: Technical communication and cultural studies* (pp. 21–24). State University of New York Press.

Newport, C. (2020, May). When technology goes awry. *Communications of the ACM, 63*(05), 49–52. https://doi.org/10.1145/3391975.

Ong, W. J. (1992). Writing is a technology that restructures thought. In P. Downing, S. Lima & M. Noonan. (Eds.), *The linguistics of literacy* (pp. 293–319). J. Benjamins. https://doi.org/10.1075/tsl.21.22ong.

Oxford University Press. (n.d.). Technology. In *Oxford English Dictionary*. Retrieved March 11, 2011, from https://www.oed.com.

Postman, N. (1993). *Technopoly: The surrender of culture to technology*. Vintage Books.

Slack, J. D., Miller, D. J. & Doak, J. (1993). The technical communicator as author: Meaning, power, authority. *Journal of Business and Technical Communication, 7*(1), 12–36. https://doi.org/10.1177/1050651993007001002.

Walton, R., Moore, K. & Jones, N. (2019). *Technical communication after the social justice turn: Building coalitions for action*. Routledge Press. https://doi.org/10.4324/9780429198748.

Williams, M. F. & Pimentel, O. (2012). Introduction: Race, ethnicity, and technical communication. *Journal of Business and Technical Communication, 26*(3), 271–276. https://doi.org/10.1177/1050651912439535.

Williams, R. (1985). *Keywords: A vocabulary of culture and society* (Rev. ed.). Oxford University Press.
Winner, L. (1978). *Autonomous technology: Technics-out-of-control as a theme in political thought*. The MIT Press.

37. Translation

Bruce Maylath
North Dakota State University

The word *translation* has long roots in Latin, and the act of translation goes back centuries further, with equivalent terms in other ancient languages. In their book *Found in Translation*, Nataly Kelly and Jost Zetzsche (2012) observe that "*Translation* comes from the Latin *translatus*, which means 'to carry over,' as across a river . . ." (p. 41). Noting the same in its etymology of the word, *The Oxford Etymological Dictionary* offers two definitions of *translate*: "A. remove from one place to another; B. turn from one language to another" (Onions, 1966, p. 937). Drawing likewise on river imagery, Kirk St.Amant (2019) describes translation as "Transfers of meaning [which] often involve bridging different systems of conveying ideas" (p. 5). While the Latin word, and its anglicized form, are rooted in a crossing, the equivalent words in English's close Germanic cousin languages provide a slightly different image, e.g., *übersetzen* in German, *oversette* in Norwegian, in both cases meaning literally to "overset" or, more idiomatically, set over.

Translation studies, as a discipline, is a relatively recent development of mainly the past half century. Drawing on earlier theorists, James Melton (2008) identifies three types of translation: 1) intra-lingual ("within a single language or sign system"), 2) inter-lingual ("from one language into another"), and 3) inter-semiotic ("from verbal signs into non-verbal sign systems"; pp. 189–190). Federica Scarpa (2019) incorporates these three types within the discipline's more expansive and comprehensive taxonomy:

> *Translation* refers to
>
> - The process of transferring meaning from an original text written in a source language to another language according to the specific socio-cultural context of that language
> - The product resulting from that process: The target (i.e., final, translated) text that should address the socio-cultural context of the intended audience reading in the target language
>
> The word *translation* can also refer to other activities and products based on criteria such as
>
> - Medium: Written, oral, audiovisual, etc.
> - Mode: Conversion of a text from one language to another including
>   - Intralingual translation: Within the same language
>   - Interlingual translation: Between different languages

DOI: https://doi.org/10.37514/TPC-B.2023.1923.2.37

- Intersemiotic translation: Between different verbal/non-verbal systems, such as from a novel to the medium of film. (pp. 19–20)

Because this volume focuses on keywords as technical communicators are likely to encounter them, the rest of this essay dwells on the most common types of translation, interlingual and intralingual.

Throughout ***history***, translation has generally and most commonly been understood to refer to transferring meaning from one human language to another, especially when written. (In the translation industry, translators work with written texts while interpreters convey oral renderings between languages.) Translation as an occupation has its earliest roots in religious texts, especially in the West, where translation focused on fidelity or faithfulness of Latin translated from Greek (Windle & Pym, 2011, pp. 8–9). In parallel, Scott Montgomery (2013) demonstrates how crucial a role translation has played in spreading scientific ***knowledge*** throughout the centuries by allowing the transfer of ideas from Greek into Syriac, Latin, and Arabic; from Arabic into Latin; from Latin into Chinese and European vernaculars (including English); and from Chinese into Japanese and other East Asian languages (p. 158). In time, as the industrial revolution took place, translation historically became viewed, especially before the era of globalization, as Jeremy Munday (2016) describes it:

> The process of translation between two different written languages involves the changing of an original written text (the source text or ST) in the original verbal language (the source language or SL) into a written text (the target text or TT) in a different verbal language (the target language or TL). (p. 8)

Demand for translation of technical documents has soared since the late 20th century as global trade surged in the wake of such trade pacts as the North American Free Trade Agreement (NAFTA, now replaced by the United States-Mexico-Canada Agreement [USMCA]) and the World Trade Organization (WTO). As technical ***documentation*** has increasingly involved translation, many technical communicators have taken on the role of translation project managers. Some have sought cross-training as translators, while even more translators have sought cross-training as technical communicators (Gnecchi et al., 2011). Thus, over time, "Technical communication researchers are increasingly pushing for a move away from thinking of translation as an afterthought to content design and development" (Gonzales, 2017, p. 96).

Simultaneously, translation theory has moved from conceptions of faithfulness of the target text to the source text and instead emphasized equivalency of meaning. Examining the history of translation studies, Sandra Halvorson (2010) notes that this move transpired by the mid-20th century. Birthe Mousten and Dan Riordan (2019) credit this move to the theorist Ernst-August Gutt, who "switches *translation* from meaning 'two language versions of the same text' to meaning 'two texts with similar purpose and understanding'" (p. 160). Thus, as Patricia Minacori (2019) puts it, "Translation is a process that relates first and foremost to meaning, as opposed to words. In that regard it is fundamentally a process focused on comprehension" (p. 39). Among the best known of translation theorists, Lawrence Venuti (2008) sums up the current theoretical stance in this way: "Translation is a process by which the chain of signifiers that constitutes the foreign text is replaced by a chain of signifiers in translating language which the translator provides on the strength of an interpretation" (p. 13).

In focusing on equivalencies of meaning, translation theorists have also increasingly acknowledged the importance of accounting for culture: "The apparent division between cultural and linguistic approaches to translation that characterized much translation research until the 1980s is disappearing," observes Susan Bassnett (2014, p. 3), "for translation is not just the transfer of texts from one language to another, it is now rightly seen as a process of negotiation between texts and between cultures" (p. 6). This new view, of translation as ***intercultural communication***, has begun to seep into the thinking of the technical communication community as well. As Josephine Walwema (2018) observes, "At its most basic level, language is intertwined with culture, which itself comes with a set of values and belief systems" (p. 24). Or, as international technical communication specialist Timothy Weiss (1997) has put it, "translation, in the broadest sense of the term . . . is the fundamental process by which we interpret and express our reading of reality" (p. 322).

All languages display a continuum of formal to informal registers—with the latter sometimes interpreted as "***plain language***"—between which speakers and writers sometimes "translate" intralingually (Lanham, 1983). However, native English speakers—so many of whom have never bothered to learn other languages—are often unaware that English is unique among major languages in the extent to which its vocabulary is largely two languages merged into one: Germanic (Anglo-Saxon/Old English) and Latinate (Latin and its offspring Romance languages, including French), a result of the Norman Conquest. (For parallels, only a few minor languages exhibit extensive dual-language vocabularies, e.g., Luxembourgish, with German and French; Maltese, with Arabic and Italian; Romansch, with German and Latin.) The result has been frequent intralingual translation between Latinate and Germanic vocabularies. In the centuries since 1066, Latinate vocabulary, typifying the jargon of the educated professions, has been translated into Germanic "everyday English" or "plain language" for the masses (Crystal, 2004). As David Corson (1985, 1995) has shown in depth, such intralingual translation is necessary because the Latinate vocabulary in English remains foreign to so many native English speakers. They encounter what he calls a "lexical bar," resulting in "lexical avoidance" and "lexical apartheid," even though Latinate vocabulary is the most accessible lexicon for English language learners whose first language is Spanish or another Romance language (Maylath, 1997, 2000; Thrush, 2001).

Lexical apartheid in English-language cultures, and the history of French being required in the ***public*** affairs of England, has held staying power much longer than the vast majority of English speakers realize. Debate in England's parliament was conducted chiefly in French for hundreds of years post-Conquest. Furthermore, cases in common-law courts were argued in French until 1731, when parliament required that they be pleaded in English (albeit with many stock French terms incorporated wholesale in the proceedings; Fisher, 1992, p. 1169). Even as late as 1892, when delivering his "Introductory Lecture" at University College, London, A. E. Housman "translated" by repeating each point twice, once in Latinate English and once in Germanic English (Lanham, 1983). In our own time, a student's use of highly Latinate vs. highly Germanic English predictably can yield highly different ***assessments*** from college-level writing instructors of the quality of students' writing (Maylath, 1996).

The Plain Language Movement, as it exists in English, rests largely on the presumption of a dual vocabulary that requires intralingual translation. In fact, the U.S. government's current plain language guidelines webpage ("Choose Your Words Carefully," 2011) quotes H. W. Fowler's 1906 rule, "Prefer the Saxon word to the Romance word." The rise of scientific and technical communication as a ***profession***, especially in English-speaking lands, can be seen as a response to the need to provide users of new technologies with intralingual translation. Carol Barnum and Saul Carliner (1993) stated so plainly as the ***profession*** blossomed:

> Technical communication is translation. Technical communicators must take complicated subject matter, easily understood by subject-matter experts, and "translate" it into a language, a format, a style, and a tone that can be easily understood by non-specialists. . . . It requires recognizing jargon—the specialized vocabulary of one group—and reducing it to terms and expressions that can be understood by those outside the group. (pp. 3–4)

We see a similar view taken toward ***science*** communicators/journalists, when Kira Dreher (2020) writes,

> the scientific paper has traditionally had a gatekeeping function, inaccessible both in terms of language, rhetoric, and restricted access (via paid journals). In the past, the public has relied primarily on translators—science communicators and journalists—to bridge this gap.

When encountering or using the term *translation*, technical communicators, especially in the United States, need to be aware that in some subfields, such as ***risk communication***, the intended meaning is intralingual, or even intersemiotic, rather than interlingual, as one can see in such risk communication literature as "Translating Risk Management Knowledge" (Maule, 2004) or *Social Media in Disaster Response* (Potts, 2014). Indeed, the meaning of the term can go well beyond language and culture in this subfield of technical communication. Such usage is especially apparent in Liza Potts' work (2014), where "participants in the social web" become "'translators' who perform 'translations'" (p. 28), as defined by Michael Callon (1986), across four stages:

1. Problematization: Establishing and defining the event
2. Interessement: Encouraging participants to accept the network definition of the event
3. Enrollment: Actors align themselves with anchor actors and accept definition of the network.
4. Mobilization: Actors assemble across the network and mobilize to validate and distribute content. (paraphrased and summarized from Potts, 2014, pp. 28–29)

Language seems to be at some remove in this rendering of translation. Language can certainly be employed in "defining an event," but even there, defining can occur through still or moving images, thus falling into intersemiotic translation. At no point is there clear reference to interlingual translation.

Employing multiple meanings of *translation* might seem innocuous, but without explicit operational definitions, their use can halt communication. Such became apparent in 2015 during technical communication conferences held in quick succession. The first, in Austria, drew participants mainly from Europe. During the concluding session, winners of a European Union grant announced that they had just received the funds to carry out groundbreaking ***research*** on translating ***social media*** messages during disasters. They explained that such messages are typically transmitted in the national language, without regard for speakers of other languages in the disaster locale. The next week, during a conference in Ireland that drew mainly Americans, a participant in the conference in Austria relayed the prior week's news. American participants objected, saying that translation had long been addressed in risk communication, as evident in Potts' recent book. Europeans in the ***audience*** were surprised but held their tongues. Sadly, not until after the conferences did anyone realize that the Americans were using a far different definition of *translation* than Europeans were accustomed to—*intra*lingual, or perhaps even intersemiotic, instead of *inter*lingual. Without explication, what could have been a fruitful discussion was squandered and lost without translation.

Will *translation* take on new meanings in the future, perhaps especially as artificial intelligence develops and spreads? Only time will tell. However, as linguists since Ferdinand de Saussure are fond of pointing out, 1) the sign is arbitrary, and 2) language is in a constant state of flux. As words are signs, their meanings are unfixed and almost inevitably evolve and multiply as living speakers alter living languages, making translation necessary even between older versions of a language (e.g., Old English, Middle English) and newer versions (e.g., Modern English, in its many varieties around the world).

References

Barnum, C. M. & Carliner, S. (1993). *Techniques for technical communicators.* Macmillan.

Bassnett, S. (2014). *Translation studies* (4th ed.). Routledge. https://doi.org/10.4324/9780203488232.

Callon, M. (1986). Some elements of a sociology of translation: Domestication of the scallops and the fishermen of St Brieuc Bay. In J. Law (Ed.), *Power, action and belief: A new sociology of knowledge?* (pp. 196–223). Routledge.

Choose your words carefully. (2011, May). Plainlanguage.gov. https://www.plainlanguage.gov/guidelines/words/.

Corson, D. (1985). *The lexical bar.* Pergamon Press.

Corson, D. (1995). *Using English words.* Kluwer Academic Publishers. https://doi.org/10.1007/978-94-011-0425-8.

Crystal, D. (2004). *The stories of English.* The Overlook Press.

Dreher, K. (2020, July 21). *Plain language in the sciences: A qualitative meta-analysis of research* [Presentation]. IEEE ProComm 2020. https://doi.org/10.1109/ProComm48883.2020.00038.

Fisher, J. H. (1992). A language policy for Lancastrian England. *PMLA, 107*(5), 1168–1180. https://doi.org/10.2307/462872.

Gnecchi, M., Maylath, B., Scarpa, F., Mousten, B. & Vandepitte, S. (2011). Field convergence: Merging roles of technical writers and technical translators. *IEEE Transactions on Professional Communication, 54*(2), 168–184.

Gonzales, L. (2017). But is that relevant here? A pedagogical model for embedding translation training within technical communication courses in the US. *Connexions: International Professional Communication Journal, 5*(1), 75–108.

Halvorson, S. (2010). Translation. In Y. Gambier & L. van Doorslaer (Eds.), *Handbook of translation studies* (pp. 378–384). John Benjamins. https://doi.org/10.1075/hts.1.tra2.

Kelly, N. & Zetzsche, J. (2012). *Found in translation: How language shapes our lives and transforms the world.* Perigee.

Lanham, R. (1983). *Analyzing prose.* Macmillan.

Maule, A. (2004). Translating risk management knowledge: The lessons to be learned from research on the perception and communication of risk. *Risk Management, 6*, 17–29. https://doi.org/10.1057/palgrave.rm.8240177.

Maylath, B. (1996). Words make a difference: The effects of Greco-Latinate and Anglo-Saxon lexical variation on college writing instructors. *Research in the Teaching of English, 30*(2), 220–247.

Maylath, B. (1997). Why do they get it when I say "gingivitis" but not when I say "gum swelling"? In D. Sigsbee, B. Speck & B. Maylath (Eds.), *Approaches to teaching non-native English speakers across the curriculum* (pp. 29–37). Jossey-Bass. https://doi.org/10.1002/tl.7003.

Maylath, B. (2000). Floods of foreign words: Building language awareness through the study of borrowed lexicon. In L. White, B. Maylath, A. Adams & M. Couzijn (Eds.), *Language awareness: A history and implementations* (pp. 33–40). Amsterdam University Press.

Melton, J. (2008). Beyond standard English: Re-thinking language in globally networked learning environments. In D. Stärke-Meyerring & M. Wilson (Eds.), *Designing globally networked learning environments: Visionary partnerships, policies, and pedagogies* (pp. 185–199). Sense Publishers. https://doi.org/10.1163/9789087904753_014.

Minacori, P. (2019). Pragmatic translation and assessment for technical communicators: Countering myths and misconceptions. In B. Maylath & K. St.Amant (Eds.), *Translation and localization: A guide for technical and professional communicators* (pp. 39–64). Routledge. https://doi.org/10.4324/9780429453670-3.

Montgomery, S. L. (2013). *Does science need a global language?* University of Chicago Press. https://doi.org/10.7208/chicago/9780226010045.001.0001.

Mousten, B. & Riordan, D. (2019). Technical communicators on track or led astray! Using quality standards in practice. In B. Maylath & K. St.Amant (Eds.), *Translation and localization: A guide for technical and professional communicators* (pp. 158–179). Routledge. https://doi.org/10.4324/9780429453670-8.

Munday, J. (2016). *Introducing translation studies: Theories and applications* (4th ed.). Routledge. https://doi.org/10.4324/9781315691862.

Onions, C. T. (Ed.). (1966). *The Oxford dictionary of English etymology.* Oxford University Press.

Potts, L. (2014). *Social media in disaster response: How experience architects can build for participation.* Routledge.

Scarpa, F. (2019). An overview of the main issues of translation. In B. Maylath & K. St.Amant (Eds.), *Translation and localization: A guide for technical and professional communicators* (pp. 19–38). Routledge. https://doi.org/10.4324/9780429453670-2.

St.Amant, K. (2019). Introduction: The dynamics of—and need to understand—translation and localization in technical communication. In B. Maylath & K. St.Amant (Eds.), *Translation and localization: A guide for technical and professional communicators* (pp. 1–15). Routledge.

Thrush, E. A. (2001). Plain English? A study of plain English vocabulary and international audiences. *Technical Communication*, *48*(3), 289–296.

Venuti, L. (2008). *The translator's invisibility* (2nd ed.). Routledge.

Walwema, J. (2018). Digital notebooks: Composing with open access. In R. Rice & K. St.Amant (Eds.), *Thinking globally, composing locally: Rethinking online writing in the age of the global internet* (pp. 15–34). Utah State University Press.

Weiss, T. (1997). Reading culture: Professional communication as translation. *Journal of Business and Technical Communication*, *11*(3), 321–338. https://doi.org/10.1177/1050651997011003005.

Windle, K. & Pym, A. (2011). European thinking on secular translation. In K. Malmkjær & K. Windle (Eds.), *The Oxford handbook of translation studies* (pp. 7–27). Oxford University Press. https://doi.org/10.1093/oxfordhb/9780199239306.013.0002.

38. User Experience (UX)

Guiseppe Getto
Mercer University

User experience, or *UX*, can be defined as the sum total of activities that need to occur during a design process to ensure a high-quality user experience is produced by that process. It is a growing focus of a diverse array of professionals, from academic researchers to technical communicators and web developers working in industry settings to specialists who focus solely on the UX process. Variously called UX designers, UX leads, UX researchers, and a host of other titles, these professionals have experienced considerable job growth in recent years due in large part to the explosion of the mobile app marketplace and the increasing need for large-scale (or "enterprise") applications developed for major corporations. At the same time, many academic researchers focused on ***technology*** have developed ***research*** agendas and courses devoted to UX, as well as full-scale majors, minors, and graduate programs.

UX is a complex term with a rich ***history*** in fields like technical communication, human-computer interaction (HCI), and ***design***. No exploration of the evolution of UX would be complete, either, without describing the important contributions of practitioners working in industry. As evidenced by the above definition, in contemporary usage, the term *UX* denotes both a design process focused on the user's experience and the experience that users have when utilizing the product of that process, be it a website, mobile application, enterprise application, or other type of technology. Closely related terms, such as *user-centered design* (*UCD*), are sometimes used as synonyms for UX and sometimes used as sub-terms.

The notion that design processes should focus primarily on user needs was first introduced to broad audiences by Don Norman in his book *The Psychology of Everyday Things*, first published in 1988 and later revised and expanded into *The Design of Everyday Things* in 2013. Norman called this notion UCD, a term he referred to earlier in his edited collection with Stephen Draper, *User-Centered System Design: New Perspectives on Human-Computer Interaction* (Norman & Draper, 1986). In these works, Norman argued that the products we use on a daily basis, even simple objects like door handles, will either succeed or fail based on how much prospective users are incorporated into the processes for designing them. Positioning users at the center of design processes would become a central attribute of UX that follows through to this day.

It is much harder to trace the etymology of the second use of the term, the experience a user has while utilizing a product. Early works such as those by Norman stressed that users have specific experiences when utilizing a product and

DOI: https://doi.org/10.37514/TPC-B.2023.1923.2.38

that these experiences matter. Another important touchstone in the evolution of UX was Jesse Garrett's 2003 book *The Elements of User Experience: User-Centered Design for the Web*. This important book invoked the term *UX* (as opposed to *UCD*) to describe both design processes and users' experiences, and it described the many dimensions, or "planes," of UX, which ranged from "the strategy plane" at the highest level to "the surface plane" at the level of the interface (pp. 31–34).

The idea that UX has not only a dual meaning but many different levels of operation and even closely related sub-terms carries through to contemporary usage. In more recent conceptions of UX, terms like *usability*, *information architecture*, *content strategy*, *visual*, and *design* often serve as sub-elements of the broader term (Buley, 2013; Garrett, 2003; Hartson & Pyla, 2012; Hoober, 2014; Morville, 2007) and are also explained as workflows that fit within the broader UX design process.

Most recently, the term *UX process* (or *UX lifecycle*) has been used to describe UX as a series of smaller workflows that represent the sum total of activities that need to occur during a design process to ensure a high-quality user experience (Hartson & Pyla, 2012, pp. 55–60). This process is typically depicted as a series of stages like the following:

1. Preliminary research
2. Prototyping
3. Usability testing
4. Maintenance

Less a linear process than a recursive and iterative one, the UX process helps practitioners make decisions when designs reach a certain threshold. A prototype (Banerjee, 2014), for instance, or "simulation of the final product," enables designers to "test whether or not the flow of the product is smooth and consistent." Similarly, preliminary research can teach designers what kind of prototype will be best to test with or what specific methods they need to deploy within the design process (Buley, 2013, p. 86). Maintenance, on the other hand, addresses what ongoing UX-related activities might look like, including when to engage in follow-up usability testing or prototyping of new features (Abercrombie, 2019). Sustainability and iteration are key concerns here, as resources are always finite, and keeping an entire design team functioning full time isn't always feasible.

Many developments in UX have been fueled, of course, by the advent of new technologies. Design processes are increasing in complexity and scope, with technologies such as ***social media*** applications, mobile applications, enterprise applications, web applications, augmented and virtual reality applications, and the numerous devices that make use of these applications. Because "we cannot consistently predict what kinds of information might be important to specific groups and in specific situations, we need methods by which we can understand the dynamic relationships between users and technologies" (Potts, 2009, p. 285). In other words, as digital technologies become more pervasive, the relationships among users and technologies become increasingly complex and increasingly unpredictable. Yet despite or perhaps because of this, "most users are involved in the design process too late to influence the final product" (Andrews et al., 2012, p. 124). This failure to account for users and their contexts "explains systems which function technically but fail because of lack of user acceptance" (Albers, 2003, p. 270). In other words, UX is only growing in importance as new challenges arise in the relationships between users and the technologies they depend on.

As these new challenges arise, a wide variety of individual UX methods have arisen. Since the publication of Jakob Nielsen's landmark *Usability Engineering* in 1993, usability testing has arguably remained the primary method for assessing the quality of a product's user experience. A method devoted to empirical observation of users while they test out an application in semi-controlled settings, usability testing enables UX experts to assess an application from the user's point of view. Typically, testers recruit users who are demographically similar to an application's intended user base. These participants are then asked to complete a series of tasks using the application or a prototype or mock-up of it. Users are then asked about why they completed the tasks the way they did to give designers a better grasp of how users navigate the application. Recently, remote, unmoderated usability testing has grown in popularity as UX experts use apps, such as UserTesting and UserZoom, to recruit, test, and record sessions with users through a combination of videoconferencing and screen-recording software. Regardless, the goal remains the same: to test a user interface for intuitiveness, usefulness, and ease-of-use.

Only a few years after Nielsen popularized usability, Hugh Beyer and Karen Holtzblatt's 1998 *Contextual Design: Defining Customer-Centered Systems* would introduce a second important method for assessing user contexts: contextual inquiry. Unlike usability testing that typically assesses user responses to an application's user interface in a semi-controlled environment, contextual inquiry is a semi-ethnographic method that seeks to observe users in their own context. Methods for contextual inquiry vary, from simple interviews with users in the setting in which they intend to use an application to fly-on-the-wall field studies in which researchers observe users conducting their daily tasks over a period of time. What unifies these variants, however, is an approach that attempts to balance the semi-controlled nature of usability testing with a more qualitative understanding of user behavior in context. Such an understanding is now agreed to be essential for designing an effective application.

While these two original methods remain important for both researchers and practitioners alike, a dizzying array of additional methods have since been developed, often by practitioners struggling to deal with the challenges of increasingly complex product development cycles. A complete catalog of UX methods is beyond the scope of this chapter, but an online list entitled *UX Design Methods & Deliverables* purports to be a continually updated collection of UX methods and associated deliverables, complete with links to fuller explanations of each method listed (UX Collective, 2016). These methods, which include persona development (Golz, 2014), competitive analysis (Withrow, 2006), and storyboarding (Little, 2013), have largely arisen due to new technological exigencies and design workflows.

One method that has cropped up largely due to the growing complexity of applications is customer journey mapping (Gibbons, 2018). This method typically pools information garnered from other methods, such as usability testing and contextual inquiry, in order to create a map of how different types of users attempt to navigate and make use of an application. The central deliverable of this method is a literal map of individual users' journeys that includes their goals, pain points, and other details important for improving their flow through the application.

For decades within the field of technical communication, scholars focused primarily on usability and how it should inform the practice and teaching of technical communication (Breuch et al., 2001; Cooke, 2010; Redish, 2010; Skelton, 1992; Sullivan, 1989). This focus remains strong in the field. However, recent work has broadened the scope of UX beyond usability (Getto & Beecher, 2016; Lauer & Brumberger, 2016; Potts, 2013; Redish, 2011; Sun, 2013). This work often seeks to identify new relationships between technical communicators and UX specialists, with many scholars arguing that these roles are beginning to blur in productive ways.

Within related fields like HCI and design, *UX* has similarly begun to take center stage over the last few decades as the predominant term for describing design processes that center users (Benyon, 2019; Bevan, 2005; Kreitzberg et al., 2019; Vermeeren et al., 2016). This shift builds on a long history of *UCD* being the predominant term—and continuing to be an important term—to describe user-focused design processes (Karat, 1997; Lazar, 2005; Silva da Silva et al., 2011).

Meanwhile, within the broad community of industry practitioners, it is almost undeniable that *UX* has taken center stage as the primary term describing work to improve user experiences. Indicative shifts include the Usability Professionals Association changing its name to the User Experience Professionals Association (UXPA) in 2012 as well as the ever-expanding list of industry-hosted conferences in UX (https://uiuxtrend.com/events/). In addition, much of the work cited in this chapter, including that from Arijit Banerjee (2014), Leah Buley (2013), Jesse James Garrett (2003), Steven Hoober (2014), and Peter Morville (2007), is from industry practitioners, all of whom seem to use *UX* as their primary term, though many still refer to the associated terms mentioned above as components of the UX umbrella. This shift can also be witnessed in important trade publications and presses such as *User Experience Magazine* (the publication of the UXPA: https://uxpamagazine.org/), *Boxes and Arrows* (https://boxesandarrows.com/), *UX Matters* (https://www.uxmatters.com/), *Rosenfeld Media* (https://rosenfeldmedia.com/), *A List Apart* (https://alistapart.com), and *Nielsen Norman Group* (https://www.nngroup.com/)—publications representing the collected ***knowledge*** of hundreds, if not thousands, of UX practitioners.

Overall, in the past several decades, *UX* has grown from a relatively novel term to an important one within a wide range of conversations and practitioner workflows. It has become the de facto descriptor for design processes that put human needs before other concerns. And it has begun to represent a discipline in its own right, a discipline devoted to improving the experiences users have when utilizing any form of technology, from a website to a household appliance. During this time, it has also permeated other, more established fields, such as technical communication, HCI, and design. And, perhaps most persuasively, it has become a kind of rallying cry for user-focused practitioners working in a variety of industry contexts who contribute to the development of the ever-broadening array of products and services we use on a daily basis.

That being said, UX is also an emerging field, given the pace at which technologies change. With new advances in augmented reality, virtual reality, wearables, and the Internet of Things, the interfaces that users use to access technologies, not to mention the organizing principles behind them, are multiplying every year. It is possible, if not probable, that UX experts will continue to specialize in the future into different applications of UX, such as conversational UX for voice-activated systems, wearable UX for items users attach to their bodies, even ***social justice***-related UX for contributing to activist causes. One thing is certain: UX will continue to grow and evolve as technologies and their attendant design processes grow and evolve. The UX we have today may very well be completely different only a few years from now. That is the exciting challenge, but also the predicament, of a field devoted to adapting new technologies to human needs.

References

Abercrombie, R. (2019, September 5). *Give usability maintenance a seat at the table.* Usability Geek. https://usabilitygeek.com/give-usability-maintenance-seat-table/.

Albers, M. J. (2003). Multidimensional audience analysis for dynamic information. *Journal of Technical Writing & Communication, 33*(3), 263–279. https://doi.org/10.2190/6KJN-95QV-JMD3-E5EE.

Andrews, C., Pohland, E., Burleson, D., Saad, D., Dunks, K., Scharer, J., Elmore, K., Wery, R., Lambert, C., Wesley, M., Oppegaard, B. & Zobel, G. (2012). A new method in user-centered design: Collaborative prototype design process (CPDP). *Journal of Technical Writing and Communication, 42*(2), 123–142. https://doi.org/10.2190/TW.42.2.c.

Banerjee, A. (2014, November 17). What a prototype is (and is not). *UX Magazine.* http://uxmag.com/articles/what-a-prototype-is-and-is-not.

Benyon, D. (2019). *Designing user experience: A guide to HCI, UX and interaction design* (4th ed.). Pearson.

Bevan, N. (2005). Creating a UX profession. In *CHI EA '05: CHI '05 extended abstracts on human factors in computing systems* (pp. 1078–1079). Association for Computing Machinery. https://doi.org/10.1145/1056808.1056820.

Beyer, H. & Holtzblatt, K. (1998). *Contextual design: Defin•ing customer-centered systems.* Morgan Kaufmann.

Breuch, L.-A. K., Zachry, M. & Spinuzzi, C. (2001). Usability instruction in technical communication programs: New directions in curriculum development. *Journal of Business and Technical Communication*, *15*(2), 223–240. https://doi.org/10.1177/1050651 90101500204.

Buley, L. (2013). *The user experience team of one: A research and design survival guide.* Rosenfeld Media.

Cooke, L. (2010). Assessing concurrent think-aloud protocol as a usability test method: A technical communication approach. *IEEE Transactions on Professional Communication*, *53*(3), 202–215. https://doi.org/10.1109/TPC.2010.2052859.

Garrett, J. (2003). *The elements of user experience: User-centered design for the web*. New Riders.

Getto, G. & Beecher, F. (2016). Toward a model of UX education: Training UX designers within the academy. *IEEE Transactions on Professional Communication*, *59*(2), 153–164. https://doi.org/10.1109/TPC.2016.2561139.

Gibbons, S. (2018, December 9). *Journey mapping 101*. Nielsen Norman Group. https://www.nngroup.com/articles/journey-mapping-101/.

Golz, S. (2014, August 6). A closer look at personas: What they are and how they work (part 1). *Smashing Magazine*. https://www.smashingmagazine.com/2014/08/a-closer-look-at-personas-part-1/.

Hartson, R. & Pyla, P. (2012). *The UX book: Process and guidelines for ensuring a quality user experience*. Morgan Kaufmann.

Hoober, S. (2014, May 5). The role of user experience in the product development process. *UXmatters*. http://www.uxmatters.com/mt/archives/2014/05/the-role-of-user-experience-in-the-product-development-process.php.

Karat, J. (1997). Evolving the scope of user-centered design. *Communications of the ACM*, *40*(7), 33–38. https://doi.org/10.1145/256175.256181.

Kreitzberg, C., Rosenzweig, E., Shneiderman, B., Churchill, E. & Gerber, E. (2019). Careers in HCI and UX: The digital transformation from craft to strategy. In *CHI EA '19: Extended abstracts of the 2019 CHI Conference on Human Factors in Computing Systems* (pp. 1–6). Association for Computing Machinery. https://doi.org/10.1145/3290607.3311746.

Lauer, C. & Brumberger, E. (2016). Technical communication as user experience in a broadening industry landscape. *Technical Communication*, *63*(3), 248–264.

Lazar, J. (2005). *Web usability: A user-centered design approach*. Addison-Wesley.

Little, A. (2013, February 8). Storyboarding in the software design process. *UX Magazine*. https://uxmag.com/articles/storyboarding-in-the-software-design-process.

Morville, P. (2007, July 23). *User experience strategy*. Semantic Studios. http://semanticstudios.com/user_experience_strategy/.

Nielsen, J. (1993). *Usability engineering*. Morgan Kaufmann.

Norman, D. (1988). *The psychology of everyday things*. Basic Books.

Norman, D. (2013). *The design of everyday things*. Basic Books.

Norman, D. & Draper, S. (1986). *User-centered system design: New perspectives on human-computer interaction.* CRC Press. https://doi.org/10.1201/b15703.

Potts, L. (2009). Using actor network theory to trace and improve multimodal communication design. *Technical Communication Quarterly*, *18*(3), 281–301. https://doi.org/10.1080/10572250902941812.

Potts, L. (2013). *Social media in disaster response: How experience architects can design for participation*. Routledge. https://doi.org/10.4324/9780203366905.

Redish, J. (2010). Technical communication and usability: Intertwined strands and mutual influences. *IEEE Transactions on Professional Communication*, *53*(3), 191–201.

Redish, J. (2011). Overlap, influence, intertwining: The interplay of UX and technical communication. *Journal of Usability Studies*, *6*(3), 90–101.

Silva da Silva, T., Martin, A., Maurer, F. & Silveira, M. (2011). User-centered design and Agile methods: A systematic review. In *2011 Agile Conference* (pp. 7–13). https://doi.org/10.1109/AGILE.2011.24.

Skelton, T. M. (1992). Testing the usability of usability testing. *Technical Communication*, *39*(3), 343–359.

Sullivan, P. (1989). Beyond a narrow conception of usability testing. *IEEE Transactions on Professional Communication*, *32*(4), 256–264. https://doi.org/10.1109/47.44537.

Sun, H. (2013). *Cross-cultural technology design: Creating culture-sensitive technology for local users*. Oxford University Press. https://doi.org/10.1093/acprof:oso/9780199744763.001.0001.

UX Collective. (2016, April 30). *UX design methods & deliverables*. https://uxdesign.cc/ux-design-methods-deliverables-657f54ce3c7d.

Vermeeren, A., Roto, V. & Väänänen, K. (2016). Design-inclusive UX research: Design as a part of doing user experience research. *Behaviour & Information Technology*, *1*, 21–37. https://doi.org/10.1080/0144929X.2015.1081292.

Withrow, J. (2006, February 27). Competitive analysis: Understanding the market context. *Boxes and Arrows*. https://boxesandarrows.com/competitive-analysis-understanding-the-market-context/.

39. Visual

Han Yu
Kansas State University

From illustrations and photographs to principles of document ***design***, visual elements are an essential part of technical communication. But what does it mean for something to be "visual," and how have theories of the visual shaped technical communication scholarship and practice?

The earliest uses of the term *visual* recorded in the *Oxford English Dictionary* include "visual beams" and "visual rays," which reflect the ancient (and incorrect) belief that we see by shooting a beam of light from our eyes—or by the eyes receiving beams emanating from objects. For example, in Nathanael Carpenter's 1625 *Geography Delineated Forth in Two Books*, "The visuall Ray wherein the sight is carried, is alwaies a right line" (Oxford University Press, n.d.). The contemporary meaning of "pertaining to sight or vision" became prevalent after the 18th century, as in "a clear and settled idea of visual beauty," from Edmund Burke's 1757 *A Philosophical Enquiry into the Origin of Our Ideas of the Sublime and Beautiful* (Oxford University Press, n.d.). It was not until well into the 19th century that we saw increasing use of the word *visual* to refer to non-physical imageries conjured up by a viewer, as in Thomas Carlyle's 1845 *Letters and Speeches*: "Let the reader try to make a visual scene of it as he can" (Oxford University Press, n.d.).

This etymology, in some ways, predicts the two major theoretical frameworks used by our field in its study of visuals. If vision is caused by physical beams that seize an object or seize the eye, then what one sees is a material reality. Studies of visuals thus become an attempt to understand how the eye—and the optical nerve and visual cortex behind it—automatically reacts to that reality. The framework employed by these studies is variably called perceptual or cognitive. On the other hand, if, instead or in addition, *visual* means the formation of an imagined, self-constructed view, then studies of visuals become an attempt to understand how individuals—replete with different experiences, ***knowledge***, and assumptions—make sense of what *they* see. The framework employed by these studies is variably called critical, social, or cultural. These two frameworks have competing—but also complementary—focuses and applications in technical communication.

The cornerstone of the perceptual/cognitive framework is the Gestalt theory. Originated from the 20th century Gestalt psychology, Gestalt is the study of visual perceptual organization—with the German word "Gestalt" translating loosely to "shape" or "pattern." The theory includes a set of principles that govern our perception. The principle of *proximity*, for example, states that visual elements close to each other tend to be perceived as belonging to one group and conveying related ***information***; by contrast, elements that are set apart are perceived as

DOI: https://doi.org/10.37514/TPC-B.2023.1923.2.39

conveying unrelated information. Other commonly applied Gestalt principles include *closure*, *similarity*, *continuation*, *enclosure*, and *figure-ground*, which are summarized in Figure 39.1.

Drawing upon or overlapping with the Gestalt theory are various other perceptual/cognitive lenses: for example, Edward Tufte's (1990) concepts of layering, separation, and small multiples; Charles Kostelnick and David Roberts' (1998) ideas of emphasis and clarity; Evelyn Goldsmith's terms of syntactic and semantic unity, location, emphasis, and text parallels (Dragga, 1992); and Stephen Kosslyn's (2006) principles of salience and discriminability.[1]

Local differences aside, the overriding goal of these perceptual/cognitive lenses is to expedite the workings of the human eye and brain, to design visuals in such ways that a viewer can derive information from them most swiftly and accurately. This goal has obvious relevance and value to technical communication, a field concerned with communicating complex information where it is expedient (and reassuring) if viewers follow a consistent process in visual processing. The process starts with viewers sensing visual stimuli (lines, colors, etc.) on the retina, which are processed by working memory where visual queries and pattern searches allow viewers to recognize the stimuli as, say, a human face.

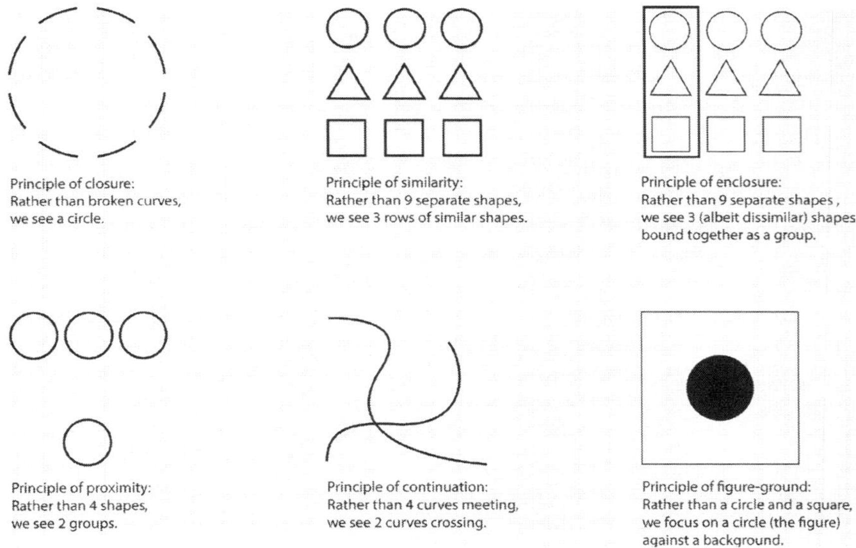

Figure 39.1. Commonly applied Gestalt principles.

1. It is important to note that most theoretical lenses are not 100 percent perceptual/cognitive or 100 percent social semiotic. Kostelnick and Roberts, for example, also emphasize visual tone and ethos, while Goldsmith speaks of pragmatic, all of which implicate factors beyond biological processing of sensory data. Even Edward Tufte is not straightforwardly positivistic (see Kimball, 2006).

With this predictable process, targeted (and thereby effective) interventions become possible. Because viewers may not optimally sense visual stimuli (due to color vision deficiency, visual impairment, or environmental conditions), technical communicators are instructed to practice universal design principles: for example, using adjustable fonts, accessible color schemes, or redundant visual cues (Chaparro & Chaparro, 2017; Chisnell et al., 2006; Wong, 2011). Even when viewers can physically sense stimuli, problems may arise at the stage of working memory, which has a low capacity and can only "hold" a few items at a time (Kosslyn, 2006; Miller, 1956; Ware, 2012). Thus, excessive visual details or failures to configure those details vis-à-vis Gestalt will confuse—even harm—viewers, deterring their comprehension of popular ***science*** visuals, for example (Yu, 2017), or failing to alert them of safety warnings (Paradis, 1991).

Despite its valuable applications in technical communication, the perceptual/cognitive framework runs the risk of espousing a positivist visual outlook, which assumes that visuals embody an objective reality and should help (universally conceived) viewers decode that reality through a transparent conduit. Ben Barton and Marthalee Barton (1985) were among the first in our field to critique this visual outlook and to emphasize visuals as contextualized, ***rhetorical*** productions subject to ideological and cultural consensus.

Since then, various critical, social, and cultural lenses have been applied to studying technical and scientific visuals: the anti-positivist, anti-hegemonic, feminist, environmental, ethical, or the more broad-ranging humanistic, which acknowledge a range of human-centered factors from emotions to lived experiences (e.g., Barton & Barton, 1993; Brasseur, 2003; Kimball, 2006; Mellor, 2009; Robles, 2018; Ross, 2008; Welhausen, 2017; Yu, 2017).

These individual lenses can be comprehended through the larger framework of social semiotics. Originated from the work of Swiss linguist Ferdinand de Saussure, semiotics is, put simply, the study of signs. A sign contains two parts: the signifier, which is originally a sound pattern (e.g., the pronunciation /pɛn/), and the signified, which is the concept denoted by the sound pattern (a writing device). "Signifier" was later broadened so that signs become "anything which 'stands for' something else" and can "take the form of words, images, sounds, gestures and objects" (Chandler, 2007, p. 2).

Social semiotics believes that in a given sign, the signifier (e.g., an image of a pen), rather than the signified (the actual pen), assumes primacy (Chandler, 2007). This is because signifiers set the stage and create the parameters for us to conceptualize, imagine, and deliberate the signified. In other words, reality is actively constructed rather than passively reflected in signs, the construction driven by sign-makers' interest tied to social-cultural histories and contexts (Kress & van Leeuwen, 2006).

These beliefs profoundly complicate the way we look at visuals. For example, the abundant rectangles in contemporary life—in the shapes of buildings and devices—are not random. Rather, with their parallel lines and controlled angles, they support and perform a rational, disciplined, and impersonal modern society (Kress & van Leeuwen, 2006). Similarly, stunning photographs of prepared dishes in *Elle* magazine—"golden partridges studded with cherries" or "a faintly pink chicken chaud-froid"—are less about cooking and foodstuffs and more about petit-bourgeois' preoccupation with gentility and ornamentation (Barthes, 1991, p. 78).

Once invested in social semiotics, we realize perceptual/cognitive principles are precisely some of the means by which visuals conceal their social/cultural values. For example, the iconic map of the London Underground depicts routes in straight lines connecting stations of homogenous distances—when, in reality, routes are meandering and stations are congested (Barton & Barton, 1993). In the name of clarity and consistency, the map, Barton and Barton (1993) argued, belies nationalist and capitalist attempts to depict urban London as orderly (when it isn't) and to persuade tourists that travel is easy (when it isn't).

The perceptual/cognitive and the social semiotic frameworks more visibly clash when scholars attempt to reveal visuals for the ideological signs that they are and to articulate individual, social-cultural, and humanistic values in established technical ***genres***. Sam Dragga and Dan Voss (2001), notably, suggested using pictorial human icons and other means to combat the absence of human emotions and lives in Cartesian graphs and technical illustrations. The suggestion invited considerable criticism from technical communicators who called the idea "off-base," "almost laughable," and "totally wrong-headed" ("Correspondence," 2000, pp. 9–10). Similar attitudes can be found in the writings of renowned information designer Edward Tufte (1990, 2001), who coined the term *chartjunk* to denigrate non-data-ink or redundant data-ink—anything from dark grid lines to pictorial elements—that does not directly contribute to perceptual/cognitive processing. Symbols of patrons and religious orders in 17th century diagrams, rather than seen as fabrics of social-cultural identities, are deemed "strident, contradicting nature's rich pattern" (Tufte, 1990, p. 21).

But nature's pattern is never truly detached from human interference. The moment a pattern is visualized—is deemed worthy of visualization—it always already is wrought with subtle or unsubtle signs of beliefs and interests. Consider the making of scientific visuals, the quintessential endeavor to portray nature. Prior to the 19th century, natural philosophers aspired to achieve "truth to nature"—by peeling away nature's spurious elements and revealing its divine and hidden truth (Galison, 1998). Thus, in Andreas Vesalius' *De Humani Corporis Fabrica*, muscle men line up for a dance of death (Hildebrand, 2004). As each layer of their bodily tissues is stripped away, Vesalius reveals the structure and purpose of the human body as the Creator intended—whilst the last muscle man collapses to his fate. Circa 1830, such artistic attempts to reveal nature's truth gave way to mechanical objectivity (Galison, 1998). In making visuals, scientists aspired to rely not on humans but machines—first, the camera lucida; later, increasingly sophisticated apparatuses from electron microscopes to DNA sequencers—with the hope of producing objective evidence. But machines are *made* and *set* by humans: Even without obvious re-touching, factors such as position, zooming, exposure time, and shutter speed can manipulate images into publishable evidence (Knorr-Cetina & Amann, 1990; Meyer, 2007). In opening the black-boxed, machine-made inscriptions (Latour, 1986, 1998), we can expect to find one visual no more strident than another and all a result of interpretation.

It is in acknowledging—and celebrating—"interpreted images" (Galison, 1998) that we can start to synthesize the perceptual/cognitive and the social semiotic, to consider viewers' information needs as well as emotional state, cultural beliefs, political ideologies, economic interests, and more. In her study of popular science visuals, for example, Yu (2017) combined perceptual/cognitive discussions of shapes, colors, and layouts with social semiotic considerations of female bodies, genetic determinism, and citizen science. Such integrated approaches are more likely to result in visuals that are persuasive, compelling, and useful—as opposed to merely easy to use (Mirel, 2002).

An integrated approach also enriches our understanding of visual ***ethics***. Many agree that misleading readers in perceptual/cognitive processing—by drawing what one didn't observe or omitting what one did observe, for example—is unethical (Dombrowski, 2003). But what about removing human bodies from accident reports (Dragga & Voss, 2003)? "Staging" experimental contexts for scientific photographs (Buehl, 2014)? Underrepresenting women and minorities in popular science images (Yu, 2017)? Selecting perfect as opposed to representative visual evidence for publication (Frow, 2012)? Or asking readers to take responsibilities in scrutinizing visuals (Dragga, 1996)? We cannot broach—or even conceive—these questions without seeing visuals as social/cultural artifacts.

As new (and old) visual types and technologies find their relevance in technical communication, we will benefit by approaching them from the interrelated domains of the perceptual/cognitive and the social/cultural. For example, comics, with their abundant pictorial images (including staple technical communication genres such as illustrations), make rich sites for multi-pronged studies (Yu, 2015; Bahl et al., 2020). Interactive visuals—from web-based 3D molecular modeling tutorials (Yu, 2017) to geovisualization ***risk communication*** tools (Stephens & Delorme, 2019)—represent another important (and underdeveloped) area for integrated studies to prioritize users' diverse needs and contexts. Ultimately, visual technical communication, whether between experts or between experts and non-experts, relies on the interplay between the perceptual/cognitive and the social/cultural to make and share knowledge, reflecting the dual empirical and humanistic values that undergird our field.

References

Bahl, E. K., Figueiredo, S. & Shivener, R. (2020). Special issue on comics and graphic storytelling in technical communication. *Technical Communication Quarterly*, 29(3). https://doi.org/10.1080/10572252.2020.1768297.

Barthes, R. (1991). *Mythologies*. Noonday Press.

Barton, B. & Barton, M. (1985). Toward a rhetoric of visuals for the computer era. *The Technical Writing Teacher, 12*(2), 126–145.

Barton, B. & Barton, M. (1993). Ideology and the map: Toward a postmodern visual design practice. In N. Blyler & C. Thralls (Eds.), *Professional communication: The social perspective* (pp. 49–78). Sage.

Brasseur, L. E. (2003). *Visualizing technical information: A cultural critique*. Baywood Publishing Company.

Buehl, J. (2014). Toward an ethical rhetoric of the digital scientific image: Learning from the era when science met Photoshop. *Technical Communication Quarterly, 23*(3), 184–206. https://doi.org/10.1080/10572252.2014.914783.

Chandler, D. (2007). *Semiotics: The basics.* Routledge. https://doi.org/10.4324/9780203014936.

Chaparro, A. & Chaparro, M. (2017). Applications of color in design for color-deficient users. *Ergonomics in Design: The Quarterly of Human Factors Applications, 25*(1), 23–30. https://doi.org/10.1177/1064804616635382.

Chisnell, D. E., Redish, J. C. & Lee, A. (2006). New heuristics for understanding older adults as web users. *Technical Communication, 53*(1), 39–59.

Correspondence: Half-baked pies, cruel cover, and anecdotal accuracy. (2002). *Technical Communication, 49*(1), 9–14.

Dombrowski, P. (2003). Ernst Haeckel's controversial visual rhetoric. *Technical Communication Quarterly, 12*(3), 303–319. https://doi.org/10.1207/s15427625tcq1203_5.

Dragga, S. (1992). Evaluating pictorial illustrations. *Technical Communication Quarterly, 1*(2), 47–62. https://doi.org/10.1080/10572259209359498.

Dragga, S. (1996). "Is this ethical?": A survey of opinion on principles and practices of document design. *Technical Communication, 43*(3), 255–265.

Dragga, S. & Voss, D. (2001). Cruel pies: The inhumanity of technical illustrations. *Technical Communication, 48*(3), 265–274.

Dragga, S. & Voss, D. (2003). Hiding humanity: Verbal and visual ethics in accident reports. *Technical Communication,* 50(1), 61–82.

Frow, E. (2012). Drawing a line: Setting guidelines for digital image processing in scientific journal articles. *Social Studies of Science, 42*(3), 369–392. https://doi.org/10.1177/0306312712444303.

Galison, P. (1998). Judgment against objectivity. In C. A. Jones, P. Galison & A. E. Slaton (Eds.), *Picturing science, producing art* (pp. 327–359). Routledge.

Hildebrand, R. (2004). Alternative images: Anatomical illustration and the conflict between art and science. *Interdisciplinary Science Reviews, 29*(3), 295–311. https://doi.org/10.1179/030801804225018864.

Kimball, M. (2006). London through rose-colored graphics: Visual rhetoric and information graphic design in Charles Booth's maps of London poverty. *Journal of Technical Writing and Communication, 36*(4), 353–381. https://doi.org/10.2190/K561-40P2-5422-PTG2.

Knorr-Cetina, K. & Amann, K. (1990). Image dissection in natural scientific inquiry. *Science, Technology, and Human Values, 15*(3), 259–283. https://doi.org/10.1177/016224399001500301.

Kosslyn, S. M. (2006). *Graph design for the eye and mind.* Oxford University Press. https://doi.org/10.1093/acprof:oso/9780195311846.001.0001.

Kostelnick C. & Roberts, D. (1998). *Designing visual language: Strategies for professional communicators.* Allyn and Bacon.

Kress, G. R. & Van Leeuwen, T. (2006). *Reading images: The grammar of visual design* (2nd ed.). Routledge.

Latour, B. (1986). Visualization and cognition: Thinking with eyes and hands. In H. Kuklick & E. Long (Eds.), *Knowledge and society: Studies in the sociology of culture past and present* (Vol. 6, pp. 1–40). Jai Press.

Latour, B. (1998). How to be iconophilic in art, science, and religion? In C. A. Jones, P. Galison & A. E. Slaton (Eds.), *Picturing science, producing art* (pp. 418–440). Routledge.

Mellor, F. (2009). Image–music–text of popular science. In R. Holliman, E. Whitelegg, E. Scanlon, S. Smidt & J. Thomas (Eds.), *Investigating science communication in the information age* (pp. 205–220). Oxford University Press.

Meyer, E. T. (2007). *Socio-technical perspectives on digital photography: Scientific digital photography use by marine mammal researchers* [Doctoral dissertation, Indiana University]. ProQuest Dissertations and Theses.

Miller, G. A. (1956). The magical number seven, plus or minus two: Some limits on our capacity for processing information. *Psychological Review, 63*(2), 81–97. https://doi.org/10.1037/h0043158.

Mirel, B. (2002). Advancing a vision of usability. In B. Mirel & R. Spilka (Eds.), *Reshaping technical communication: New directions and challenges for the 21st century* (pp. 165–187). Lawrence Erlbaum.

Oxford University Press. (n.d.). Visual. In *Oxford English Dictionary*. Retrieved June 2, 2021, from https://www.oed.com.

Paradis, J. (2004). Text and action: The operator's manual in context and in court. In C. Bazerman & J. Paradis (Eds.), *Textual dynamics of the professions: Historical and contemporary studies of writing in professional communities* (pp. 256–278). The WAC Clearinghouse. https://wac.colostate.edu/books/landmarks/textual-dynamics/ (Originally published 1991 by University of Wisconsin Press)

Robles, V. D. (2018). Visualizing certainty: What the cultural history of the Gantt Chart teaches technical and professional communicators about management. *Technical Communication Quarterly, 27*(4), 300–321. https://doi.org/10.1080/10572252.2018.1520025.

Ross, D. G. (2008). Dam visuals: The changing visual argument for the Glen Canyon Dam. *Journal of Technical Writing and Communication, 38*(1), 75–94. https://doi.org/10.2190/TW.38.1.e.

Stephens, S. & Delorme, D. (2019). A framework for user agency during development of interactive risk visualization tools. *Technical Communication Quarterly, 28*(4), 391–406. https://doi.org/10.1080/10572252.2019.1618498.

Tufte, E. R. (1990). *Envisioning information*. Graphics Press.

Tufte, E. R. (2001). *The visual display of quantitative information*. Graphics Press.

Ware, C. (2012). *Information visualization: Perception for design* (3rd ed.). Morgan Kaufmann.

Welhausen, C. (2017). Visualizing science: Using grounded theory to critically evaluate data visualizations. In H. Yu & K. Northcut (Eds.), *Scientific communication: Practices, theories, and pedagogies* (pp. 82–106). Routledge. https://doi.org/10.4324/9781315160191-5.

Wong, B. (2011). Points of view: Color blindness. *Nature Methods*, *8*(6), 441. https://doi.org/10.1038/nmeth.1618.
Yu, H. (2015). *The other kind of funnies: Comics in technical communication*: Routledge. https://doi.org/10.4324/9781315231266.
Yu, H. (2017). *Communicating genetics: Visualizations and representations*. Palgrave Macmillan. https://doi.org/10.1057/978-1-137-58779-4.

Afterword: Diversity, Equity, and Inclusion Through Citational Practice

Kristen R. Moore
University at Buffalo

Lauren E. Cagle
University of Kentucky

Nicole Lowman
University at Buffalo

Up until the late 2000s, if not later, the very idea of diversity as a central concept for the field of technical communication (TC) would have been laughable. Now, in 2023, diversity, defined as representation of multiple populations across race, class, gender, sexuality, ability, and other identity markers, is understood as "necessary but insufficient" for achieving an inclusive field with ethical and equitable practices at its center. Pursuing equity and inclusion in the field of technical communication comprises a range of practices that consider how our work—described throughout this collection—might contribute to and/or combat the systems of oppression that do harm to particular groups of people. As Natasha N. Jones, Kristen R. Moore, and Rebecca Walton (2016) articulate, the ***social justice*** turn has emerged from these pursuits. This afterword considers a narrow slice of the field's attempt to address equity and inclusion: how our citation and writing practices amplify and suppress particular perspectives. More specifically, this afterword takes up the meta-analytic question of what the citations in this very book you're reading right now tell us about TC's nonlinear movement towards establishing itself as a diverse, equitable, and inclusive field.

When I (Kristen) was accepted into this collection, I wondered how a book like this, with its focus on identifying and defining the field's keywords, might become a tool for either the amplification or suppression of ideas that emerge from groups who have been historically marginalized in our field and beyond. It's not a stretch to think that such collections are not just descriptive, but also normatively definitional. Johndan Johnson-Eilola and Stuart Selber's (2004) *Central Works in Technical Communication*, as an example, serves as a key text in many TC graduate courses (Faris & Wilson, 2022), providing TC students with perhaps their first overview of the field. Although the text provides some of the more forward-looking texts from the field, it presents TC as a field informed primarily by white scholars, though we know from Edward A. Malone's entry in this volume on the ***history*** of technical communication that this is hardly accurate. Without attention to whose story of TC the keyword collection is telling, then,

DOI: https://doi.org/10.37514/TPC-B.2023.1923.3.2

there is a major risk of reifying and committing to an exclusionary, limited story about the field and its contributors.

Storytelling is a collective political act, and limiting the stories we tell about ourselves and the fields we belong to is a political act of exclusion. In a project like this *Keywords* collection, our citation practices can function as a proxy for understanding mechanisms of a story's exclusion. So even as the "social justice turn" has been widely celebrated in the field of technical communication, close attention to citational practices can reveal that our everyday scholarly politics and practices, such as citing the same central homogeneous canon by default, have not caught up.

This close attention is an example of what Walton et al. (2019) call "recognizing" in their book *Technical Communication After the Social Justice Turn*. Recognizing is the first of four steps they recommend for addressing injustice; the remaining three are Reject, Reveal, and Replace. At times, recognizing can be an anticipatory move; rather than recognizing where inequity is already entrenched, we might strive to recognize where inequity threatens to creep in. This anticipatory recognition allows us to address inequity *before* harm is done. Knowing this, we undertook these four steps, beginning with recognizing the role of diverse citations in inclusionary field-building, as an anticipatory move to push for this collection to tell an inclusive story about TC. Here are some specific examples of how our process followed the 4Rs steps:

- **Recognize:** A text like this has the potential to amplify particular voices that have been silenced;
- **Reveal 1:** Kristen reveals to Han Yu and Jonathan Buehl her concern;
- **Coalitional Rejection:** Han and Jonathan confirm that they *recognize* the concern and *accept* Kristen's offer to consider representation and amplification as a part of the editorial process;
- **Reveal 2:** Kristen reveals to a coalition of scholars Han and Jonathan's response;
- **Coalitional Rejection 2:** Cagle *recognizes* and *agrees* this is a potential harm that needs to be anticipated and agrees to help Kristen consider opportunities for amplification.

The following citation audit consisted of further iterations of recognizing and revealing, and after we completed it, the ball was then in the editors' and authors' courts to decide if and how to reject and replace any of their own potentially exclusionary citational practices.

An Imperfect Methodology

To address the potential for harm in the citation and writing practices in the collection, we developed an imperfect methodology that draws on the accountability framework used in Catherine D'Ignazio and Lauren Klein's (2020) *Data Feminism*. In their book, the authors (two white women) hold themselves accountable for considering intersectionality by establishing quantifiable metrics and criteria for the projects and authors they cite. In the afterword of their book, they include both their metric table and an audit, which was conducted by Isabel Carter "in the interest of remaining accountable to the values statement for this book" (D'Ignazio & Klein, 2020, p. 223).

Using their heuristic as a starting place, we began reviewing early drafts of keyword entries using three major questions to guide our reading: 1) *Who* did the authors cite? 2) *How* did the authors write about others? and 3) *What* themes or examples revealed a commitment to or acknowledgement of the need for diversity, equity, and/or inclusion (broadly construed) in TC? The second question was fairly easy to assess: We found that most entries were inclusive in the way the authors wrote about others, using inclusive language.

The first question proved tricky: Unlike with *Data Feminism*, the authors weren't instructed from the outset of their drafting to purposefully construct an inclusive entry or strive to be accountable to an explicit value statement. Perhaps some authors (this is true of Kristen, for example) considered the politics of citation and amplification in the drafting, but given the recentness of the turn towards seeing diversity, equity, and inclusion as an integral part of the field, it seems likely that other authors did not build reference pages with an inclusive imperative. Additionally, few constraints were placed on authors as they constructed their keyword entries in order to (we presume) enable academic freedom and support authorial autonomy. Finally, we reviewed only the initial drafts of the keyword entries, which varied considerably in their level of completeness.

As a result, using citation metrics as a proxy for inclusivity was complicated by the astonishing variation simply in the total number of citations across entries. The spread of total raw citation count for a single entry ranged from four to nearly 150 citations (initial drafts with extremely low citation counts increased their citations in final drafts). Therefore, "counts" were only useful in the context of an individual entry.

Even trickier was the difficult project of deciding how to "count" authors; indeed, our own experience reflected Carter's difficulty in *Data Feminism*. She warns, "Future attempts to replicate this audit should take seriously the difficulty of clearly establishing these identity categories without formally consulting with those who are being referenced and therefore classified" (as cited in D'Ignazio & Klein, 2020, p. 224). Further, although intersectional scholars (like us) resist the idea that marginalizing identity characteristics can be disarticulated, the act of auditing the citation practices of authors left us to do this very thing: to count how many total women, women of color, etc.

Our method attempted to account for these two intractable challenges, but it did so imperfectly. We began by pulling out the references from each keyword entry's draft to create a set of citation lists sorted by entry. We also built a comprehensive list of citations for the entire collection, in order to identify and manage duplicate citations of the same work across multiple entries. Having these different datasets to work with allowed us to analyze both the diversity of citations within any given entry and the diversity of citations across the entire collection. For the cross-collection citation diversity, we were interested both in whether the authors being cited represented the true diversity of the field and in how many different publications by marginalized or multiply marginalized and underrepresented (MMU) scholars were cited across the collection. In other words, hypothetically speaking, each entry might cite at least one article by a Black woman, but if each entry's citation is of the *same* article by a Black woman, then the appearance of diversity across the full collection would be more tokenization than inclusion.

We put each citation list for individual entries through three analytic phases. In each phase, we used a different tool to determine how and if an entry amplified voices via citation of those who have traditionally been marginalized or are MMU scholars: Phase One relied on our personal knowledge to identify scholars across race, gender, sexuality, etc.; Phase Two relied on pre-existing lists of MMU scholars; and Phase Three sought out "knowable" information by conducting a public search of authors through their faculty pages, personal/professional websites, social media bios, or other sites of online presence. Table 40.1 provides an overview of these phases and their imperfections.

Because the collection focuses on technical communication, we benefited from three established lists of *self-identified* MMU and Black, Indigenous, and People of Color (BIPOC) scholars:

1. Chapter 7 of Rebecca Walton, Kristen R. Moore and Natasha N. Jones' *Technical Communication after the Social Justice Turn: Building Coalitions for Action* (Walton et al., 2019).
2. Cana Uluak Itchuaqiyaq's *MMU Scholar List* (Itchuaqiyaq, 2020).
3. Jennifer Sano-Franchini, Sweta Baniya, and Chris Lindgren's "Bibliography of Works by Black, Indigenous, and People of Color in Technical and Professional Communication" (Sano-Franchini et al., 2021).

After coding the author identities for each citation in each entry draft using each of these tools, we tallied the numbers. To align our findings with a more intersectional approach, we also added an MMU marker. We tabulated the percentages by dividing the numbers by the total number of citations in the chapter. We additionally made a notation for authors who clearly failed to include MMU scholars in their citation list.

In addition to quantifying diversity and inclusion via the citation count, we attempted to answer our third research question through a more holistic approach. We read each of the entries multiple times and offered suggestions about missed opportunities to create a more inclusive entry. For example, some entries missed the opportunity to amplify the work of MMU scholars, and we used Sano-Franchini et al.'s list along with our own knowledge of work in the field to suggest additions to the citation lists. We also tried to note where neutrality was assumed as a part of the entry and, where appropriate, provide suggestions for acknowledging the role of power differentials and/or oppression in the treatment of the keyword. In doing so, we followed Cecilia Shelton's (2020) call for TC instructors (along with practitioners and researchers) to "shift out of neutral" by giving explicit attention to inequities related to TC.

Table 40.1. Overview of the Three Analytic Phases

	Phase One: Author Knowledge	Phase Two: MMU Lists	Phase Three: Knowable Information
What We Did	Cagle and Kristen identified any cited authors in terms of gender, race, and class.	Cagle and Kristen cross-checked all citation entries with three MMU lists (details below). Using the MMU lists, we marked authors who self-identified on the lists into a separate category: MMU.	A research assistant, Nicole, searched all unknown citations using Google and Twitter. If the author self-identified as a member of a minoritized group, Nicole marked them as such; if the author was clearly marked, Nicole labeled them as marked.
Example	Cagle and Kristen know that Rebecca Walton is a white woman and were able to mark her as such; in a different way, Cagle and Kristen both know that Dorothy Winsor self-identifies as a woman in her work, and so we were able to mark her as a woman.	Although Cagle and Kristen do not know all scholars personally, we were able to mark them as MMU based upon these lists.	Nicole saw that "Wegner, D." was unmarked, searched for their name, and found that their bio uses she/her pronouns. These pronouns are then taken as a proxy for gender identity, which is itself of course imperfect, as she/her pronouns may be used by cis women, trans women, nonbinary people, and others.
Why It Is Imperfect and Flawed	We don't know everyone. And even with those we do know, we aren't necessarily keen on assuming that we are privy to how they self-identify.	The MMU lists don't differentiate among marginalized and multiply marginalized scholars. Additionally, these lists are limited in their inclusion of scholars outside the field of TC and prior to the most recent generation of writers.	So many imperfections here: We cannot actually know anyone's identity by looking at them. Our objective here was to be as inclusive as possible, so we wanted to give the benefit of the doubt to authors and amplify as many choices to include women and MMU scholars as possible.

What We Found and Did

Our methodology is imperfect. Resources such as the three lists of self-identified MMU and BIPOC scholars we mentioned previously can be an asset for inclusionary citation practices, but these lists are imperfect. So too is performing internet searches and attempting to determine whether a person is marginalized or MMU based on appearance and what their online bios might say. One's sexuality and gender identity can go unmarked, as can their disability status and their race. In other words, someone might "look like" a cisgender, heterosexual white female but might use they/them pronouns. One's economic background is also unmarked; it's difficult to tell from a picture whether someone is a first-generation student, for example. It's also true that an MMU scholar might choose not to self-identify to avoid the material effects of exclusionary hiring, tenure, publication, and citation practices. There are these and more issues with trying to determine a person's identity based on internet presence, so what we "found" is also imperfect. However, we want to offer here some observations about the *drafts* we feel confident in noting:

- White women authors were well cited among the authors in the collection; scholars of color were not. In the drafting stage at which we reviewed, for example, white women comprised at least 20 percent of more than 25 entries' citations; scholars of color, on the other hand, were only prominent (more than 20%) in two entries.
- Men were less likely than women to cite MMU scholars. For example, although men were responsible for 49 percent of the total citations and 60 percent of the entries, their entries accounted for only 34 percent of the citations of MMU scholars.
- The numbers of MMU scholars we counted represent only a select few authors, not a wide range of MMU scholarship. Five scholars (Natasha Jones, Miriam Williams, Godwin Agboka, Angela Haas, and Huiling Ding) are repeatedly cited.

The citation numbers suggest that most entries could meaningfully engage with more MMU authors, even by simply consulting the lists of MMU scholars we referred to in Phase Two. This additional step may be beyond some of our traditional research practices, but David L. Wallace (2006) reminds us of our duty to frame

> our arguments with a new awareness, a multiplicity that acknowledges and transcends what has been taken as normative, that gets beyond the presumption that the way we have always done things is more or less neutral and well enough informed to be adequately inclusive. (p. 503)

That is, it is incumbent upon each of us to reconsider how we are making our arguments and who we are citing to support our claims, and this may require a bit of extra work.

In reading the early drafts of the keyword entries, we noted missed opportunities in multiple entries, and we created a table that offered concrete suggestions for topical or scholarly inclusion for these authors. After we collated our data and analyzed it, we met with the editors to discuss our findings and shared a brief report. The editors were enthusiastic about recommending more inclusive practices to authors. From there, the editors provided our individual feedback to respective authors in addition to recommending that all authors consider additional citations and the integration of MMU scholars.

While we heard from one author as a follow up, aside from that author, we don't know how or if authors integrated our suggestions for more inclusive entries. In the course of writing and revising this afterword, we have learned that many authors seriously considered suggestions, implemented changes, and used the feedback to shape their projects.

An Invitation to Readers

We have elected not to conduct a second audit on the final version of this collection. The point of such an audit lies in its relevance to the revision process; an audit doesn't serve our goals of creating a more inclusive narrative of the field when conducted after the fact on a final, fixed text. But as we close our afterword, we invite you as a reader to engage with this collection through the lenses we brought to our mid-process audit: Whom and what does this text, entry, collection include? Whom does it amplify? And whose knowledge does it suppress?

As relatively early career scholars, we acknowledge that we are junior to many, if not most, of the well-established authors in this text. You might be, too. But we hold that the social justice turn in TC empowers readers to consider how and if texts acknowledge systems of power and oppression and represent difference and diversity. We have agency as readers, and we can push back in our own reading practices, in our review practices, and in our willingness to accept the limits of particular narratives. Moreover, we *should* push back. An audit such as ours is one way to push back; others include methods such as antenarrative (Jones et al., 2016) and counterstory (Martinez, 2020), both of which center the questioning of and writing against established narratives as critical knowledge-making practices.

What's lovely about the story of this audit is that it's incomplete: We hand this story off to you. Our invitation for you is to *not* see these keywords as the whole story of the field. When you read anything that claims to be essential about a field, it is crucial to know that that claim is always coming from a particular place, always shaped by power, and always subject to amendment. Even as we finalize this afterword and reflect on the process, we recognize our own positions of privilege and the limits of what we can and can't know about authors' decisions. We see this work as a part of the long-term work of coalition-building in the field, and we hope this flawed effort provides a generative roadmap for interrogating our writing and reading practices in technical communication.

References

D'Ignazio, C. & Klein, L. F. (2020). *Data feminism*. The MIT Press. https://doi.org/10.7551/mitpress/11805.001.0001.

Faris, M. J. & Wilson, G. (2022). Mapping Technical Communication as a Field: A Co-Citation Network Analysis of Graduate-Level Syllabi. In J. Schreiber & L. Melonçon (Eds.), *Assembling critical components: A framework for sustaining technical and professional communication* (pp. 69–115). The WAC Clearinghouse; University Press of Colorado. https://doi.org/10.37514/TPC-B.2022.1381.

Itchuaqiyaq, C. U. (2020, August 5). *MMU scholar list*. Cana Uluak Itchuaqiyaq. https://www.itchuaqiyaq.com/mmu-scholar-list.

Johnson-Eilola, J. & Selber, S. A. (Eds.). (2004). *Central works in technical communication*. Oxford University Press.

Jones, N. N., Moore, K. R. & Walton, R. (2016). Disrupting the past to disrupt the future: An antenarrative of technical communication. *Technical Communication Quarterly*, *25*(4), 211–229. https://doi.org/10.1080/10572252.2016.1224655.

Martinez, A. Y. (2020). *Counterstory: The rhetoric and writing of critical race theory*. National Council of Teachers of English.

Sano-Franchini, J., Baniya, S. & Lindgren, C. A. (2021). *Bibliography of works by BIPOC in TPC*. Retrieved June 16, 2021, from https://docs.google.com/document/d/1ybStErUgIQcE5toof_1hGuOVh5olCx08oqBcXnH2nf8/edit?usp=sharing&usp=embed_facebook.

Shelton, C. (2020). Shifting out of neutral: Centering difference, bias, and social justice in a business writing course. *Technical Communication Quarterly*, *29*(1), 18–32. https://doi.org/10.1080/10572252.2019.1640287.

Wallace, D. L. (2006). Transcending normativity: Difference issues. *College English*, *68*(5), 502–530. https://doi.org/10.2307/25472168.

Walton, R. W., Moore, K. R. & Jones, N. N. (2019). *Technical communication after the social justice turn: Building coalitions for action*. Routledge. https://doi.org/10.4324/9780429198748.

Contributors

Rebekka Andersen is Associate Professor in the University Writing Program at the University of California, Davis, where she teaches courses in professional and technical communication and serves as the Associate Director for Professional Writing. Her research focuses on strategies for building stronger connections between academia and industry as well as on implications of digital transformations, particularly around content, for education and research in professional communication. She serves on the Advisory Council for the Center for Information-Development Management (CIDM) and is an Associate Editor for *IEEE Transactions on Professional Communication*.

Elizabeth L. Angeli is Associate Professor of English at Marquette University and a spiritual director. A leading expert in prehospital healthcare communication, Liz has partnered with clinicians and educators to research and improve writing training and practice. Her first book, *Rhetorical Work in Emergency Medical Services: Communicating in the Unpredictable Workplace*, won the 2020 NCTE CCCC Best Book in Technical and Scientific Communication Award. Liz's work has appeared in *JEMS: Journal of Emergency Medical Services*, *Wisconsin Medical Journal*, *Jesuit Higher Education: A Journal*, and *Presence: An International Journal of Spiritual Direction*. As a spiritual director, Liz accompanies people on their personal and professional journeys, teaches discernment-based writing classes and workshops, and serves on retreat leadership teams.

Tatiana Batova is Associate Professor of Business Administration in the Communication area at the University of Virginia's Darden School of Business, where she teaches Leadership Communication, Storytelling with Data, and User Experience (UX). Her research focuses on cross-cultural communication with applications to business and healthcare; data visualization; user and customer experience; social psychology; content strategy; and rhetoric of technology. Her articles appeared in the *International Journal of Business Communication*, the *Journal of Business and Technical Communication*, *IEEE Transactions on Professional Communication*, *Technical Communication Quarterly*, *Technical Communication*, and *Substance Use and Misuse*. Alongside numerous academic conferences, she presented her research at practitioner-oriented venues such as UXPA, CIDM, and Congility. She is the recipient of the 2010 Frank R. Smith Outstanding Journal Article Award from the Society for Technical Communication.

Ann M. Blakeslee is Professor of English and Director of Campus & Community Writing at Eastern Michigan University. Blakeslee coordinates the University Writing Center, WAC, the Eastern Michigan Writing Project, and YpsiWrites. She has served on the executive committees of AWAC and ATTW and is Associate Publisher for Books for the WAC Clearinghouse. She has published a book on writing for audiences in physics and a textbook on qualitative research methods.

She has also published articles and book chapters on disciplinary and workplace writing, learning transfer, and community writing centers. She has been recognized for her scholarly achievements with the Society for Technical Communication Ken Rainey Award for Excellence in Research in Technical Communication and with the Association of Teachers of Technical Writing Fellows Award.

Pam Estes Brewer is Professor of Technical Communication in Mercer University's School of Engineering. Brewer is a Fellow in the Society for Technical Communication (STC) and a recipient of STC's Jay R. Gould Award for Excellence in Teaching, the STC's President's Award, and Mercer's Vulcan Award for Innovation in Teaching. She researches and trains on remote/hybrid teaming, and her book entitled *International Virtual Teams: Engineering Global Success* was published by Wiley in 2015. With George Hayhoe, she published the 2nd edition of *A Research Primer for Technical Communication* with Taylor & Francis.

Tracy Bridgeford is Professor of Technical Communication and editor of *Technical Communication Quarterly*. She has published *Teaching Professional and Technical Communication: A Practicum in a Book* with Utah State University Press and four co-edited collections, including *Teaching Content Management in Professional and Technical Communication*, *Academy-Industry Relationships: Perspectives for Technical Communicators*, *Sharing Our Intellectual Traces: Narrative Reflections from Administrators of Professional, Technical, and Scientific Programs*, and *Innovative Approaches to Teaching Technical Communication*. She has contributed chapters to *Editing in the Modern Classroom*, *Resources in Technical Communication: Outcomes and Approaches,* and *Teaching Writing with Computers: An Introduction* (awarded the 2003 *Computers and Composition* Distinguished Book Award).

Jonathan Buehl is Associate Professor of English at The Ohio State University, where he teaches courses on research methods, rhetoric, and technical and professional communication. He is the author of *Assembling Arguments: Multimodal Rhetoric and Scientific Discourse* and the co-editor of *Science and the Internet: Communicating Knowledge in a Digital Age*. His essays have appeared in such venues as *College Composition and Communication, Technical Communication Quarterly, The Routledge Handbook of Scientific Communication,* and *Landmark Essays on Archival Research*. As a consultant and trainer, he has worked with teams of writers in organizations ranging from small nonprofits and biotech startups to business consulting firms and multinational insurance companies.

Lauren E. Cagle is Associate Professor of Writing, Rhetoric, and Digital Studies and Affiliate Faculty in Environmental and Sustainability Studies and Appalachian Studies at the University of Kentucky (UK). She is the co-founder and Director of the Kentucky Climate Consortium, a multi-institutional network of climate teachers and researchers in Kentucky higher education. Cagle teaches scientific, environmental, and technical communication, and her research focuses on overlaps among digital rhetorics, research ethics, and scientific, environmental, and technical communication, frequently in collaboration with local and regional environmental and technical practitioners such as the Kentucky Division for Air Quality, the Kentucky Geological Survey, the UK Recycling Program, and The Arboretum, State Botanical Garden of Kentucky. Cagle's work has been published in *Technical Communication Quarterly*, *Rhetoric Review*, and *Computers & Composition*.

Kelli Cargile Cook is Professor and Founding Chair of the Professional Communication Department at Texas Tech University. Previously, she served as Professor of Technical Communication and Rhetoric at Texas Tech. Her scholarship focuses on online education, program development and assessment, and user-experience design. Most recently, she co-edited *User Experience as Innovative Academic Practice* (2022) with Kate Crane. She also co-edited two collections on online education: *Online Education 2.0: Evolving, Adapting, and Reinventing Online Technical Communication (2013)* and *Online Education: Global Questions, Local Answers (2005)*. She is a past president of the Association of Teachers of Technical Writing and the Council for Programs in Technical and Scientific Communication. She is currently a member of the International Association of Business Communicators Professional Development Committee.

Huiling Ding teaches technical communication at North Carolina State University. Her research focuses on intercultural professional communication, technical communication, risk communication, responsible AI, and epidemic communication. Her recent projects have been exploring the connections between artificial intelligence, communication technologies, labor market analytics, job screening, risk communication, and social justice.

Angela Eaton is the owner of Angela Eaton & Associates, LLC. She was previously Associate Professor of Technical Communication and Rhetoric at Texas Tech University, where she taught grant writing, quantitative research methods, and technical editing. She designed the Certificate in Grant and Proposal Writing there. She also co-authored the 5th edition of *Technical Editing* with Carolyn Rude.

Norbert Elliot is Professor Emeritus of English at New Jersey Institute of Technology. A specialist in writing assessment, his final academic book was a co-edited collection with Diane Kelly-Riley—*Improving Outcomes: Disciplinary Writing, Local Assessment, and the Aim of Fairness* (Modern Language Association, 2021). In 2021–2023, he completed a series of articles with Mya Poe, Jessica Nastal, Maria Elena Oliveri, David Slomp, and other colleagues on fairness and justice in assessment. He remains on the Editorial Board of *Assessing Writing*.

Carlos Evia is Professor of Communication and Director of the Academy of Transdisciplinary Studies at Virginia Tech, where he is also Associate Dean for Transdisciplinary Initiatives, and Chief Technology Officer in the College of Liberal Arts and Human Sciences. Carlos worked in the intersection of information technology and the humanities as database designer and technical writer. In his academic career, he has been a Professor of English (Professional and Technical Writing) and Communication (Digital Publishing and Content Strategy), and award-winning researcher of transdisciplinary Technical Communication and Content Operations. He authored *Creating Intelligent Content with Lightweight DITA*, edited *Content Operations from Start to Scale*, and worked on the development of the Darwin Information Typing Architecture (DITA) and Lightweight DITA (LwDITA) standards for digital content.

David Farkas began teaching technical communication as a graduate student at the University of Minnesota in 1975. After completing his doctorate in British literature, he taught technical communication at Texas Tech, West Virginia University and, for most of his career, the University of Washington. He transitioned to emeritus status in 2014 but remains active in the field. He is an STC Fellow. He received the Society's Ken Rainey Award for research and Jay Gould award for teaching. He served as a Fulbright Senior Scholar in Egypt. He has published on many topics, but his focus in recent years has been developing reading environments (QuikScan) and techniques that can overcome the increasing resistance of modern readers to long non-fiction documents.

Erin Clark Frost is Associate Professor in the Department of English at East Carolina University, where she teaches technical communication and rhetoric with a focus on intersectional feminist issues. She has published in *Computers and Composition, Journal of Business and Technical Communication, Technical Communication Quarterly, Programmatic Perspectives,* and *Peitho: Journal of the Coalition of Feminist Scholars in the History of Rhetoric & Composition*, and her book *Feminist Technical Communication* is due out at the end of 2023.

Guiseppe Getto is Associate Professor of Technical Communication and Director of the M.S. in Technical Communication Management at Mercer University. His research focuses on utilizing user experience (UX) design, content strategy, and other participatory research methods to help people improve their communities and organizations. Read more about him at: http://guiseppegetto.com.

William Hart-Davidson is Professor in the Department of Writing, Rhetoric, and American Cultures and Associate Dean for Research and Graduate Education in the College of Arts & Letters at Michigan State University.

Brent Henze is Associate Professor of English at East Carolina University, where he coordinates the internship program and serves as graduate advisor for technical and professional communication. His research focuses on the rhetoric of scientific disciplines and the engagement of novice and lay practitioners in scientific and technical activity.

Johndan Johnson-Eilola is Professor and Chair of Communication, Media & Design at Clarkson University, where he teaches courses in design. In addition to more than fifty book chapters and journal articles, he has written, co-written, or co-edited books including *Datacloud*, *Writing New Media* (with Anne Wysocki, Cindy Selfe, and Geoff Sirc), *Central Works in Technical Communication*, and *Solving Problems in Technical Communication* (both co-edited with Stuart Selber). His work has won awards from the National Council of Teachers of English, *Computers & Composition*, *Technical Communication Quarterly*, *Kairos*, and the National Council of Writing Program Administrators.

Richard Johnson-Sheehan is Professor of Rhetoric, Composition, and Professional Writing at Purdue University. He researches rhetoric of science, rhetoric of health and medicine, ancient rhetorics, and science and medical writing. He is the author of *Writing Proposals*, 3e, *Technical Communication Today*, 7e, and *Writing Today*, 5e, among other books.

Natasha N. Jones is a technical communication scholar and co-author of the book *Technical Communication after the Social Justice Turn: Building Coalitions for Action* (winner of the 2021 CCCC Best Book in Technical or Scientific Communication). Her research interests include social justice, narrative, and technical communication pedagogy. She holds herself especially accountable to Black women and marginalized genders and other systemically marginalized communities. Her work has been published in *Technical Communication Quarterly*, *Journal of Technical Writing and Communication*, and *Journal of Business and Technical Communication*. She has received national recognition for her contributions and currently serves as the President for the Association of Teachers of Technical Writing (ATTW). She is Associate Professor at Michigan State University in the African American and African Studies department.

Steven B. Katz is Pearce Professor Emeritus of Professional Communication, and Professor Emeritus of English, at Clemson University. He has published several books (*The Epistemic Music of Rhetoric* [1996] and *Writing in the Sciences* [with Nancy Penrose] 1st-4th editions [2020]), and has several new books forthcoming, including *Plato's Nightmare* (Parlor Press, 2024). Katz also has published numerous articles on scientific and technical writing, medical communication, and ethics. "The Ethic of Expediency: Rhetoric, Technology, and the Holocaust" was the recipient of the National Council of Teachers of English Award for Best Article on the Theory of Scientific and Technical Communication (1993), and has been reprinted in different anthologies, most notably in *Central Works in Technical Communication* edited by Stuart Selber and Johndan Johnson-Eilola (Oxford UP, 2004).

Miles A. Kimball, Professor of Communication and Media at Rensselaer Polytechnic Institute, is interested in the relationships between technology and humanity, particularly in terms of technical communication. His concept of "tactical tech comm" highlights the unrecognized ubiquity of technical communication in our society. He has published broadly on e-portfolio pedagogy, information design, digital humanities, and the history of data visualization. Kimball is the coeditor of the SUNY Technical Communication book series. He is also a longtime member of the Society for Technical Communication, which named him an Associate Fellow in 2020.

Charles Kostelnick is Professor at Iowa State University, where he has taught technical communication and a graduate and undergraduate course in visual communication in business and technical writing. He has published several articles and book chapters on visual communication as well as authored *Humanizing Visual Design: The Rhetoric of Human Forms in Practical Communication* (2019), co-edited *Visible Numbers: Essays on the History of Statistical Graphics* (2016), and co-authored *Shaping Information: The Rhetoric of Visual Conventions* (2003) and *Designing Visual Language: Strategies for Professional Communicators* (second edition, 2011).

Chris Lam is Associate Professor of Technical Communication at the University of North Texas. He studies communication in team projects and examines the literature on professional and technical communication and its impact on the profession.

Benjamin Lauren is Associate Professor and Chair of the Department of Writing Studies at the University of Miami. His work focuses on the intersections of learning, professional writing, rhetorical theory, and creative-critical methods of inquiry, such as design, songwriting, and soundwriting. His work has been published in journals such as *Kairos*, *Technical Communication*, and *Reflections*. His first book *Communicating Project Management* was published by Routledge in the ATTW Series. For more information about his work, please visit http://benlauren.com.

Bernadette Longo recently retired from her position as Associate Professor in the Department of Humanities and Social Sciences at New Jersey Institute of Technology. She is the author of *Spurious Coin: A History of Science, Management, and Technical Writing* (SUNY Press, 2000), *Edmund Berkeley and the Social Responsibility of Computer Professionals* (ACM Press, 2015), and *Words and Power: Computers, Language, and U.S. Cold War Values* (Springer Press, 2021). She is co-editor of *Transnational Research in Technical Communication: Stories, Realities and Reflections* (SUNY Press, 2022) and *Critical Power Tools: Technical Communication and Cultural Studies* (SUNY Press, 2006), as well as co-author of *The IEEE Guide to Writing in the Engineering and Technical Fields* (IEEE Press, 2017). She currently enjoys life by a small lake in New Jersey.

Nicole Lowman teaches technical communication in the Departments of Engineering Education and English at the University at Buffalo. Their research focuses on rhetorics of race, legal humanities, and contemporary American culture. Their work has been published by *The Journal of Contemporary Rhetoric* and *The New Americanist* and is forthcoming in *African American Literature in Transition: 2000–Present*.

Edward A. Malone is Professor of Technical Communication at Missouri University of Science and Technology (Missouri S&T), where he serves as Assistant Chair for Graduate Studies in his department and teaches courses in the history of technical communication, technical editing, and layout and design.

Bruce Maylath is Professor Emeritus of English at North Dakota State University, where he directed the university's program in Upper-Division Writing and taught courses in linguistics and international technical writing. He is the author of many articles and the co-editor of eight books, the most recent of which is *Translation & Localization* (Routledge, 2019). A Fellow of the Association of Teachers of Technical Writing, he is the recipient of the IEEE Professional Communication Society's Ronald S. Blicq Award for Distinction in Technical Communication Education, the Society of Technical Communication's J. R. Gould Award for Excellence in Teaching, the Council for Programs in Technical and Scientific Communication's Distinguished Service Award, and

NDSU's Faculty Lectureship recognizing "sustained professional excellence in teaching, scholarly achievement, and service."

Kristen R. Moore is Associate Professor of Technical Communication in the Departments of Engineering Education and English at the University at Buffalo. Her research explores the role of mundane injustices in technical projects and the academy and has been published in a range of journals, including *Technical Communication Quarterly, IEEE Professional Communication, Technical Communication*, and *The Journal of Business and Technical Communication*, among others. Her award-winning, co-authored book *Technical Communication After the Social Justice Turn* and subsequent studies provide an applied theory of addressing inequities that she uses regularly in her work as the chair of Justice, Equity, Diversity, and Inclusion initiatives in the School of Engineering and Applied Sciences at UB.

Kathryn Northcut serves as Professor of English and Technical Communication at the Missouri University of Science and Technology. She teaches courses on proposal writing, visual theory, and technical communication to undergraduate and graduate students. Among her scholarly works are two edited collections: *Scientific Communication: Practices, Theories, and Pedagogies* (2018), co-edited with Han Yu, and *Designing Texts: Teaching Visual Communication* (2013), co-edited with Eva Brumberger. She is fascinated with the interplay of science, technology, and text.

Sushil K. Oswal is Professor of Human-Centered Design in the School of Interdisciplinary Arts and Sciences and CREATE Faculty at the Center for Research and Education on Accessible Technology and Experiences at the University of Washington. The broad focus of his HCI research is on the employment of technology in the knowledge industry. His research has encompassed human-computer interaction design issues in medical devices, distributed web environments, digital library databases, self-service kiosks, and learning management systems. His current projects include informational access about preparedness for wildfires and flash floods in climate change scenarios and the accessibility of healthcare information for blind users in pandemic conditions. He consults in the areas of HCI, technology design, and digital accessibility of work spaces.

James E. Porter is Professor of Rhetoric and Professional Communication at Miami University, where he holds a joint appointment in the Departments of English and Emerging Technology in Business & Design. He has been teaching and/or administering programs in the field of technical/professional communication since 1982. His recent research focuses on human-machine teaming, rhetorical intelligence, and the ethics of AI-based writing systems, an inquiry that began with his co-authored 2017 book (with colleague Heidi McKee), *Professional Communication and Network Interaction: A Rhetorical and Ethical Approach* (Routledge).

Liza Potts is Professor in the Department of Writing, Rhetoric, and Cultures at Michigan State University. Her research interests include networked participatory culture, social user experience, and digital rhetoric. She has published books and articles focused on disaster response, user experience, and participatory memory.

Dirk Remley is Professor of English at Kent State University, where he teaches writing courses that include technical writing and professional writing. He has authored several books, chapters, and articles on topics related to multimodality and multimodal rhetoric in technical, professional and leadership communication contexts.

Gerald Savage is Emeritus Professor of Technical Communication and Rhetoric at Illinois State University. He is co-editor of *Technical Communication & Social Justice*, an online open-source journal (https://techcommsocialjustice.org/index.php/tcsj). His articles have appeared in *TCQ*, JTWC, JBTC, *Programmatic Perspectives*, and elsewhere. He is co-editor with Han Yu of *Negotiating Cultural Encounters: Stories in Intercultural Engineering and Technical Communication*, with Teresa Kynell-Hunt of *Power and Legitimacy in Technical Communication, Volumes 1 & 2*, and with Dale Sullivan of *Writing a Professional Life: Stories of Technical Communicators On and Off the Job*. He is a Fellow of ATTW, and has received the CPTSC Distinguished Service and the STC Excellence in Teaching awards.

J. Blake Scott is Professor of Writing & Rhetoric at the University of Central Florida. His technical and professional communication (TPC) scholarship has focused primarily on advancing cultural, community-based, and social justice-oriented pedagogical approaches. His scholarship in the rhetoric of health and medicine (RHM)—which has included studies of HIV/AIDS risk rhetorics, rhetorical stigma in clinical healthcare setting, and arguments about transnational pharmaceutical risk conflicts—has been driven by the goal of advancing more just and effective public health policy arguments and efforts. He is the former founding co-editor of the journal *RHM*.

Stuart A. Selber is Professor of English and Director of Digital Education at Penn State University, where he directs the Penn State Digital English Studio and the Program in Writing and Rhetoric. His latest book, *Institutional Literacies: Engaging Academic IT Contexts for Writing and Communication* (University of Chicago Press), won the Distinguished Book Award from *Computers and Composition*. Selber is a past president of the Association of Teachers of Technical Writing and the Council for Programs in Technical and Scientific Communication.

Clay Spinuzzi is a Professor of Rhetoric and Writing at the University of Texas at Austin. His research interests include workplace studies, qualitative research methodology, activity theory, actor-network theory, and genre theory. Spinuzzi has conducted multiple workplace studies, resulting in several articles and books.

Jason Swarts is Professor of Technical Communication in the Department of English at North Carolina State University. He regularly teaches courses on technical document design, networks, and discourse analysis. His research focuses on interrelated areas of genre studies, computer-mediated communication, networks, knowledge work, and knowledge communities.

Christa Teston is the Andrea Lunsford Designated Associate Professor in Rhetoric, Composition, and Literacy in the Department of English at Ohio State University. She mobilizes multiple methods to study how people navigate uncertainty in technoscientific and biomedical contexts. Her first book, *Bodies in Flux: Scientific Methods for Negotiating Medical Uncertainty*, was published by University of Chicago Press in 2017 and won two national best book awards. Her second book, *Doing Dignity: Ethical Praxis and the Politics of Care*, is forthcoming from Johns Hopkins University Press and draws on analyses of three case studies about how in/dignities emerge in contemporary caretaking contexts. Teston also directs Ohio State's business, professional, and technical writing courses and is Vice Chair of the Writing, Rhetoric, and Literacy Program.

Michael Trice received his Ph.D. in Technical Communication and Rhetoric from Texas Tech University. He is currently a Lecturer II with the Writing, Rhetoric, and Professional Communication (WRAP) Program at MIT. His professional career includes work for Apple Computer, Hart InterCivic, and Wizards of the Coast. His work has appeared in *IEEE Transactions on Professional Communication* and *Present Tense: A Journal of Rhetoric in Society*. His research interests include usability issues related to public deliberation in digital platforms and the ways that knowledge of system theory influences participant behavior within digital systems.

Rebecca Walton is an associate dean in the College of Humanities and Social Sciences at Utah State University, Professor of Technical Communication and Rhetoric in the Department of English, and editor of the journal *Technical Communication Quarterly*. She researches how people intervene for justice in their workplaces. Her co-authored scholarship has won multiple national awards, including awards for best book, best collection of essays, best theory article, and best empirical research article. Her research has informed implicit bias training, policy revision, and curriculum development at multiple universities.

Russell Willerton is a Technical Writer/Editor for Po'okela Solutions, LLC, which contracts with federal clients. He graduated from Texas Tech University with a Ph.D. in technical communication and rhetoric. Before returning to industry, he spent two decades in higher education, earning the rank of Professor at Boise State University and Georgia Southern University. His book *Plain Language and Ethical Action: A Dialogic Approach to Technical Content in the Twenty-First Century* (Routledge, 2015) is part of the ATTW Series in Technical and Professional Communication.

Han Yu is Professor of English at Kansas State University, where she teaches technical and scientific communication. Her research interests include writing assessment, intercultural technical communication, visual communication, and popular science communication. Han has published many books, edited collections, and articles on these topics. She is the co-editor of *Negotiating Cultural Encounters: Narrating Intercultural Engineering and Technical Communication* and *Scientific Communication: Practices, Theories, and Pedagogies*. Her books on visual communication include *The Other Kind of Funnies: Comics in Technical Communication* and *Communicating Genetics: Visualizations and Representations*. Her latest work includes two public-facing popular science books titled *Mind Thief: The Story of Alzheimer's* and *The Curious Human Knee*.